# Fundamentals of
# Linear Control

# FUNDAMENTALS OF LINEAR CONTROL

**Prof. Sanjay Kumar Choudhary**

ARCLER
PRESS

www.arclerpress.com

**Fundamentals of Linear Control**

*Prof. Sanjay Kumar Choudhary*

**Arcler Press**
4164 Lakeshore Road
Burlington, ON L7L 1A4
Canada
www.arclerpress.com
Email: orders@arclereducation.com



ISBN: 978-1-77956-980-6 (Hardcover)

# ABOUT THE AUTHOR

**Prof. Sanjay Kumar Choudhary** is a senior academician and experienced university administrator with a long and distinguished career in higher education. Born on 4 February 1964, he currently serves as the Vice Chancellor of Lalit Narayan Mithila University (LNMU), a position he has held since 24 January 2024. Over the years, he has played an important role in strengthening academic systems, governance, and research culture in Indian universities.

Prof. Choudhary has a strong academic background in Physics and Material Science. He holds an M.Sc. in Physics and Material Science, an M.Tech. in Surface Science and Engineering, and two Ph.D. degrees—one in Physics and Material Science and another in Electronics. He has also qualified the GATE examination in Material Science, reflecting his solid technical foundation. His main area of specialization is Material Science and Engineering.

Before joining LNMU, Prof. Choudhary served as Vice Chancellor at Munger University and T.M. Bhagalpur University, where he contributed significantly to academic planning and institutional development. He has also worked as Principal and Professor at institutions such as J.P. University, Sanjay Institute of Engineering and Management, and Bengal College of Engineering and Technology. Earlier in his career, he served as a Lecturer and Research Fellow at NIT Jamshedpur and B.I.T. Mesra, including work on an ISRO-sponsored project.

Prof. Choudhary has published 26 research papers, guided two Ph.D. scholars, and presented his work at international conferences, including events held in Germany. His academic contributions have been recognized with honors such as the Shiksha Ratna Puraskar (2010) and the Rastriya Gaurav Award (2011).

# TABLE OF CONTENTS

# LIST OF FIGURES

# PREFACE

Control is the engineering of dynamic behavior. From auto cruise control to satellite attitude control, from reactor temperature control to noise reduction in audio equipment, we engineer signals to drive systems to desired directions and maintain them there in the presence of uncertainty. The key issues are specific and pragmatic: What model most accurately describes the dominant dynamics? Where must the poles of the closed loop be placed to get a rapid but well-damped response? How do we trade off tracking accuracy for disturbance rejection and noise amplification? Linear control has a strong first set of answers to these questions. By modeling systems near operating points and describing behavior in terms of linear differential equations, we acquire tools that provide understanding, enable prediction, and facilitate sound design in fields as diverse as robotics, aeronautics, power electronics, and process engineering.

Just as powerful is the feedback principle—measure the output, compare it to a reference, and utilize the difference to adapt future actions. Feedback can stabilize a plant that would otherwise be unstable, decrease sensitivity to parameter drift, and condition transients to achieve performance objectives. The mathematics underlying these advantages is within reach once the foundation ideas are established: time and frequency responses, transforms, transfer functions, and state models. With these, designers examine step responses for settling time and overshoot, trace loci as gains are varied, estimate bandwidth and margins, and think about what is going to happen before a prototype is ever constructed. This book approaches these techniques as a unified toolbox for constructing reliable systems that satisfy specifications with restraint and clarity.

*Fundamentals of Linear Control* is structured to develop judgment through a consistent progression of concepts, examples, and design problems. The book begins with the foundations of signals, systems, and feedback loops, progresses to analysis methods in the time and frequency domains, and concludes with synthesis: how to select controllers that satisfy specified requirements under realistic constraints.

Chapter 1 introduces the topic of control systems, stimulating feedback with common and industrial examples. It establishes open- and closed-loop arrangements, defines the concepts of setpoint, disturbance, and error, and states the objectives of design—stability, accuracy, speed, and robustness. The chapter also sketches the modeling assumptions underlying linear time-invariant (LTI) analysis and identifies when such approximations are valid.

Chapter 2 concerns dynamic systems and their mathematical models. Mass-spring-damper mechanics, RLC circuits, and basic thermal models offer tangible beginnings.

Students learn to derive differential equations, use the Laplace transform, create transfer functions, and interpret the significance of poles and zeros. The chapter ends with the notion of canonical forms and the connection between physical parameters and dynamic characteristics.

Chapter 3 describes time-domain analysis as a performance lens. Impulse and step responses are employed to introduce transient specifications like rise time, peak time, overshoot, and settling time, and steady-state specifications like error constants. Root-locus techniques illustrate how closed-loop poles shift with a changing gain so that time-domain requirements can be satisfied with straightforward compensators.

Chapter 4 extends frequency-domain analysis to insight and design. Sinusoidal steady-state, frequency response, and the design and interpretation of Bode, Nyquist, and Nichols plots are presented with solved examples. Bandwidth, resonant peak, and phase characteristics are connected to disturbance rejection and noise attenuation, establishing an intuitive connection between plot characteristics and system behavior.

Chapter 5 defines stability and how to check it. Internal versus BIBO stability is separated, and methods like the Routh–Hurwitz criterion, root-locus tests, and Nyquist stability are discussed. The chapter concentrates on margins—gain and phase—and demonstrates how to extract them from plots, why they are important to uncertainty, and their relation to damping in the time domain.

Chapter 6 covers block diagrams and signal flow graphs as concise notations for interconnections. Block-diagram reduction rules, the interpretation of summing junctions and pick-off junctions, and application of Mason's gain formula to calculate overall transfer functions are where readers become knowledgeable. The goal is mastery at restructuring complicated structures without altering their fundamental input–output behavior.

Chapter 7 considers the analysis of feedback in detail. Sensitivity and complementary sensitivity functions are introduced and employed to represent trade-offs between tracking, disturbance rejection, and shaping of noise. The impacts of actuator saturation, sensor noise, and model uncertainty are addressed, together with simple techniques—prefilters, two-degree-of-freedom design, and loop-shaping heuristics—that enhance performance in practice without excessive complexity.

Chapter 8 is dedicated to controller design using PID control. Proportional, integral, and derivative actions are described with their individual contributions: proportional to provide responsiveness, integral for removing offset, derivative for predictive damping. Tuning methods—from trial rules like Ziegler–Nichols to specification-based tuning based on desired pole locations or frequency targets—comparing them is presented. Implementation issues range from derivative filtering, bumpless transfer, anti-windup, to discrete-time realization.

Chapter 9 elaborates state-space representation as an alternative, and usually more versatile, framework. Readers build up state models using physical principles, estimate controllability and observability, and relate state poles to eigenvalues. State-feedback design by pole placement and use of observers (such as the Luenberger observer) are illustrated at a practical level, demonstrating how state techniques carry over easily to multivariable systems and constraints.

Chapter 10 deals with robustness and performance in linear control. Uncertainty descriptions of gain variation, unmodeled dynamics, and time delays are related to classical margins and straightforward parametric checks. Classical loop-shaping guidelines, weighted specs, and basic $H\infty$ concepts are handled at an introductory level in the chapter for model-based design. The audience learns to specify performance envelopes that trade speed, accuracy, and insensitivity to disturbances and modeling errors.

This book is for undergraduate and first-year graduate students in electrical, mechanical, aerospace, and chemical engineering, as well as working engineers who require a brief yet comprehensive path from principles to reliable design. Familiarity with differential equations and transforms is useful, and reminders of main mathematical tools are provided where necessary. The material sustains a one-semester course focused on classical techniques, or a two-module program that inserts state-space and robustness with lab or simulation exercises. The overall objective is the development of practical fluency: comprehending how to model, analyze, and design linear systems so that performance requirements are achieved economically and confidently.

CHAPTER 1

# INTRODUCTION TO CONTROL SYSTEMS

## Contents



## INTRODUCTION

Control systems play a critical role in both engineering applications and everyday life by regulating the behavior of various systems. They can be broadly classified into two categories: open-loop and closed-loop systems. Open-loop systems operate without feedback, meaning that they execute commands without monitoring the output. A classic example of this is a simple toaster, which heats bread for a predetermined duration regardless of the final state of the bread itself. Conversely, closed-loop systems utilize feedback to constantly adjust their operation based on the output. This feedback mechanism enables the system to self-correct, enhancing accuracy and effectiveness. One prominent example is a modern thermostat in a heating system. It continuously measures the ambient temperature and adjusts the heating accordingly to maintain the desired temperature, showcasing the importance of feedback in achieving a controlled environment. In engineering, control systems are integral to the design and functionality of many complex systems, optimizing performance while ensuring reliability. For instance, the autopilot feature in aircraft functions as a closed-loop control system; it continuously assesses various parameters like altitude and speed to maintain the aircraft's stability during flight. By making real-time adjustments based on sensor input, it enhances safety and operational efficiency. Open-loop control systems can also be found in engineering, such as in conveyor belts that run at a constant speed, where adjustments are not needed based on output conditions.

**Figure 1.1**. *Control systems.*

*Source: https://media.geeksforgeeks.org/wp-content/uploads/20231101182309/Block-diagram-of-CONTROL-SYSTEM-660.webp.*

The contrast between these two systems highlights the importance of feedback in enhancing performance and managing variability in different engineering scenarios, ultimately contributing to the development of smarter, more responsive technology. Understanding the definition and functionality of control systems is essential for grasping their significance in various applications. These systems determine how inputs related to a desired output or performance objective are managed and manipulated, making them vital for automation and system regulation. The principles underlying control systems can be applied across a wide range of fields, from robotics to process industries, showcasing their versatility and importance. The increasing integration of control systems in everyday technology, such as smart home devices that adjust lighting and temperature based on user preferences, exemplifies how control systems are shaping modern life. Furthermore, as industries are pushed toward greater efficiency and sustainability, control systems will play an indispensable role in developing innovative solutions that address both technical and environmental challenges. Thus, an in-depth understanding of control systems is critical for future advancements and their implications (Figure 1.1).

## *Explanation of Control Systems in Engineering*

In engineering control systems, the open-loop and closed-loop distinction is critical in comprehending how devices work. Open-loop systems do not use feedback; they run through a pre-programmed process using specified inputs, generating outputs without compensating for any deviation in the environment or system behavior. One such common example is a washer running on a timer. After it has been started, the machine cycles through regardless of the cleanliness of items contained within. Closed-loop systems, or feedback control systems, by contrast, use feedback to make continuous adaptations in their operation. For example, a sophisticated temperature control system on a thermostat continuously monitors room temperature and makes actual-time compensations in heating or cooling output to set the desired temperature. This inherent disparity exemplifies how control systems can be adapted to suit diverse operation requirements and optimizations in engineering and daily usage. Another look at control systems illustrates their significant implications in engineering fields, where accuracy and efficiency are paramount. Closed-loop control systems are becoming more widely used in industrial automation in order to increase production accuracy. For instance, robotic arms in assembly lines depend on closed-loop feedback to make sure that movements and operations are regulated according to the precise position and orientation of the components. Conversely, for everyday use, the working of devices like microwave ovens can be seen as another illustration of an open-loop system. As users establish the timer and power level, the microwave works without quantifying the food's actual temperature, showing an absence of feedback. The equilibrium between these two kinds of systems highlights their applicability and versatility to certain tasks and contexts and therefore emphasizes the significance of choosing the right control strategy to obtain favorable results. The applications of control systems extend beyond simple mechanical functions to impact technology integration into wider areas of society and industry. As artificial intelligence and machine learning continue to advance rapidly, there is an emerging move towards incorporating old-school control systems with improved capabilities. For example, adding AI into a closed-loop system can enhance decision-making processes through evaluating higher amounts of data more quickly and accurately. As noted in the analysis of generative AI technologies, such as ChatGPT, the inclusion of intelligent systems within engineering systems raises ethical concerns—ranging from data protection to automated decision bias—the implication being that control system development must also include not only technical effectiveness but ethical accountability. Therefore, as control system landscapes transform with emerging technologies, the paradigm upon which they function will stretch far beyond typical engineering boundaries, an expression of complex interaction between innovation and social influence (Yogesh et al., 2023; Berg et al., 2020).

## *Significance of Feedback in Control Systems*

Feedback is a critical mechanism in the control system that differentiates open-loop and closed-loop systems. Open-loop systems do not involve any feedback and use only pre-programmed instructions to attain a desired result. A simple example is a washing machine programmed for a particular cycle; once started, it executes the cycle purely based on its own internal timer and preprogrammed parameters without any real-time feedback regarding the dirtiness of the clothes. Closed-loop systems, however, include feedback to make their operations better with each iteration. This interactive mechanism enables the system to respond to condition changes, enhancing accuracy and efficiency. A classic illustration of this would be a thermostat-based heating system, one that varies the temperature based on instantaneous readings, achieving the best possible use of energy and comfort. The vital function of feedback is further emphasized when comparing the advantages and uses of closed-loop systems. Through the use of feedback, these systems are able to naturally compensate for errors and prevent oscillations that can be caused by disturbances or changed conditions. As an example, in car cruise control systems, feedback is employed to keep a constant speed regardless of terrain changes or power fluctuations. A vehicle so equipped will automatically modulate its throttle position as per instantaneous speed data, making for a smooth ride all the time. This dynamic modulation not only improves customer experience but also overall road safety. Such uses demonstrate how feedback systems are fundamental not just in engineering applications but also in everyday life, highlighting their relevance in different areas. Further, the relevance of feedback carries over into more advanced technological frontiers, like reconfigurable intelligent surfaces (RISs) used in wireless communication networks. These networks, which are aided by decision-making functions supplemented by feedback loops, can dynamically change configurations to maximize transmission. As highlighted in current research, RISs represent the harmony between advanced theories of control and real-world applications, enabling responsive adaptation to environmental adjustments for enhancing clarity and quality of connections in wireless networks (Renzo et al., 2020). In the same way, feedback is also becoming increasingly critical in machine learning networks, with the algorithms' complexity requiring interpretability and verifiability. The collaboration between feedback mechanisms and smart technologies points to an overall trend toward integrating feedback into system design, toward environments that can adapt freely to the needs of users and operating challenges (Diogo et al., 2019).

## *Overview of System Dynamics*

In control systems, systems are divided into open-loop and closed-loop systems to explain various working mechanisms and applications. Open-loop systems do not operate in conjunction with feedback; they execute orders based on inputs only, yielding predictable outcomes. A simple example of this is an open irrigation system which makes water function depending on a timer rather than on soil moisture. These systems work well for straightforward tasks where external conditions remain unchanged. They are not capable, however, of self-correcting when they run into unforeseen variations and so do not work well in dynamic environments. Thus, it is imperative to know the limitations of open-loop systems in engineering usage as well as in everyday life situations where there is a need for accuracy, such as in basic domestic appliances that utilize fixed parameters but do not adjust according to actual results. Closed-loop systems, however, incorporate feedback systems that allow continuous monitoring and adjustments utilizing real-time data. This cyclical process is the best representation of reactivity in order to achieve effective control. For instance, a domestic heating system with a thermostat exemplifies this concept: it measures ambient temperature and adjusts heating levels according to it to maintain desired temperature. Closed-loop systems are integral to engineering applications, such as automation, for instance, in robotics where precise control is critical. The ability to change outputs in response to feedback not only enhances performance but also increases efficiency and precision, such that closed-loop systems are particularly valuable where operating environments are dynamic. Generally speaking, appreciation of the intrinsic differences between these two forms of system is essential for engineers and specialists involved in developing effective control systems. System dynamics are not just applicable as they are technically defined, but have relevance in areas such as environmental science and economics. For example, the pyramidal engineering structures in energy systems emphasize the incorporation of closed-loop designs to enhance efficiency and reliability in uses such as sodium-ion batteries, as presented in Liu et al. (2022). Generally, educational technologies, and in particular Artificial Intelligence in Education (AIEd), have begun leveraging system dynamics to promote individualized learning but with very little regard for ethical implications, as Zawacki et al. (2019) ascertain. This is to illustrate that learning system dynamics is vital not only for technological development but for resolving societal issues. Thus, ongoing development of control systems, particularly through system dynamics, has a pivotal part in driving innovation in many fields, underscoring its importance in engineering as well as broader societal use.

## 1.1. TYPES OF CONTROL SYSTEMS

The functioning of control systems is primarily based on their mechanism and structure, which primarily divide them into two types: open-loop and closed-loop systems. In an open-loop system, output is generated from a pre-decided set of inputs without any feedback system to change or rectify the output. For instance, a time-cycled washing machine is an open-loop system; it runs its predetermined cycle whether the clothes are clean or not. Closed-loop systems, on the other hand, have feedback to continuously monitor and adjust their process, thus providing accuracy and stability. A great example of a closed-loop system is a thermostat on a heating system. It takes the temperature reading and regulates heating based on attaining or maintaining a specified set point. This distinction not only emphasizes the extent of operation proficiency but also sheds light on the application of these systems in technical and practical contexts. The understanding of open-loop and closed-loop control systems is significant in engineering due to the fact that both systems serve distinct uses based on the application requirements. Open-loop systems tend to be employed where speed and simplicity are most significant, so tend to be best for conditions where exactness is not actually necessary. For example, in certain industrial processes, a conveyor belt may function without constant feedback. Closed-loop systems, on the other hand, serve a vital purpose in applications where exactness and flexibility are needed. The role of feedback is linked with improved disturbance rejection and system stability, which are important in robotics, aerospace, and automotive engineering. The ability to select the two types of systems appropriately can make or break technological solutions and is a basic issue in control systems engineering. The choice of implementing control systems can also be influenced by external factors, specifically environmental variability and operating dynamics. For instance, in agriculture, increased global food demand calls for efficient use of resources, and this can be achieved using closed-loop control systems. The systems can efficiently optimize water consumption based on the soil moisture levels and therefore avoid wastage but produce as much crop as possible—a required factor as noted in current studies dealing with food security concerns (Charles et al., 2010). Conversely, open-loop systems have the potential to be constrained by different conditions, without the adaptability inherent in closed-loop systems. In summary, the type of control system chosen is based on the specific needs of the application, along with the general context it operates within, and that helps to underscore their pivotal role within modern engineering and daily life.

## 1.1.1. Open-Loop Control Systems

The control system dynamics are crucial to different fields, where the fundamental distinction between open-loop and closed-loop systems dictates the operation and design of numerous applications. Open-loop control systems do not provide any feedback; once an input has been provided, the system carries out the pre-defined output without any provision for changing or altering the actions based on the resulting effects later. In engineering, an illustration of an open-loop system would be the use of a washing machine with a preset sequence of operations once it is initiated, without consideration to how clean the clothes actually are. This kind of lack of feedback can generate inefficiencies in which conditions will differ in unforeseen ways, therefore creating the necessity of learning about the limitations of such systems in giving optimum performance. In everyday use, open-loop control systems are ubiquitous and highly effective though the limitation may not be easy to detect. A toaster provides an illustration of an open-loop system: once a user has adjusted the heat and engaged the toasting action, it will operate for an appointed duration without sensor feedback to ascertain the toasting level of bread. Such systems are usually less complex and more inexpensive compared to their closed-loop counterparts, which utilize feedback loops in monitoring and correcting themselves constantly.

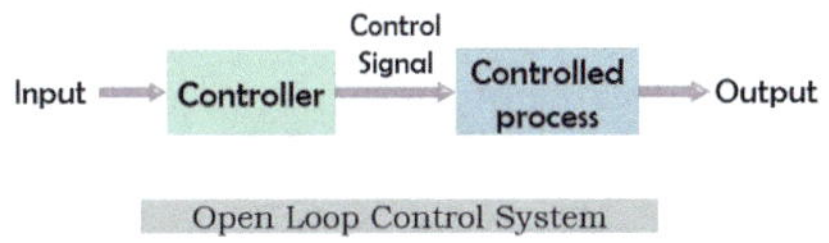


**Figure 1.2**. *Open-loop control systems.*

*Source: https://electronicscoach.com/wp-content/uploads/2019/11/open-loop-control-system-1.jpg.*

As simple as open-loop systems are to design and maintain, they are unable to react to changing circumstances, hence leading to inconsistent results, thus making proper deliberation in engineering as well as daily life a necessity. By way of contrast, closed-loop control systems leverage feedback to improve accuracy and responsiveness, illustrating the benefits of dynamic adjustment. Closed-loop systems monitor output on an ongoing basis and make corrections as required using feedback to deliver desired performance. An example is a thermostat-regulated heating system. It measures the room's actual temperature and controls the heat output to the level necessary to maintain a predetermined temperature, thus the conditions are maintained equal and within the specifications. With evolving technology, the use of both forms of control systems is inevitable, particularly in advanced applications such as the Digital

Twin technology, which illustrates the significance of establishing control systems and procedures to ensure successful implementation (Jones et al., 2020) and counteract the problems caused by their working structures (Guidotti et al., 2018) (Figure 1.2).

### 1.1.2. Closed-Loop Control Systems

In various domains, the distinction between open-loop and closed-loop systems is pivotal in understanding control dynamics. While open-loop systems operate on a predefined sequence of operations without feedback to correct performance, closed-loop systems integrate feedback mechanisms that allow real-time adjustments based on actual output. Consider a simple example: a toaster operates as an open-loop system, where it simply heats the bread for a set time, regardless of the browning achieved. In contrast, a thermostat functions as a closed-loop control system by continuously measuring the room temperature and adjusting the heating or cooling output accordingly to maintain a desired climate. This feedback element not only enhances efficiency but also ensures the systems performance aligns with the established goals, thus emphasizing the critical role of feedback in effective control mechanisms. The practical applications of closed-loop control systems extend across engineering, automation, and everyday scenarios, showcasing their versatility and effectiveness. In the realm of engineering, closed-loop control is essential in processes like automated manufacturing, where feedback from sensors helps maintain optimal production rates and quality standards. For instance, in a robotic assembly line, sensors provide positional data to adjust the movements of robots based on real-time evaluations. This feedback greatly reduces errors and improves overall productivity. In daily life, closed-loop systems can be witnessed in various forms, such as adaptive cruise control in vehicles, which continuously monitors speed and adjusts acceleration to maintain a safe distance from other cars. Such examples underline the importance of closed-loop systems in ensuring safety, efficiency, and precision across multiple sectors. Moreover, the development and integration of technologies into closed-loop control systems continue to evolve, particularly in high-stakes environments such as healthcare and smart grid systems. In medical applications, closed-loop systems are employed in insulin pumps that automatically adjust insulin delivery based on continuous glucose monitoring, thereby optimizing patient care and enhancing health outcomes. Similarly, smart grid technologies utilize closed-loop control to dynamically manage electricity distribution, responding to real-time consumption data to improve energy efficiency and reliability. With advancements in artificial intelligence and machine learning, the capacity for closed-loop systems to adapt and learn from feedback is growing, promising even more sophisticated applications in the future. As outlined in recent literature, including studies on robust closed-loop automation systems, the future holds significant potential for enhanced operational efficiencies that can impact various facets of life and industry (John et al., 2020; Yang et al., 2019) (Figure 1.3).

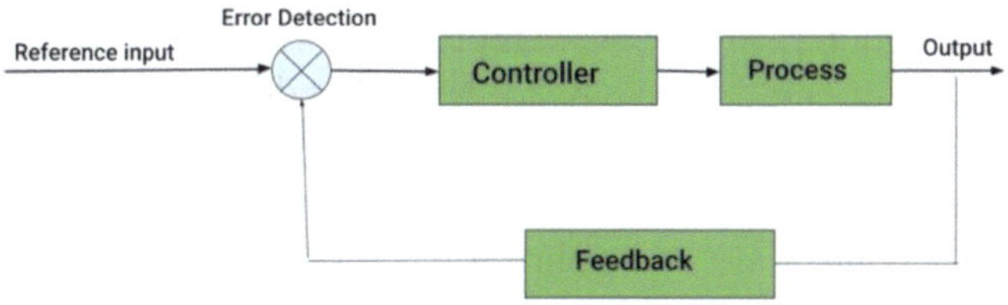


**Figure 1.3**. *Closed loop control system.*

*Source: https://media.geeksforgeeks.org/wp-content/uploads/20231109204652/closed-loop-control-system-660.jpg.*

## 1.1.3. Comparison Between Open-Loop and Closed-Loop Systems

In many engineering applications and everyday situations, separate control systems are employed to produce certain effects. Open-loop systems, characterized as one-way information flow systems, operate without the delivery of feedback from the output. It means that the initial input directs the systems' behavior regardless of any adaptation based on the outcome. For example, a basic toaster is an open-loop system; once the toast is in and the timer turned, the toaster works regardless of the toasts state, potentially leading to unwanted outcomes like charred bread. This absence of feedback can be useful where speed is critical but can sacrifice accuracy and flexibility, highlighting an important limitation of open-loop setups. Closed-loop systems, on the other hand, employ feedback to measure their performance and adjust inputs in light of it, creating a mechanism for self-control. In this arrangement, the output is monitored and compared with a predetermined reference, allowing the system to take corrective action. A thermostat-regulated heating system is a typical illustration of a closed-loop control system. Here, the thermostat constantly monitors room temperature against the set point and adjusts for fluctuations with heating levels. This live control enables higher precision and efficiency, particularly where spaces are complex and variables change rapidly. Through feedback, closed-loop systems can immediately rectify errors, and therefore, represent a solution of preference in applications requiring responsiveness and reliability. In examining the implications of these two control measures, it can be seen that choosing open-loop or closed-loop systems depends on specific requirements of operations. Open-loop systems, even though less complex and easier to establish, can create inefficiencies and potential breakdowns when conditions suddenly change. Meanwhile, the complexity of closed-loop systems creates design and implementation complexities but encourage more reliability and accuracy. This compromise is typical of the fundamental principles governing control systems in general and becomes a foundation for engineers and decision-makers to utilize when selecting the best configuration for their projects. Essentially, awareness of these differences provides priceless insight into how control systems spur on the forward pace of technology and daily life and shows the applicability in the evolving face of engineering.

## 1.2. COMPONENTS OF CONTROL SYSTEMS

A deep understanding of control systems ultimately relies on the recognition of their various components, namely open-loop and closed-loop systems. Open-loop systems operate without feedback, in the sense that they deliver an output based only on a pre-defined input, relying on the assumptions that the environment remains constant. For instance, a simple toaster is a typical example of an open-loop control system; it responds solely on the basis of user-programmed timer, regardless of the actual degree of toasting. Though unimportant, these kinds of systems abound in engineering and day-to-day life and provide efficiency when precision is not a requirement. However, they are limiting because they do not account for internal changes or external interference and will give unsatisfying outcomes if it matters that they are controlled precisely, thus the introduction of feedback mechanisms to provide expectations. Closed-loop systems utilize feedback to allow precision, continually monitoring output to accordingly modify input. This feedback system is essential so that a system can respond to alterations and, in that way, become more capable of maintaining set objectives. A highly intuitive example of a closed-loop system is a house heating system with a thermostat. The thermostat senses room temperature and adjusts the heating output as a function of the difference between desired temperature and actual temperature, hence a comfortable temperature. Closed-loop systems are a basic feature of most engineering applications such as aviation control systems whose continuous adjustment gives them stability in flight. The success of closed-loop control also translates to improved performance but highlights the fundamental distinction between holistic control mechanisms and their open-loop counterparts, with zero adaptability. Finally, the distinction between open-loop and closed-loop systems is a reflection of the fundamental principles of control theory. All forms of system have their respective merits and drawbacks, making them suited to different areas and applications. Open-loop systems can be helpful for basic processes where precision of control is not quite significant, while closed-loop systems are required in complex scenarios that demand constant adjustments. This differentiation not only shed light upon the theoretical framework of control systems but also upon their applicative consequences in everyday technology and engineering procedures, thereby validating the necessity of effective control mechanisms in achieving operational effectiveness and reliability. Awareness of these factors not only deepens one's grasp of control systems but also emphasizes their significance in theoretical as well as practical contexts in engineering disciplines (Figure 1.4).

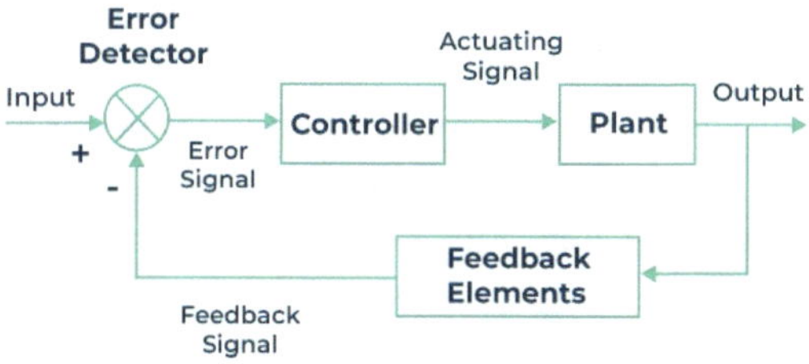


**Figure 1.4**. *Components of control systems.*

*Source: https://media.geeksforgeeks.org/wp-content/uploads/20240124124222/OL-(1).webp.*

## 1.2.1. Sensors and Their Role in Feedback

The intricate control system design is strongly based on feedback mechanisms, the system performance of which depends on various applications. Sensors are especially important elements in closed-loop and open-loop systems. Basic washing machines function without feedback; such operate on a predefined sequence based solely on user command without reflecting on variation in conditions. Compared to this, closed-loop systems like modern thermostats utilize sensors to monitor temperature levels in real-time. In case the environment deviates from the set point, the sensor provides real-time feedback that triggers the adjustments, thereby keeping the system at its desired output. The use of sensors in closed-loop systems not only enhances accuracy but also enables adaptive modification of responses to changing conditions, further establishing their fundamental role in overall efficiency and functionality in engineering fields. Apart from mechanical systems, application of sensors permeates everyday life, influencing a broad array of sectors such as industrial automation and home automation. In intelligent homes, sensors inform on various aspects like occupancy, humidity, and light. Sensors are used to adjust heating, lighting, and security systems in real time, illustrating the feedback mechanism that promotes energy efficiency and comfort for the user. In manufacturing use, sensors are core to DT (Digital Twins) technology, which is real-time modeling of physical assets and therefore allows performance monitoring and predictive maintenance. Smart manufacturing and Industry 4.0 are, to a great extent, dependent on sensor feedback to establish process enhancement and improved decision-making. All of them not only highlight the capacity of the sensor but also emphasize the significance of feedback mechanisms towards goal-oriented operational results in both individual and industrial uses (Tao et al., 2019). As cities are transforming towards smart applications of technology, the role of sensors cannot be overlooked, particularly in optimizing urban services and infrastructure. The concept of smart cities is greatly dependent on the integration of sensors and linked feedback systems, revolutionizing traditional urban living. For instance, traffic management systems with sensor-integrated vehicles can monitor vehicle movement and adjust signals dynamically, thereby reducing congestion and enhancing transportation efficiency. This illustrates the manner in which sensor data feedback can induce real-time optimizations toward enhancing the overall quality of life in urban areas. Further, such integration calls for embracing potent feedback mechanisms to address the complexities of smart city development. As cities strive to redefine their working systems, the intelligence derived from sensor data will surely play an important role in designing sustainable and responsive cities (Albino et al., 2015) (Figure 1.5).

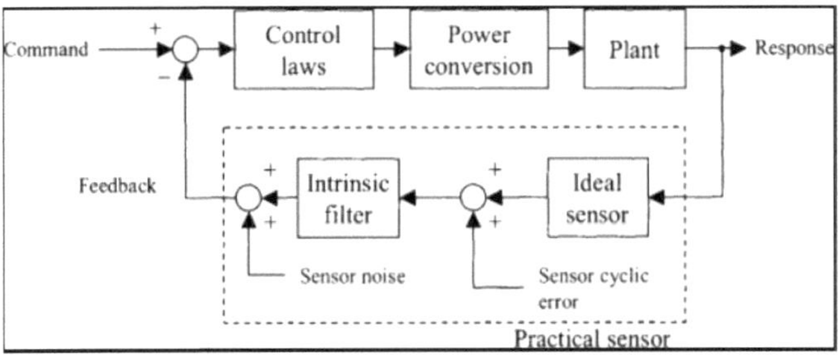


**Figure 1.5**. *Feedback sensor.*

*Source: https://ars.els-cdn.com/content/image/3-s2.0-B9780122374722500031-f02-05-9780122374722.jpg.*

## 1.2.2. Controllers and Their Functions

In control systems, the open-loop versus closed-loop systems difference is critical in deciding the way controllers function. An open-loop system is a system that does not employ feedback; therefore, the controller sends commands to the actuator or process without feedback on its output. The method is straightforward and typically effective for processes where input and output relationship is very well established and constant. A standard case of an open-loop system is a microwave oven, which runs for a certain amount of time without adjusting for the temperature within the food. Closed-loop systems, however, have feedback, which allows them to adjust based on output responses. An everyday example is in a thermostat-heating system, where the temperature is continually monitored and adjusted as needed to maintain the level of temperature desired, which shows flexibility and precision in closed-loop control systems. The functionality of controllers is at the center of the efficiency of open-loop and closed-loop systems. Controllers, both for open and closed systems, are designed to regulate a given process so that it operates in accordance with specified performance criteria. Controllers for closed-loop systems utilize feedback to minimize error by continuously modifying the input in light of a difference between desired and actual output. For instance, load frequency control (LFC) in industrial application is an important attribute in power systems where system frequency is controlled to within permissible ranges by managing generators and loads in an optimal way (Latif et al., 2021). This adaptive adjustment makes closed-loop systems best suited to operating conditions where the environment is dynamic, ensuring consistency and reliability in their performance over time. In open-loop control systems, even though the lack of feedback may introduce constraints in handling disturbances, they remain pivotal in engineering and domestic applications. For example, agricultural automatic irrigation systems are invaluable in optimizing water utilization; such systems operate based on schedules programmed beforehand without continually observing soil water contents. Alternatively, in robotics, recent developments have made use of closed-loop systems, as in minimally invasive surgery, where controllers serve as the medium for transferring human manipulation experience to robots through teaching by demonstration. Sophisticated control methods have made possible surgical robots to respect constraints and optimize operation efficiency, thus demonstrating the flexibility and effectiveness of controllers in both open-loop and closed-loop systems (Su et al., 2021). Since the technologies continue to develop, compatibility between controllers and their functions will remain essential in achieving intended outcomes under different applications.

### 1.2.3. Actuators and Their Importance in System Execution

The functioning of control systems is primarily based on the operation of actuators, that is, devices by which control signals are transformed into physical activities. In achieving the dual nature of control systems, one can distinguish between open-loop and closed-loop systems. Open-loop systems, wherein there is no feedback, demonstrate the basic operation of devices like a simple electric fan that works at a constant speed regardless of room temperature. In such instances, actuators respond immediately to command inputs, taking actions regardless of outcome.

Closed-loop systems, however, utilize feedback to self-adjust performance dynamically, as in the case of a heating system controlled by a thermostat. In such an example, actuators not only execute commands but also adapt their output by adjusting it according to temperature readings, further outlining actuators' contribution in the overall process of achieving desired performance outcomes and system stability, further proving their importance in successful implementation of systems. Actuators are applicable in various uses in engineering and in everyday life, providing the ability to carry out advanced control processes. In robotics, for instance, actuators find a major use by making it possible for robots to move accurately and reconfigure their activities in real-time depending on sensor data.

In manufacturing, pneumatic and hydraulic actuators play a crucial role in automating machinery, enhancing precision, and enhancing productivity. Also, in the car industry, actuators regulate numerous functions like throttle control and braking systems, which ensure safety and efficiency. The utilization of advanced control algorithms specifically further enhances actuators' performance, like in systems where feedback mechanisms are used for accurate control, leading to extensive enhancement in efficacy as well as response. The versatility and difference in functionality of actuators, thus, highlight their core contributions toward the success of modern control systems. The union of actuator technology with control techniques is extremely apparent in the creation of model-free control methods, providing new paradigms for system operation. Utilizing input-output information rather than complicated internal models, this new method simplifies control implementation and enhances flexibility. For example, systems founded on such techniques can remain operational even under varying actuator execution frequencies, which is key in dynamic environments. This adaptability is of essential significance not only in industry but also in emerging technologies such as autonomous vehicles. Bounded stability by continuous adaptation of control inputs enhances operational performance and safety. Lastly, the integration of actuator performance and modern control systems explains the groundbreaking impacts across various fields and highlights the importance of healthy actuator systems in sound control realization, further asserting their role in system performance improvement (Al-khafaji, 2024; Zhou et al., 2024).

## 1.3. APPLICATIONS OF CONTROL SYSTEMS IN ENGINEERING

In most engineering applications, the distinction between open-loop and closed-loop control systems is essential to achieve desired outcomes. Open-loop systems lack feedback; they execute commands without assessing the results of the operations. An everyday example is a washing machine on a time-of-day program; it completes its washing cycle regardless of the efficiency of the detergent or the cleanliness of the clothes. Closed-loop systems, in contrast, employ feedback systems whereby adjustments can be made continuously based on the output states. A good example is the thermostat of a heating system, which is checking the temperature in the room continuously and modulates the heating operations to maintain the required temperature accordingly. The perception of these two classes of systems is essential to engineers as it determines the design and implementation choices that affect efficiency, response time, and performance in different fields of engineering practice, referencing the pivotal role control systems play in engineering practice. The application of control systems is not exclusive to industrial settings to day-to-day life, affecting the way we interact with many other different technologies. In car engineering, for example, adaptive cruise control is a closed-loop system where sensors collect information about car speed and distance between cars, enabling automatic speed adjustment. The system enhances security and driving comfort by keeping optimal distance. On the other hand, a basic microwave oven is an open-loop control system; it works from pre-programmed timing without any adjustments depending on user feedback or food heating. These daily examples not only validate the technical uses of control systems but also recognize the necessity for engineers to adequately select and employ the proper control practices that are right for specific business applications in industrial and consumer environments, emphasizing the engineering generalizability of control systems. Additionally, the significance of control systems is observed in their ability to improve efficiency and dependability in processes and products. An area where this is more pronounced is mining, where intelligent mining systems are increasingly being applied to enhance operational security and environmental sustainability. Industry studies point out that intelligent mining technologies for hard-to-mine coal seams demonstrate innovative control system applications that facilitate key operations such as the identification of risks, scheduling equipment, and managing manpower (Fan, 2024). Such advancements are aimed at minimizing the environmental impacts while maximizing production efficiency to achieve a sustainable mining atmosphere. Similarly, computer graphics uses control systems through image morphing methods, as witnessed in the Beier-Neely technique, where changes can be made from feature control points (Bangal, 2024). This reflects how control systems aid advanced engineering applications as well as advanced technological advancements, underpinning their significance in many sectors (Figure 1.6).

**Figure 1.6**. *Application of control systems.*

*Source: https://media.licdn.com/dms/image/v2/D5612AQH7LlWcy4SwSw/article-cover_image-shrink_600_2000/article-cover_image-shrink_600_2000/0/1702129105785?e=2147483647&v=beta&t=RI7n14-5duz4RW2sebUPqzj4YApdpIW-407PtTe9a15k.*

## 1.3.1. Automation in Manufacturing Processes

One of the key aspects of modern manufacturing processes is the application of control systems, which can further be classified into two broad types: open-loop and closed-loop systems. Open-loop systems consist of a basic mechanism in that control action is solely on the basis of input without the facility for feedback for adjustment, similar to an on/off switch. One such example in engineering is a simple conveyor belt system that operates for a predetermined time period without measuring the actual output. In contrast, closed-loop systems have feedback mechanisms, monitoring output continuously and making changes to inputs to maintain desired performance levels. One such good example of this is automatic temperature control systems used in industrial ovens, which alter heating elements with respect to real-time temperature measurements. These are significant in manufacturing since they dictate automation plan decisions and system complexity, impacting efficiency and product quality. Deep into exploring the implications of control systems on manufacturing, the revolutionary potential of integrating automation becomes clear. The drive for greater precision and efficiency has a tendency to bring about the implementation of advanced technologies, such as vision systems, that are capable of significantly improving assembly line operation. A nice illustration appears in the automotive sector, where implementation of processes like Process and DOK Improvement in Engine Shop by Using Vision System aims to solve complex production problems, avoiding defects that arise during components assembly (Gangadhar, 2025). Through such implementations, manufacturers manage to maintain rigid quality standards while gaining advantages due to increased overall productivity. The move to mechanized systems acknowledges heightened requirements for precision and reliability in an era increasingly dominated by short production runs and customized output. Furthermore, the improvement of artificial intelligence with Industry 4.0 represents a significant leap in the capability of control systems. By adding decision-making with prediction analysis, AI can optimize processes and maximize assets on a manufacturing level. For instance, AI applications in energy consumption forecasting enable producers to better organize their resources, leading to sustainable production procedures. These technologies not only boost efficiency in operations but also resolve difficulties associated with complexity in modern manufacturing settings. The need for smart systems to be capable of balancing efficiency with demand responds to a broader move toward increasingly smarter production, with everything from parts to equipment to process being interconnected and capable of adjusting on their own (Iriakuma et al., 2024). Therefore, as automation technology progresses, the use of advanced control systems will remain at the heart of the evolution of manufacturing processes (Figure 1.7).

**Figure 1.7**. *Automation in manufacturing processes.*

*Source: https://modula.us/wp-content/uploads/2023/02/automation-in-production-robotic-arm-1.jpg.*

### 1.3.2. Robotics and Control Systems

The complexities of control systems lend themselves to two general classifications: open-loop and closed-loop systems. Open-loop systems control without feedback, through a series of actions beforehand guided by an input, and are naive in nature. A simple lawn sprinkler operating on a timer is an example, an open-loop system because it water sprays according to a pre-programmed schedule regardless of environmental conditions such as rain or ground wetness. Closed-loop systems, however, have feedback mechanisms to continuously monitor and control processes. A good example can be found in temperature control systems, where a thermostat measures the temperature around it and adjusts the output of heating or cooling accordingly. This dual characteristic of control systems is important in robotics, where they control the precision and flexibility of robot operation in various applications, making them more applicable across industries from manufacturing to health care. In robotics, control systems play a pivotal role in providing the intended performance and functional effectiveness. A closed-loop control system, for example, for mobile robots can be capable of making real-time adjustments through sensory feedback, enabling the possibility of better interaction with the dynamic environment. Algorithms of robotic arm control are also a good example; they require constant feedback so that they can work at high accuracy rates when processes like surgery or assembly are in question. The use of feedback is meant to minimize error, hence enhancing robotic dependability in industrial processes. Lastly, open-loop system concepts are applied when designing less sophisticated robotic systems where feedback is not a concern. The kind of control system that has been selected will directly influence the whole design, operation, and effectiveness of robotic solutions and thus hints at the need to be aware of such systems in developing effective robotic technology. With the advancement of robotics, the addition of advanced control systems is increasingly important, particularly in the area of automation and artificial intelligence. End-user development (EUD) of software and hardware technologies has introduced a new dimension of how control systems can be designed and utilized by anyone who is not an expert. It is all about introducing user-centric interfaces that can enable a person to tailor functionalities according to individualized requirements (Fogli et al., 2022). The convergence point between user empowerment and control systems is one that can transform the way robot solutions are developed and interacted with every day, bridging the gap between advanced engineering design and usability. Beyond that, as shown with the application of electromagnetic converters in flexible manufacturing systems, the importance of the role control system accuracy plays is growing and impacts production as well as operational use of robots in most industries

(Mammadov et al., 2025). Therefore, recognizing the intersection of robotics and control systems is the key to achieving innovation in technological as well as practical applications (Figure 1.8).

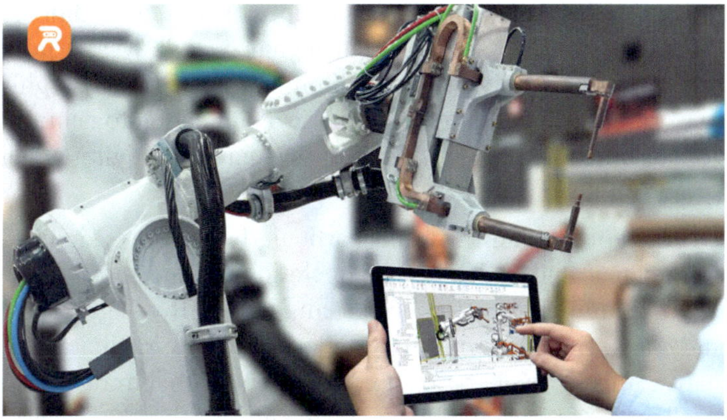

**Figure 1.8**. *Robotics and control systems.*

*Source: https://rozitek.com/wp-content/uploads/2023/12/robot-simulation.webp.*

### 1.3.3. Aerospace Applications and Flight Control

Aerospace control systems' complexity is a reflection of their position at the center of ensuring flight control mechanisms are both effective and safe. Minimizing their applications into functional systems might involve the employment of both open-loop and closed-loop control mechanisms. Open-loop systems neither have feedback nor is their output monitored or corrected in relation to performance. This is exemplified in basic aircraft altitude hold systems that command a fixed altitude without always adjusting for such external forces as turbulence. Closed-loop systems, conversely, usually more sophisticated, employ sensors to give feedback that serves to adjust the performance of the system in real time. For instance, autopilot control systems in newer aircraft have the capability to automatically adjust control surfaces based on the altitude, speed, and atmospheric conditions of the aircraft, reflecting the enhanced accuracy and responsiveness of closed-loop systems in accomplishing optimal flight behavior. The value of effective control systems in aerospace is both relevant to component-level processes as well as integrated system design. The design and realization of the electro-hydraulic servomechanism give a candid description of the way control systems manage flight operations needs requires. The systems exhibit the intrinsic characteristics of closed-loop control's flexibility as well as responsiveness by utilizing feedback from multiple sensors to activate flight control surfaces in a precise manner. Such mechanisms are essential in supplying precise position servo control to promote maximum aircraft safety and efficiency. The simulation of such systems is significant since it allows engineers to test various operating conditions, analyze potential risks, and develop solutions enhancing overall system performance and reliability. As previously argued, Alimhillaj et al. (2024) highlights the importance of rigorous modeling and numerical simulation to achieve realistic snapshots of such complex electro-hydraulic systems, highlighting their role in core aerospace engineering tasks.

The evolution of hydraulic systems particularly illustrates significant gains in energy efficiency and reliability in operation on the move in aerospace application. The traditional hydraulic system is typically plagued with energy dissipation by leakage and inertial friction, appearing as inefficiencies that can have a negative impact on operational performance. The introduction of digital hydraulic technology is a paradigm change to enhance these systems efficiency with accompanying high energy savings. As reported in Nostrani et al. (2023), new designs featuring multi-chamber hydraulic actuators, controlled by on/off valves and digital pumps, promise to be an innovation step. The systems can generate different output velocities while saving energy consumption, demonstrating the synergetic benefit of integrating control theory with modern engineering approaches. Effective hydraulic actuator function not only powers vital aircraft systems, such as control surfaces and landing gear, but also acts as a model for further control system innovation across many areas of engineering.

## 1.4. USES OF CONTROL SYSTEMS IN EVERYDAY LIFE

In the study of control systems, the distinction between open-loop and closed-loop systems yields deeply insightful information regarding their applications in various disciplines. Open-loop systems follow a straightforward rule: the input is established, and the output is generated without the use of feedback control. A typical example of this is a washing machine on a particular cycle; after it is started, it cycles through its process without any adjustments during runtime depending on the soiling of the clothes. In contrast, closed-loop systems involve feedback loops whereby adjustments are made in response to the output. A practical example is a heating system based on a thermostat, where output is adjusted in response to feedback from temperature sensors in a room. The success of closed-loop systems lies in their ability to react to altering conditions, and therefore exhibiting particular value where accuracy and responsiveness are crucial. Such characteristics mirror the basic rules that control system design and operation are governed by. Furthermore, the pervasiveness of control systems in common life says a lot regarding the importance of control systems in themselves, aside from the academic stratosphere. Automotive engineering, for instance, employs both types of systems extensively; cruise control is an open-loop system to maintain a vehicle at a set speed, whereas anti-lock brake systems (ABS) are quintessential closed-loop control in modulating brake pressure in response to present wheel speed data. Environmental control has its systems to regulate aquarium water temperature as examples of closed-loop practice, achieving optimal conditions for aquatic life while responding to change.

These depictions draw attention to the discreet application of control systems in different spheres of life that make operations more efficient, safe, and comfortable. As their applications increase, it is increasingly significant to understand their fundamental methods and mechanisms. The significance of control systems extends beyond fields touching cultural transmission and communal culture. As societies evolve, so does the method whereby indigenous knowledge, such as agricultural methods and environmental awareness, is passed on. Mediation by digital technology is becoming more prevalent in this process, enabling societies to document and pass on cultural myths and environmental behavior through apps. For example, in Cikarawang village, digital platforms enable the construction of community knowledge regarding local agriculture methods with consideration of environmental policy for sustainable development. Here, control systems play a central role by solidifying feedback mechanisms that support ongoing improvement and adjustment in community and environmental settings (Touitou et al., 2024; Lawanda, 2019). The integration of current technology with traditional methods is evidence of the versatility and relevance of control systems in the development of educated and culturally rich communities (Figure 1.9).

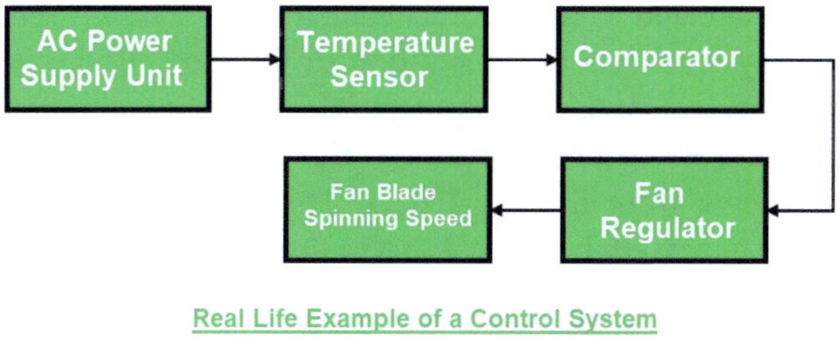


**Figure 1.9**. *Applications of control systems in daily life.*

*Source: https://images.theengineeringprojects.com/image/main/2020/04/Introduction-to-Control-Systems-1.jpg.*

### 1.4.1. Home Heating and Cooling Systems

Heating and air conditioning systems in homes are prototypical examples of open-loop and closed-loop control systems and illustrate the dynamic application of control theory. There is no feedback in open-loop systems, performing what has been programmed with no concern for present feedback. A basic example is a basic thermostat-regulated heating system that activates when a certain temperature is achieved but does not adjust according to present temperature feedback. In contrast, these systems can't self-adjust or respond to environmental changes and may generate inefficiencies and discomfort. This distinction is important to homeowners in selecting HVAC systems because the absence of feedback in open-loop systems can result in excessive energy consumption and unstable indoor climates. Thus, between these types of systems, this choice can play a phenomenal role in comfort as well as energy efficiency in living spaces. Closed-loop systems, unlike open systems, possess feedback loops within them that adjust the working of the systems in real-time, according to prevailing environmental conditions. A fine example in home heating and cooling would be a smart thermostat, which monitors house temperature and humidity continuously, and adjusts the HVAC systems output in response. These systems offer greater comfort by offering the capacity to control very specifically indoor areas with the outside weather changes, having temperatures maintained regardless of the outside weather variations. Closed-loop systems save energy because they restrict unnecessary heating or cooling since they can respond fast to sensed changes with their sensors. These developments in control systems engineering not only optimize home comfort but also promote sustainability, as reduced energy consumption is directly associated with reduced greenhouse gas emissions and homeowner energy costs. Underlining the significance of control systems to heating and cooling the home brings their inherent value to engineering as well as everyday existence into perspective. The use of sophisticated feedback mechanisms in heating, ventilation, and air conditioning systems is a demonstration of the transformative potential of closed-loop control systems. A case in point is newer technologies that can utilize the potential of machine learning algorithms to predict heating and cooling requirements from historical data. This continuous improvement of control systems is necessary because it serves the growing demands for lowered energy consumption and environmental friendliness. Through the use of smart systems that reduce waste and optimize performance, residents are now capable of embracing more energy-efficient homes. Thus, control systems move beyond theoretical abstraction and offer tangible applications that govern our daily lives and address the bigger challenges like climate change and energy efficiency (Figure 1.10).

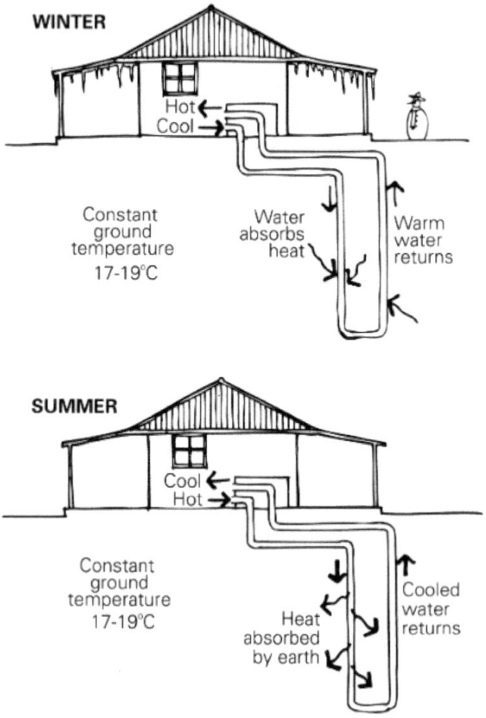


**Figure 1.10**. *Heating and cooling.*

*Source: https://www.yourhome.gov.au/sites/default/files/inline-images/E_Heating_Cooling_4.jpg.*

## 1.4.2. Automotive Control Systems

In theory of control systems, there exist two elementary kinds of systems that control the operation of systems: open-loop and closed-loop systems. Open-loop systems do not employ feedback; they execute programmed actions on input without adapting to the states of output. An excellent illustration is found in many household appliances, for example, a microwave oven. Here, the timer regulates the cooking time with no regard for the state of the food item being cooked. Closed-loop systems, by their feedback facility, realize real-time control by adjusting as a function of output responses. A typical example here is the domestic heating system that is equipped with a thermostat. This system continuously measures the room temperature and adjusts its operation to maintain the set temperature, demonstrating how closed-loop systems effectively enhance performance through continuous monitoring and adjustment. In motor control systems, these are eminently evident. For instance, open-loop systems are utilized in automobiles' basic operations, such as the control of external lights or windshield wipers, operating only on the command of the driver. Alternatively, critical systems such as Anti-lock Braking System (ABS) show how closed-loop mechanisms work. ABS modifies brake pressure during hard braking to prevent wheel lock-up by using real-time feedback from wheel speed sensors. The corrective action after occurrence shows how critical closed-loop mechanisms are in enhancing vehicle safety and performance. Through continuous monitoring of situations and adaptation of responses, closed-loop automotive control systems are the epitome of how good engineering can immediately impact user satisfaction and security. Growing integration of control systems in automotive engineering marks a revolution in car technology. With an ever-growing integration of advanced control systems into vehicles, an understanding of open-loop versus closed-loop design is imperative. An excellent example can be found in adaptive cruise control, wherein closed-loop concepts are employed to control car speed in proportion to the other car's spacing. Not only does this illustrate the practical uses of these systems but also their utilization of sophisticated algorithms and sensors now commonly utilized in automobiles. Both engineering society and everyday life more and more demonstrate the importance of these systems, where the combination of advanced technology and consumer-oriented design can significantly advance working efficiency, security, and convenience in modern cars (Figure 1.11).

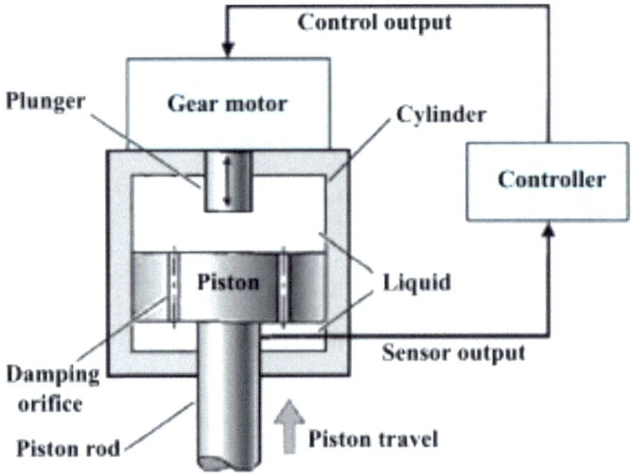


**Figure 1.11**. *Automotive control system.*

*Source: https://ars.els-cdn.com/content/image/3-s2.0-B9780128145432000011-gr024.jpg.*

### 1.4.3. Consumer Electronics and Smart Devices

The everyday interactions individuals have with consumer electronics illustrate the principles of control systems. A good example is an open-loop system, which operates based on a pre-designed input-output relationship without the feedback loop to modify its actions. A simple example of an open-loop system is a simple toaster, which gets switched on for a fixed amount of time to toast bread. Once the heat time has elapsed, the toaster switches off irrespective of the toasts actual brown color. In contrast, closed-loop systems exhibit a feedback mechanism that continuously adjusts the process based on the input received. A thermostat-controlled heating system is an ideal example, as it constantly monitors room temperature and makes adjustments to maintain the desired level. This interaction is of paramount importance in consumer electronics and smart devices, where efficiency and accuracy define user experience and comfort significantly, revealing the essence of control systems in day-to-day applications. The analysis of the integration of control systems in smart devices brings to light additional perspectives into their revolutionary impact. For instance, intelligent home appliances such as smart speakers and thermostats often utilize closed-loop systems making real-time user preference-based control feasible. The appliances are able to process a variety of inputs, such as voice input and temperature, in order to adapt their output accordingly in a bid to maximize performance and user experience. Consumer drones serve as another relevant example where advanced sensors and feedback control help maintain stability and trajectory during flight. The role of such control systems in consumer electronics becomes apparent as they enhance functionality and usability, eventually influencing consumer trends and consumer behavior. Examining the role of open-loop and closed-loop control systems in these applications demonstrates the need for continuous evolution in engineering practices. The advancements in control systems facilitate the advancement of more complex consumer electronics to integrate well into contemporary lifestyles. Energy management systems that utilize blockchain technology, for instance, illustrate the new use of closed-loop technology with decentralized ledger functionality to enable users to monitor and fine-tune their energy consumption. Such technologies offer the potential to transform consumers' interfaces with appliances and firms in energy markets, creating new avenues for efficiency and sustainability. In addition, the intersection of the metaverse and intelligent devices, as detailed by existing studies, gives an insight into the future in which consumer electronics enhance human-to-human interactions and experiences in the virtual world (Yogesh et al., 2022). The impending trend is not only an expression of the importance of intelligent control systems but also depicts the possibility to provide significant socio-economic advantages, promoting the ongoing research and development process in the field (Andoni et al., 2018).

## 1.5. IMPORTANCE OF CONTROL SYSTEMS IN MODERN TECHNOLOGY

Control systems play a very important role in engineering and everyday use applications where they help regulate desired outputs and guarantee optimal performance. Control systems are open-loop and closed-loop systems, both serving different purposes. Open-loop systems lack feedback and execute set instructions regardless of output conditions. An example here would be a washing machine on a pre-programmed cycle, which runs without varying depending on the current state of the clothes' cleanliness. Closed-loop systems, however, incorporate feedback mechanisms that allow correction in real time depending on output discrepancies. An example would be a thermostat in a heating appliance, which senses variations in heat and responds by varying the heat produced accordingly. This distinction highlights how control systems mature to meet challenging requirements, increasing efficiency and reliability across various applications, ranging from industrial automation to domestic appliances. The applicability of control systems reaches into such central areas as transportation and water management. In transport infrastructure, control systems ensure safe and efficient vehicle and infrastructure operation. With the help of dynamic models, the engineers can develop adaptive controls to manage the control variables like speed, direction, and traffic flow in order to enhance safety and reduce congestion. Similarly, in modern water management, effective control systems have become inevitable. As indicated in recent research, sophisticated systems equipped with wireless communication technology can optimally utilize water resource management through the application of Programmable Logic Controllers (PLC) along with sophisticated sensors for precise monitoring. These innovations enable real-time decision-making and increase the general system effectiveness and resource sustainability in a broad array of applications, such as agricultural application and urban infrastructure (Deokar, 2025). Furthermore, finding control systems indicates that they are at the core of addressing modern challenges. With an ever-growing reliance on automated procedures, it is critical to ensure that systems have the capacity to reverse the disruptions caused by wear and tear, environmental factors, and dynamic loads. Research has established that engineers are able to diagnose deviations' causes and accordingly make necessary adjustments to ensure desired performance levels due to understanding the dynamic properties of control systems (Artemyev et al., 2025). Application of control systems in industries, apart from enhancing accuracy in operations, also leads to innovation in developing cost-effective and sustainable solutions. This route represents a growing importance of control systems in furthering present technologies and allowing potential future technological advances, rendering them necessary across diverse industries (Figure 1.12).

**Figure 1.12**. *Modern control systems.*

*Source: https://icaincusa.com/wp-content/uploads/2023/03/Untitled-design-768x432.png.*

## 1.5.1. Enhancing Efficiency and Performance

In control systems, understanding open-loop and closed-loop mechanisms is crucial to enhance efficiency and performance in numerous applications. Open-loop systems lack feedback; they obey commands from input given, which at times may lead to inefficiency resulting from fluctuations in the external environment. For instance, a simple microwave oven accomplishes this where the time to cook is indicated without consideration of the actual food temperature. Closed-loop systems that are typically thought to be better, however, employ feedback to adjust their activities in real-time. For instance, a modern thermostatically regulated heating system monitors the current temperature and responds in real time to maintain a chosen level of environmental comfort and thereby optimize performance and energy utilization. Thus, these variations emphasize the significance of feedback to achieve efficiency in engineering and everyday use applications. The application of feedback in closed-loop systems enhances not only stability but also responsiveness to changed conditions, which is important in industrial and consumer devices. In the automotive sector, for example, adaptive cruise control systems assess a vehicle's speed and distance from other vehicles and adjust throttle and brakes accordingly. This integration significantly reduces fuel consumption and improves safety, depicting how closed-loop mechanisms may perform better than open-loop mechanisms. Similarly, home automation thermostats employ machine learning algorithms to anticipate user behavior and adjust cooling or heating accordingly, leading to considerable energy savings. Application of closed-loop control systems is therefore a crucial step towards maximizing performance and efficiency, proving their central place in modern technology applications. The transition toward more sophisticated control systems is an indication of the necessity to incorporate more evolved technologies that enhance performance in various disciplines. In underwater robotics, for instance, propulsion system optimization is a key factor due to challenging operating environments. Through methods such as bionic design and adaptive control, researchers attempt to maximize propulsion efficiency while breaking durability and environmental adaptation barriers. Moreover, electric cars are another area in which advanced motor drive systems play a crucial role. The incorporation of latest materials like silicon carbide (SiC) in power electronics significantly reduces energy losses and thus leads to increased efficiency and overall performance. All these examples demonstrate that integrating cutting-edge design and control systems into underwater robotics and EVs is fundamental to addressing the complex challenges of sustainability and performance enhancement, which are at the core of their future development (Chen, 2025; Khan, 2025).

### 1.5.2. Reliability and Safety in Safety-Critical Systems

Reliability and safety management are critical in the control systems, which can be classified as open-loop and closed-loop systems. Open-loop systems operate without any feedback, relying purely on programmed inputs to provide expected outputs. This can be observed in processes such as a washing cycle that cycles through its pre-programmed phases regardless of the hygiene of clothes. Although uncomplicated, open-loop systems can generate dangerous results if initial conditions do not fall appropriately or are disturbed by environmental fluctuations, as they are unable to adjust to altered environmental conditions. This constraint underscores the crucial importance of dependability in systems where safety is a prime concern (Saberi et al., 2018). The potential for malfunction highlights the need for implementing facilities that ensure operational integrity when there is no real-time feedback and adjustment capability. Closed-loop systems, however, consist of feedback mechanisms continuously monitoring output and adjusting input accordingly in a bid to achieve specified operational states. A typical example can be observed in modern temperature control systems, for example, those used in HVAC setups. These systems utilize sensors to gauge indoor temperature, allowing them to regulate heating or cooling function on the basis of actual real-time information. Dynamic adjustment enhances efficiency, and by far enhances reliability, particularly in systems where there are inherent safety risks. Closed-loop control's feedback process enables quick reaction to drift from desired behavior, proving its advantages in situations where safety is of utmost concern (Consuelo, 2018). Thus, closed-loop systems show the way feedback can create robust mechanisms in providing safety as well as reliability. Despite the clear benefits of closed-loop systems, trust and safety concerns continue to dominate in their application. The introduction of new technologies, such as blockchain, into closed-loop systems ensures enhanced security and transparency, silencing future fear regarding reliability failures. These technologies are able to carry key safety information from node to node, facilitating a state of mutual responsibility among stakeholders. However, as in the field of public involvement in research collaboration, there is a mutual relationship between safety and trust. It is important that the users are assured of both the technology and the processes behind the technology to allow best function (Consuelo, 2018). Thus, there has to be a thorough investigation of safety practice and reliability management measures in an effort to develop a trusting environment and hence promote safer and more reliable control systems resistant to public scrutiny and normal challenges (Figure 1.13).

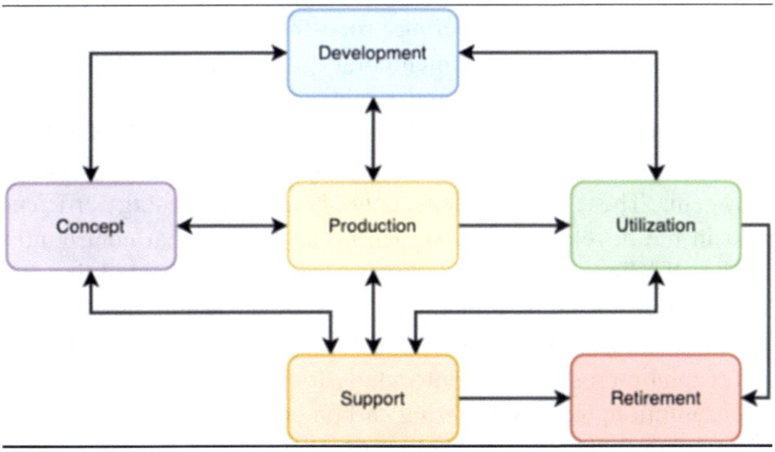


**Figure 1.13**. *Security and reliability of safety.*

*Source: https://media.springernature.com/lw685/springer-static/image/art%3A10.1007%2Fs42979-021-00753-y/MediaObjects/42979_2021_753_Fig1_HTML.png.*

### 1.5.3. Innovations Driven by Control System Advancements

Control systems, an essential part of many engineering applications, can be classified broadly into open-loop and closed-loop systems. Open-loop systems do not use feedback, instead depending on pre-specified inputs to give outputs. An example is a toaster: after the user has set the timer, the toaster acts alone, irrespective of the actual color of the toasts. Conversely, closed-loop systems possess feedback loops for realignment of outputs when desired levels of performance are not being met, typically enhancing precision and accuracy. An example is a room thermostat in an HVAC system regulating room temperature and adjusting heating or cooling dynamically to the setpoint. This feedback feature is especially critical wherever conditions can rapidly change and the necessity for the improvement of control systems is underscored in order to optimize function and efficiency in everyday application. The integration of control system innovation has driven significant developments in various industries, most significantly manufacturing and public infrastructure. The application of automatic inspection systems, powered by sophisticated control technologies, has revolutionized the way industries ensure product quality. In the car sector, for example, projects like Process and DOK Improvement in Engine Shop by Using Vision System illustrate how control systems use vision technology to rectify assembly errors and enhance overall production efficiency as well as quality control (Gangadhar, 2025). With continued growth, their ability to improve manufacturing processes without the loss of accuracy makes them focal points in contemporary industry standards, enhancing easy incorporation of efficiency and quality into production. Another good illustration of the influence of control system innovations is the field of urban infrastructure, which has undergone revolutions in the way of automatic public lighting systems. These advancements not only maximize energy efficiency but also assist in the development of stylish urban layouts that contribute to the health and well-being of citizens. Urban centers can modify lighting intensity in real-time based on climate or crowd conditions by implementing sensor-based controls and adaptive light algorithms (Daniel and Swathi et al., 2024). Dynamic adaptation fosters environmentally friendly urban spaces by reducing energy consumption, thereby lowering carbon emissions. As cities continue to evolve, the role of advanced control systems in helping them grow becomes increasingly important, their relevance in the quest for wiser, cleaner city living that is good for society at large.

## 1.6. CONTROL SYSTEMS DESIGN CHALLENGES

The complexities involved in the design of control systems extend to the fundamental distinctions between open-loop and closed-loop arrangements, both of which present unique challenges. Open-loop arrangements, characterized by a lack of feedback, operate on a pre-set input to create an output without self-correction. An excellent example of an open-loop system is a basic toaster: the operator inserts a timer, but no provision is available to adjust this timing based on the actual crispiness of the toast. In engineering applications where the degree of precision matters, these types of systems can produce unwanted outcomes, particularly when exposed to outside disturbances. This rigidity makes open-loop systems inappropriate for use in applications requiring dynamic response, noting a fundamental issue in the delivery of reliability and performance in control design. Closed-loop systems, by contrast, utilize feedback to enhance accuracy and stability. Closed-loop systems are coded to monitor continuously their output and make instant real-time adjustments in the event of deviation from a set point. A household thermostat is a closed-loop system example; it senses ambient temperature and controls heating or cooling in order to maintain the setting. Despite their merits, closed-loop systems also have their own design problems, namely stability and responsiveness. For instance, over feedback can lead to oscillation or instability, so sophisticated design techniques need to be employed in order to balance responsiveness against stability. The process of achieving this equilibrium usually compels engineers to employ very mathematical methods and simulations, which can enhance system performance but increase design time and cost. The process of developing control systems that function entails not only theoretical issues but real-world implementation where small modification can have significantly effective outcomes. Environmental influences, component variations, and potential system limitations have a big effect on the controllability of closed-loop and open-loop systems. Also, since the world is increasingly becoming dynamic with the advancing technology and environments, as brought to light in Carroll et al. (2024), the demand for flexible control systems continues to grow. All of this makes it even more difficult for designers to create systems that are robust under fluctuating conditions. Hence, engineers need to navigate through a complex web of parameters, trying to design systems that are not only functional but also immune to variations in the internal and external operation environments. Consequently, the issues in control system design illustrate the fine tuning between theory and practice, stipulating the manner in which engineers operate in the dynamic control engineering environment.

### 1.6.1. Complexity of System Modeling

Control system design contributes intrinsic complexity that is often compounded by distinctions between closed-loop and open-loop systems. Open-loop systems function without feedback under pre-programmed conditions to produce their output. One common example is in a basic washing machine, which cycles through a programmed routine regardless of how dirty the clothes are. This uncomplicated approach minimizes complexity at the cost of limiting responsiveness to new situations or unexpected events. Closed-loop systems, however, have feedback loops, allowing immediate adaptation depending on the measurement of output. A common example is a thermostat-controlled heater, wherein the ambient temperature at any instant is constantly being matched with the desired set point so that the system is able to respond in the right way when it changes. The inclusion of feedback in closed-loop systems complicates modeling, as mathematical expressions for it have to account for time delays, noise, and nonlinearities that arise in real applications. As the systems become more intricate, their dynamics being accurately represented becomes increasingly important. The intricacy involved in modeling can sometimes imply the use of complex mathematical frameworks and computational aids. For instance, for a complex system like that of a city's traffic management system, the interaction among various components like traffic lights, car sensors, and real-time computation introduces numerous variables that have to be modeled in tandem. Such systems demonstrate the intricacy involved in defining proper state variables and creating accurate mathematical models because any component can influence others in an unpredictable manner. Furthermore, model errors or simplifications can lead to suboptimal operating choices, which stresses the need for subtle and adaptable standard approaches to modeling. System complexity and the large number of operating parameters complicate achievement of specified performance levels without wasting resources. Recent work in modeling techniques, including application of neutrosophic fuzzy sets, illustrates a move towards more in-depth handling of uncertainties typical of complicated systems. This new approach has proven highly effective in cases where indeterminacy and contradictory information exist, for example, the inventory control systems exemplified in Chandrasekaran (2025). By employing methods that facilitate not merely membership but indeterminacy and non-membership, such modeling methods provide flexible frameworks that are required to facilitate disentangling the complexities of actual implementation. Further, the merit of tapping into electronic health information, as presented in research on elderly groups and hospitalization risk, underscores the difficulty of working with a high number of variables in efforts to minimize adverse outcomes (Norstedt et al., 2024). As seen in these settings, the utilization of sophisticated modeling methodology gives system engineers and decision-makers the knowledge to optimize system performance, with the added benefit of more efficient and effective operations throughout various spheres (Figure 1.14).

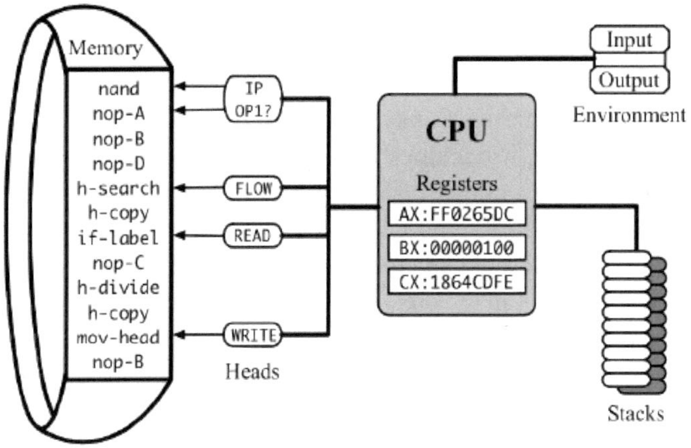


**Figure 1.14**. *Complexity of system modelling.*

*Source: https://en.algorithmica.org/hpc/complexity/img/cpu.png.*

## 1.6.2. Stability and Robustness Issues

The relationship between stability and robustness in control systems is often described through the dualistic paradigm of open-loop and closed-loop systems. Open-loop systems do not involve feedback, executing a pre-programmed set of instructions; yet they are susceptible to disturbance and uncertainty and therefore tend to be less stable in dynamic environments. For instance, a timer-activated irrigation system that does not consider unexpected weather fluctuations can overwater or under-water the plant. Closed-loop systems, on the other hand, possess feedback mechanisms that allow real-time adjustment and therefore stabilize the system. A clear example in engineering is the furnace temperature regulation system, which continuously monitors the actual temperature and adjusts the heating output to maintain the set point desired. Hence, the inherent design of closed-loop systems ensures improved robustness against external disturbances in most settings. Stability and robustness in control systems are a prime focus in engineering and technology these days. This need for robust control is underpinned by the emergence of advanced control methods, such as data-driven model predictive control (MPC). Theoretical justification in such instances is paramount to ensure closed-loop systems are stable and robust across various operating conditions, a precondition for use in reality (Berberich et al., 2024). These approaches utilize historic data to predict system response, hence enabling control actions which are not merely reactive but also insensitive to potential perturbations. For instance, in the automotive industry, adaptive cruise control systems implement such ideas to adjust automobile speed based on prevailing traffic patterns. By the use of predictive capability, these systems enhance users' reliance on automation without causing harm in unexpected driving scenarios. For systems with fractional-order dynamics, such as in complex systems, it becomes very challenging to keep them stable and robust. Fractional-order control systems are complex in their non-integer order dynamics, such that analysis and design processes for stabilization techniques are more complicated. Recent advances aim at developing necessary conditions for robust D-stabilization in the presence of complexity, focusing on a number of parameters governing system behavior Mohsenipour et al. (2022). Such studies therefore concentrate on determining constraints on uncertain parameters so that even with variation, such systems are capable of remaining stable. For example, in aerospace engineering, flight control systems utilize such methods to ensure that aircraft respond correctly to a vast number of factors and hence remain reliable and safe to operate. By this, continuous improvements in control system theory continue to enhance their stability and robustness in a myriad of applications.

### 1.6.3. Real-Time Processing and Control

Successful monitoring and control of dynamic systems require sophisticated methodologies that are capable of dealing with real-time information, thus ensuring improved decision-making processes. Such an operative process centers on open-loop and closed-loop systems, with each of them responsible for individual functions in the spectrum of control processes. Open-loop systems function without feedback, producing outputs based on inputs supplied. For example, an irrigation system that hydrates the garden for a specific period of time is a quintessential example of this approach. In contrast to closed-loop systems, which utilize feedback to adjust their outputs according to desired versus actual performance differences, an everyday example is a heating system with a thermostat regulating the room temperature by monitoring the difference between the current temperature and the set temperature at all times. This blend of control methods highlights the importance of real-time processing in terms of system accuracy and responsiveness, which is fundamental to guaranteeing stability within both engineering and everyday applications. Real-time processing is particularly important in closed-loop systems where the conditions could change swiftly, necessitating immediate correction actions. In engineering applications, this can be observed in autonomous vehicles depending on real-time data via various sensors to drive safely and navigate around obstacles. The ability to perform rapid computation of information on speed, distance, and direction provides authority to respond instantaneously to these vehicles, enabling safety as well as efficiency. Similarly, in automation of factories, robot systems use real-time monitoring to learn from alterations in the manufacturing process such that it is possible to manufacture without a hitch and automatically inspect quality. These applications underscore the significant role of real-time control mechanisms to fine-tune working effectiveness, thereby influencing productivity and safety in complex environments directly. The combination of rapid data processing and control techniques moves to highlight the role of real-time systems in addressing the demands of today's technology. Additionally, the impacts of real-time processing and control extend beyond technical realms into daily life, demonstrating its prevalence. Familiar devices such as home automation systems employ closed-loop control mechanisms to optimize user experience and efficiency in energy use. For instance, smart thermostats control heating and cooling based on occupant behavior and external environmental changes to provide comfort while conserving energy. Another instance is real-time traffic control systems, which show how processing can be applied to optimize urban mobility through dynamically controlling traffic signals to mitigate congestion and improve flow. The use of real-time processing is in many disciplines, justifying its importance in creating responsive and adaptive systems. Conclusions made from developments in hardware-in-the-loop technology and real-time simulation further indicate the promising future of control systems as they become better, showing their ability to maximize efficiency as well as safety on complex applications, such as recent studies on automatic vehicle testing and environmental monitoring (Chen, 2024).

## SUMMARY

The use of studying control systems is made relevant by their prevalence in various industries like engineering, automation, and everyday life. Among the topics in control systems, the distinction between open-loop and closed-loop systems is a first principle concept. The open-loop system is one in which there is no feedback process in it, thereby having a control action on the basis of previous input only, producing repeatable output but no provision for adjusting for alterations in the environment. For instance, a simple electric toaster is an open-loop system; it will toast bread for a preset time regardless of the actual toasting achieved. Closed-loop systems, however, are feedback to adjust operations in response to output, enhancing stability and performance. An excellent example is that of a thermostat-heated system, which constantly monitors temperature to adjust to maintain at some desired set-point, demonstrating a main feature of modern control theory. Much more in-depth research confirms the practical application of such systems goes beyond theoretical contexts, into industrial procedures and into daily life. Closed-loop systems show their effectiveness in many engineering applications, from automated assembly processes. A programmable logic controller (PLC) adjusts real-time based on feedback from multiple sensors so that assembly lines work without a hitch even under various stress conditions. At the same time, open-loop system operation continues to be of relevant use in simple operations where precise control is not so critical, such as watering plants using a timer. By knowing how each type of system operates, practitioners and engineers may meaningfully select and use the most efficient control techniques in given situations, ultimately improving system performance and reliability. Briefly, a good understanding of control systems, particularly the variations between open-loop and closed-loop designs, is key in engineering and beyond. As explained, open-loop systems offer convenience and predictability but can lack the flexibility necessary for complex operations, while closed-loop systems offer a dynamic system control method that can change according to current feedback. Their application in everyday devices, such as thermostats and factory control, testifies to their universal relevance to many fields of study. With advancements in technology, further integration of control systems is likely to increase efficiencies and function, underscoring their critical role in promoting innovation and living standards. The need to produce more sophisticated control systems and methods, as evidenced by much of the recent innovation, will continue to shape the future's engineering and technological advance.

In most industries, the structure of a control system is paramount to the effectiveness and reliability of processes. Open-loop systems, for instance, operate without utilizing feedback to control their outputs. An everyday example can be the wash machine using automatic means, where a predetermined cycle runs without changing its mode based on any fluctuation in the washing environment. Whereas basic, open-loop systems are less responsive to unexpected variables. In contrast, closed-loop systems include feedback systems which actually monitor outputs and adapt accordingly to inputs. This integration provides greater stability and performance under variability. A home thermostat's temperature-controlling system is one such closed-loop system that continuously keeps the temperature under observation and makes adjustments to maintain it at a set level, an incredible facility bestowed upon control systems to react dynamically to environmental factors. The significance of control systems is not only technical engineering-oriented, but they are omnipresent in day-to-day life through various applications employed to enhance efficiency and offer safety features. For example, modern cars have advanced closed-loop control systems that regulate engine performance to optimize fuel economy and emissions management according to driving conditions. Sensor-control algorithm synergy provides real-time revision to improve vehicle responsiveness as well as safety. Likewise, manufacturing facilities employ controlled processes with control systems to optimize production lines for minimizing waste and maximizing throughput. These systems are a wonderful demonstration of how effective control systems, apart from driving performance, also result in cost-effectiveness as well as environmental responsibility. The significance of learning these systems therefore cannot be highlighted enough, considering the expanding roles they have in dictating performance measures within different industries. The fundamentals of control systems, particularly the open-loop and closed-loop systems, are essential concepts that power developments in industries. Under the context of globalization and technology incorporation, they can trigger efficacies in operations required to maintain competitiveness. For instance, as it is noted by Labin et al. (2025), control systems are the foundation for the operation of international transport corridors facilitating logistics across borders. Such controls are in need of precise control in order to be capable of facilitating free movement of goods and services. Moreover, analytical frameworks utilized in control systems set the stage for advancements in automation and robotics and transform potential in manufacturing, transportation, and even the healthcare sector. Given our dependency on advanced control systems, an intense grasp of how they work and the importance they hold can create wider ramifications for process maximization throughout modern technology and economic fields.

# REFERENCES

1. Al-khafaji, I. M. (2024). Abdalameer. Development of autonomous mobile robots control systems. Vestnik Komp'iuternykh i Informatsionnykh Tekhnologii. https://www.semanticscholar.org/paper/1efcc7cd3d3564761cea710c7bf3a7e9aa8a32b9 (accessed on 19th September 2025).
2. Albino, V., Berardi, U., & Dangelico, R. M. (2015). Smart cities: Definitions, dimensions, performance, and initiatives. Journal of Urban Technology, 22, 3–21. https://doi.org/10.1080/10630732.2014.942092 (accessed on 19th September 2025).
3. Alimhillaj, P., Lamani, L., & Dalla Vedova, M. D. (2024). Simulation of an aerospace electrohydraulic servomechanism, with different Coulomb friction models. Journal of Physics: Conference Series, 2716. https://www.semanticscholar.org/paper/b7d68481bf79f2d61be3eb17126c32d5e69ba58a (accessed on 19th September 2025).
4. Andoni, M., Robu, V., Flynn, D., Abram, S., Geach, D., Jenkins, D., McCallum, P., et al. (2018). Blockchain technology in the energy sector: A systematic review of challenges and opportunities. Renewable and Sustainable Energy Reviews, 100, 143–174. https://doi.org/10.1016/j.rser.2018.10.014 (accessed on 19th September 2025).
5. Artemyev, V. S., & Mokrova, N. V. (2025). Development of a method for assessing indicators of discrete control systems. 2025 Systems of Signals Generating and Processing in the Field of on-Board Communications. https://www.semanticscholar.org/paper/1eadd02f9c566bf71ece10ac5205fe6869362537 (accessed on 19th September 2025).
6. Babushkin, A. (2024). The place and importance of operational investigative control of internal affairs bodies over persons prone to commit crimes in the state crime prevention system. *Vestnik Advanced Training Institute of the MIA of Russia*. https://www.semanticscholar.org/paper/ba85b0cf46e74980b55e548a0448d8dbae59720b (accessed on 19th September 2025).
7. Bangal, S. (2024). Image morpher. International Journal of Scientific Research in Engineering and Management. https://www.semanticscholar.org/paper/6ebfb1b45849bedc522c0821a5409a962f8daff5 (accessed on 19th September 2025).
8. Berberich, J., & Allgöwer, F. (2024). An overview of systems-theoretic guarantees in data-driven model predictive control. Annual Review of Control, Robotics, and Autonomous Systems, 8, 77–100. https://www.semanticscholar.org/paper/0cdd1a10dee3630b0a0bcf3b13170dea7e1116bd (accessed on 19th September 2025).

9. Berg, G., Rybakova, D., Fischer, D., Cernava, T., Champomier Vergès, M. C., Charles, T. C., Chen, X., et al. (2020). Microbiome definition re-visited: Old concepts and new challenges. *Microbiome, 8*. https://doi.org/10.1186/s40168-020-00875-0 (accessed on 19th September 2025).
10. Bharadwaj, A., El Sawy, O. A., Pavlou, P. A., & Venkatraman, N. (2013). Digital business strategy: Toward a next generation of insights. *MIS Quarterly, 37*, 471–482. https://doi.org/10.25300/misq/2013/37:2.3 (accessed on 19th September 2025).
11. Carroll, E., Tan, C., Hayes, S., Mordang, S., Rizzo, G., Zaia, V., Montagna, E., et al. (2024). Implementing high-value, cost-conscious care: Experiences of Irish doctors and the role of education in facilitating this approach. *BMC Medical Education, 24*. https://www.semanticscholar.org/paper/e1c32a3e806726683fec14686626d3a19542764f (accessed on 19th September 2025).
12. Carvalho, D. V., Pereira, E. M., & Cardoso, J. S. (2019). Machine learning interpretability: A survey on methods and metrics. *Electronics, 8*, 832–832. https://doi.org/10.3390/electronics8080832 (accessed on 19th September 2025).
13. Chalasani, N., Younossi, Z. M., Lavine, J. E., Charlton, M., Cusi, K., Rinella, M. E., Harrison, S. A., et al. (2017). The diagnosis and management of nonalcoholic fatty liver disease: Practice guidance from the American Association for the Study of Liver Diseases. Hepatology, 67, 328–357. https://doi.org/10.1002/hep.29367 (accessed on 19th September 2025).
14. Chandrasekaran, E. (2025). Revealing advanced inventory management approaches through Python-based visualization and modelling of neutrosophic and fuzzy mathematical models for enhanced decision-making and optimization. *Advances in Nonlinear Variational Inequalities*. https://www.semanticscholar.org/paper/31de7736d21c4b15a9bcb710e8150187b51afa8c (accessed on 19th September 2025).
15. Charles, H., Beddington, J. R., Crute, I. R., Haddad, L., Lawrence, D. P., Muir, J., Pretty, J., et al. (2010). Food security: The challenge of feeding 9 billion people. Science, 327, 812–818. https://doi.org/10.1126/science.1185383 (accessed on 19th September 2025).
16. Chen, G. (2025). The design and optimization technology of propulsion systems for operative underwater robots. Applied and Computational Engineering. https://www.semanticscholar.org/paper/ab2f8ac4d6f695c4d6acf3d833a3f38b79fea301 (accessed on 19th September 2025).
17. Chen, J. (2024). A survey of real-time simulation and hardware-in-the-

loop technology. Transactions on Computer Science and Intelligent Systems Research. https://www.semanticscholar.org/paper/1b14b3a923c3d357d79d620ea2bc9075ecd6c073 (accessed on 19th September 2025).

18. Daniel, M. P., Keerthana, V. S., & Vijay Kumar, N. B. S. (2024). Efficiency and elegance: Exploring automated solutions for public lighting. *International Journal for Modern Trends in Science and Technology*. https://www.semanticscholar.org/paper/df3aa95fc88e1e0fb00bd4f0afc298e492a016bb (accessed on 19th September 2025).
19. Deokar, A. (2025). Wireless level measurement and control systems using PLC. *International Journal for Research in Applied Science and Engineering Technology*. https://www.semanticscholar.org/paper/bc869dbb0cd7edf21a79e24dfad8becde4095a0a (accessed on 19th September 2025).
20. Di Renzo, M., Zappone, A., Debbah, M., Alouini, M. S., Yuen, C., de Rosny, J., & Tretyakov, S. (2020). Smart radio environments empowered by reconfigurable intelligent surfaces: How it works, state of research, and the road ahead. IEEE Journal on Selected Areas in Communications, 38, 2450–2525. https://doi.org/10.1109/jsac.2020.3007211 (accessed on 19th September 2025).
21. Dukare, G. D. (2025). Problem solving and DOK improvement in engine shop "A"–Skoda Auto Volkswagen India Pvt Ltd. *International Journal of Scientific Research in Engineering and Management*. https://www.semanticscholar.org/paper/e4d20957b403975a2a5d0c2c0f1dfffed281d163 (accessed on 19th September 2025).
22. Dwivedi, Y. K., Hughes, L., Baabdullah, A. M., Ribeiro-Navarrete, S., Giannakis, M., Al-Debei, M. M., Dennehy, D., et al. (2022). Metaverse beyond the hype: Multidisciplinary perspectives on emerging challenges, opportunities, and agenda for research, practice and policy. International Journal of Information Management, 66, 102542–102542. https://doi.org/10.1016/j.ijinfomgt.2022.102542 (accessed on 19th September 2025).
23. Fan, C. (2024). Key technologies for intelligent mining of difficult-to-mine coal seams. *Journal of Civil Engineering and Urban Planning*. https://www.semanticscholar.org/paper/da3f4e5832103f592608d255e471e98703db0356 (accessed on 19th September 2025).
24. Fogli, D., & Tetteroo, D. (2022). End-user development for democratizing artificial intelligence. *Behaviour & Information Technology, 41*, 1809–1810. https://www.semanticscholar.org/paper/15b00dc33c2495bc00f2a8055f5236de3bf8af53 (accessed on 19th September 2025).

25. Govindan, K., Soleimani, H., & Kannan, D. (2014). Reverse logistics and closed-loop supply chain: A comprehensive review to explore the future. European Journal of Operational Research, 240, 603–626. https://doi.org/10.1016/j.ejor.2014.07.012 (accessed on 19th September 2025).

26. Guidotti, R., Monreale, A., Ruggieri, S., Turini, F., Giannotti, F., & Pedreschi, D. (2018). A survey of methods for explaining black box models. ACM Computing Surveys, 51, 1–42. https://doi.org/10.1145/3236009 (accessed on 19th September 2025).

27. Ingale, K., Kadam, S., Kadam, V., Kale, A., & Korde, V. (2025). A smart mining safety system: Real-time environmental sensing and rapid emergency response. Journal of Information Systems Engineering and Management. https://www.semanticscholar.org/paper/566807fd07edce65e216d72337c8f21d8a592273 (accessed on 19th September 2025).

28. Iriakuma, E. K., Ukela, O. N., & Uche, A. P. (2024). Forecasting energy consumption with AI: A review for sustainable energy management. *International Journal of Latest Technology in Engineering, Management & Applied Science*. https://www.semanticscholar.org/paper/2dab4b104f8a1f9e91018009225ca9fca75b763a (accessed on 19th September 2025).

29. Jones, D., Snider, C., Nassehi, A., Yon, J., & Hicks, B. (2020). Characterizing the digital twin: A systematic literature review. *CIRP Journal of Manufacturing Science and Technology, 29*, 36–52. https://doi.org/10.1016/j.cirpj.2020.02.002 (accessed on 19th September 2025).

30. Khan, M. Y. A. (2025). Enhancing electric vehicle performance: A case study on advanced motor drive systems, integration, efficiency, and thermal management. Control Systems and Optimization Letters. https://www.semanticscholar.org/paper/964c2805c5e5d6f4bd8f86a7693a26779cfaa2c2 (accessed on 19th September 2025).

31. Labin, D. K., & Rzayev, R. G. (2025). New legal opportunities for the implementation of the "North–South" international transport corridor project in the context of a new global multipolarity. *Moscow Journal of International Law*. https://www.semanticscholar.org/paper/a689e4135ba498384b6de58427f66bd2542355d9 (accessed on 19th September 2025).

32. Laborde, Q., Roques, A., Robert, M. P., Armougum, A., Vayatis, N., Bargiotas, I., Oudre, L., et al. (2025). Vision toolkit part 1. Neurophysiological foundations and experimental paradigms in eye-tracking research: A review. Frontiers in Physiology, 16. https://www.semanticscholar.org/paper/865633c596871da9ccf133d0519a05f50c49467f (accessed on 19th September 2025).

33. Latif, A., Hussain, S. M. S., Das, D. C., Ustun, T. S., & Iqbal, A. (2021). A review on fractional order (FO) controllers' optimization for load frequency stabilization in power networks. *Energy Reports, 7*, 4009–4021. https://doi.org/10.1016/j.egyr.2021.06.088 (accessed on 19th September 2025).

34. Lawanda, I. (2019). The importance of information access of cultural values to the principles of sustainable development in climate change. 69, 9–19. https://www.semanticscholar.org/paper/fdc6f2ce775fc68ec2a3a5200e89c944078eb5b8 (accessed on 19th September 2025).

35. Leslie, H. A., van Velzen, M. J. M., Brandsma, S. H., Vethaak, D., García-Vallejo, J. J., & Lamoree, M. H. (2022). Discovery and quantification of plastic particle pollution in human blood. Environment International, 163, 107199–107199. https://doi.org/10.1016/j.envint.2022.107199 (accessed on 19th September 2025).

36. Liu, P., Yi, H., Chen, X., Zhu, K., Li, Z., Sun, Z., & Jiao, L. (2022). Hierarchical engineering for high-energy-oriented sodium-ion batteries. Accounts of Materials Research, 3, 672–684. https://doi.org/10.1021/accountsmr.2c00013 (accessed on 19th September 2025).

37. Mammadov, J., Nikolov, N., Mammadova, S., Safarova, T., & Ganjaliyeva, G. (2025). Research and application of information-energy indicators of analog type electromagnetic transmitters in the different industry fields. EUREKA: Physics and Engineering. https://www.semanticscholar.org/paper/90c4e57352b2813ea9841db51a1f30ce303bd6de (accessed on 19th September 2025).

38. Mitchell, J. K., Soga, K., & O'Sullivan, C. (2025). Fundamentals of Soil Behavior. https://doi.org/10.1002/9781119832348 (accessed on 19th September 2025).

39. Mohsenipour, R., & Jegarkandi, M. F. (2022). Robust D-stabilization analysis of fractional-order control systems with complex and linearly dependent coefficients. IEEE Transactions on Systems, Man, and Cybernetics: Systems, 52, 1823–1837. https://www.semanticscholar.org/paper/ee3aed638b9d6946d0aa90e7b92d915788cdacb7 (accessed on 19th September 2025).

40. Norstedt, I., Thorell, K., & Halling, A. (2024). Importance of kidney function, number of chronic conditions and medications for hospitalization in elderly in Blekinge County, Sweden: A case–control study. BMJ Open, 14. https://www.semanticscholar.org/paper/d693edffbf096b9b11a18771fbc535d8cc588b5d (accessed on 19th September 2025).

41. Nostrani, M. P., Raduenz, H., Dell'Amico, A., Krus, A. P., & Negri, V. (2023). Multi-chamber actuator using digital pump for position and velocity control applied in aircraft. International Journal of Fluid Power. https://www.semanticscholar.org/paper/33e91c3921364625d9a91401b05738ece45715c2 (accessed on 19th September 2025).
42. Panteleev, A., & Yakovleva, A. (2025). Synthesis H∞ of an observer of the state of a nonlinear continuous dynamical system, linear in control and disturbance. *Modelling and Data Analysis*. https://www.semanticscholar.org/paper/942af09b2d34c4724167dbc2ee597c2c6aaa6a0b (accessed on 19th September 2025).
43. Patil, D. S., & Patel, G. R. (2025). Energy management strategies for microgrids: A review. *International Journal of Scientific Research in Engineering and Management*. https://www.semanticscholar.org/paper/c01ecbdc8b3e12d4c2d40095af6401b62ff72150 (accessed on 19th September 2025).
44. Rastinehad, A. R., Anastos, H., Wajswol, E., Winoker, J. S., Sfakianos, J. P., Doppalapudi, S. K., Carrick, M. R., et al. (2019). Gold nanoshell-localized photothermal ablation of prostate tumors in a clinical pilot device study. *Proceedings of the National Academy of Sciences, 116*, 18590–18596. https://doi.org/10.1073/pnas.1906929116 (accessed on 19th September 2025).
45. Rogers, J. A., Chen, X., & Feng, X. (2020). Flexible hybrid electronics. Advanced Materials, 32. https://doi.org/10.1002/adma.201905590 (accessed on 19th September 2025).
46. Saberi, S., Kouhizadeh, M., Sarkis, J., & Shen, L. (2018). Blockchain technology and its relationships to sustainable supply chain management. International Journal of Production Research, 57, 2117–2135. https://doi.org/10.1080/00207543.2018.1533261 (accessed on 19th September 2025).
47. Su, H., Mariani, A., Ovur, S. E., Menciassi, A., Ferrigno, G., & De Momi, E. (2021). Toward teaching by demonstration for robot-assisted minimally invasive surgery. IEEE Transactions on Automation Science and Engineering, 18, 484–494. https://doi.org/10.1109/tase.2020.3045655 (accessed on 19th September 2025).
48. Tao, F., Zhang, H., Liu, A., & Nee, A. Y. C. (2019). Digital twin in industry: State-of-the-art. *IEEE Transactions on Industrial Informatics, 15*, 2405–2415. https://doi.org/10.1109/tii.2018.2873186 (accessed on 19th September 2025).
49. Touitou, Y., Cermakian, N., & Touitou, C. (2024). The environment and

the internal clocks: The study of their relationships from prehistoric to modern times. Chronobiology International, 41, 859–887. https://www.semanticscholar.org/paper/8668046e3c0f086cc269a3045efca02309de4c26 (accessed on 19th September 2025).

50. Wilkins, C. H. (2018). Effective engagement requires trust and being trustworthy. *Medical Care, 56*, S6–S8. https://doi.org/10.1097/mlr.0000000000000953 (accessed on 19th September 2025).
51. Yang, Z., Wang, Y., Xing, L., Yin, B., & Tao, J. (2019). Relay protection simulation and testing of online setting value modification based on RTDS. IEEE Access, 8, 4693–4699. https://doi.org/10.1109/access.2019.2963228 (accessed on 19th September 2025).
52. Zawacki-Richter, O., Marín, V. I., Bond, M., & Gouverneur, F. (2019). Systematic review of research on artificial intelligence applications in higher education–Where are the educators? International Journal of Educational Technology in Higher Education, 16. https://doi.org/10.1186/s41239-019-0171-0 (accessed on 19th September 2025).
53. Zhou, Y., Chang, J., Chen, W., & Dai, H. (2024). Interval compression-based model-free control algorithm for reducing actuator execution frequency. IET Control Theory & Applications. https://www.semanticscholar.org/paper/1e547a9ef89d2d6a6e89bffb70488e22dd53eb37 (accessed on 19th September 2025).
54. Zhu, J. Y., Park, T., Isola, P., & Efros, A. A. (2017). Unpaired image-to-image translation using cycle-consistent adversarial networks. https://doi.org/10.1109/iccv.2017.244 (accessed on 19th September 2025).

# CHAPTER 2

# DYNAMIC SYSTEMS

## Contents



## INTRODUCTION

Dynamic systems, which encompass a broad range of phenomena in many fields such as engineering, biology, and social sciences, are intricate interactions that change with time. They are characterized by their dynamic nature, in which the state of the system is a function of several variables and how these variables are related. Examining these interactions requires a holistic analytical approach combining mathematical modeling, computational simulation, and experimental verification. The intricate nature of dynamic systems allows for an examination of linear and nonlinear behaviors, culminating in the development of improvements in a number of applications. For instance, integrating feedback principles in inverse simulation has the potential to make models significantly more resilient, as they are used in engineering. Through the analysis of these dynamic interactions, researchers are able to derive conclusions that lead to improved designs and solutions in actual contexts, making the study of dynamic systems fundamental in contemporary research. Dynamic system modeling extends from its conventional engineering domains to embrace biological phenomena such as cell interactomics. Recent advances in the field suggest that quantum automata may be utilized to model biomolecular interactions along complex signaling networks, illustrating a novel framework describing cell behavior. Through the development of cellular dynamics in modular frameworks, researchers can identify discrepancies in cellular processes associated with normal and diseased states, such as cancer. The differentiation between healthy somatic cells and genomically unstable cancer cells necessitates new approaches in modeling methods, particularly when dealing with the dynamic nature of cancer interactomics.

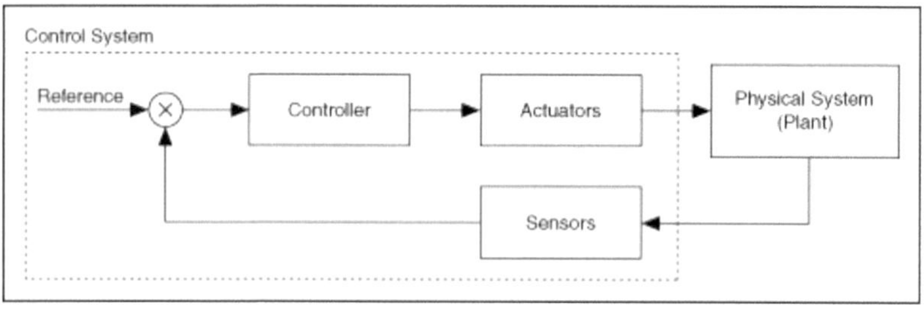


**Figure 2.1**. *Dynamic system models.*

*Source: https://docs-be.ni.com/bundle/labview-control-design-and-simulation-module/page/lvsimconcepts/loc_eps_closed_loop.gif?_LANG=enus.*

This shift in the paradigm of understanding dynamic systems at the molecular level not only sheds light on basic biological processes but also has consequential meaning for medical science and cancer therapy, exposing the central interplay between dynamic systems and health sciences. In total, dynamic systems research is a multi-disciplinary endeavor of immense significance across a wide variety of disciplines. From the engineering principles that govern the design of stable feedback loops in simulations, outlined in the use of inverse simulation (Anonymous et al., 2011), to the quantum interactomics research that investigate the complexity of cellular processes and what they hold for the explanation of human consciousness and the onset of cancer (Baianu et al., 2010), the synergy between these disciplines is evident. As researchers continue to develop more sophisticated analysis tools and methods, the role of dynamic systems in illuminating our understanding of complex phenomena will continue to increase. A thorough investigation of dynamic systems is therefore necessary for the promotion of innovation and the formulation of efficacious strategies for various scientific and engineering fields (Figure 2.1).

## *Definition of Dynamic Systems*

Dynamic systems theory comprises an extensive variety of interacting components that evolve throughout the passage of time with interactivity between themselves within the system and with external environments. In essence, a dynamic system is marked by its ability to change based on various inputs, exhibiting features such as feedback loops, state changes, and non-linearity. Though static properties have been assigned to conventional systems, dynamic systems are characterized by their ability to change and develop. Such flexibility allows them to be stable under changing conditions, a characteristic that lies at the core of resilience. As long-rumored in relevant literature, resilience occurs as a dynamic process in that it determines how systems present behavior towards environmental stimuli (Brandenburger et al., 2015). Therefore, the understanding of a dynamic system not only entails learning about its components but also the intricate interactions that lead to emergent behaviors and the capacity of the systems to handle complex situations. In systems theory language, the description of dynamic systems includes considerations of various theoretical models that help elucidate their working dynamics. An intrinsic understanding calls for awareness of meta-theoretical assumptions that guide explanation of interactions and behaviors (Ale et al., 2014). As illustrations, resilience and safety management models propose the graduated complexity that such dynamic systems might entail. By examining how different elements influence each other in terms of the progression of time, researchers can infer important information regarding such systems' operation, particularly in actual operations such as organizational safety or environmental control. Differences, for example, in applying systems theoretical presumptions when regulating safety point to the need for a comprehensive framework to tackle properly problems in dynamic environments (Ale et al., 2014). Ultimately, these results justify that a well-defined dynamic system would have to be in agreement with both their structure and with those processes regulating their development. Secondly, the definition of dynamic systems is best put in terms of their capacity for achieving resilience, a concept seminal to describing behavior when in the state of stress or change. Resilience is not an unchanging attribute, but an emergent property that arises from the interactive dynamics between the systems configuration and its operating environment (Brandenburger et al., 2015). The dynamic interaction produces the emergence of such attributes as elasticity, change tolerance, and even antifragility—terms that capture the broad array of manners in which systems can withstand or change disruptions. Therefore, dynamic systems suggest an intricate game of balance, trying to reconcile internal processes against external pressures. Through the exploration of such characteristics, greater insight into the fundamental nature of dynamic systems is gained, which makes them central to assisting us in comprehending a broad spectrum of disciplines from engineering to ecology and social sciences.

## *Importance in Various Fields*

The dynamic systems theory has its significance extend to numerous fields, offering groundbreaking theories and practical concepts. In environmental science, to take one example, what the dynamics of ecosystems as complex systems can tell us is the complexity of interactions between various species and their habitats. This is significant, particularly with regard to the destruction of habitats and global warming, since this information generates more accurate models for ecological forecasting. With the use of principles of dynamic systems, researchers are likely to develop interventions that promote sustainability and preservation of biodiversity. The paradigm shift in considering ecosystems from the viewpoint of uncurtailed dynamic complexity emphasizes more intricate approaches in managing ecological systems (Kirilyuk et al., 2009). In technology fields, the relevance of dynamic systems is evident when considering the evolution of information technology as well as communications systems. The complexities that abound in globalized technological infrastructures necessitate a paradigm shift in how data management and system effectiveness should be approached. Application of complexity-oriented IT systems, as noted in ongoing studies, can lead to monumental strides in information processing and utilization. In intelligent communication networks, for instance, the capability to discern dynamic interactions is what makes it possible to easily create context-based systems that are adaptive to user needs in real-time. Such innovation not only improves end-user experience but also the reliability of operation of key services towards a more integrated and more efficient technological framework (Kirilyuk et al., 2009). The system dynamics also find their interpretation heavily within the field of natural sciences, particularly in the modeling of fluid dynamics and phase transitions. Coarsening of droplets in salad dressing gives a paradigmatic case study of inherent dynamic processes shared by a variety of natural phenomena. Recent studies have discredited long-existing conventional wisdom by demonstrating that droplet mobility is radically governed by hydrodynamic forces rather than strictly stochastic behavior. This redirection of dynamic interaction among immiscible liquid mixtures contributes to insight into a myriad of applications from industry to bio-systems. Comprehending these complex dynamics can lead to more effective designs and systems that inherently respond to fluid behavior, ultimately leading to innovation in numerous fields.

## 2.1. MATHEMATICAL FOUNDATIONS

Mathematical underpinnings form the foundation of dynamic systems, providing tools and structure for describing complex interactions across various disciplines. Matrices, calculus, and differential equations are a few of the most important principles that allow practitioners to describe system change over time. As an example, matrices are crucial in state-space representations, which give a structured mechanism for describing linear system dynamics. The dynamic interaction of variables generally has nonlinear characteristics, and consequently, an in-depth understanding of these mathematical structures is necessary to adequately model behaviors and make forecasts. This foundation is not theoretical; its implications extend far into disciplines ranging from engineering through economics, where dynamic interactions define success or failure in processes. Lacking mathematically rigorous foundations, the study of dynamic systems would be inaccurate, preventing advancement in theoretical and applied fields. The mathematical basis of dynamic systems is also crossed by the study of computability theory. Having emerged as a response to challenges in intuitionism by Hilbert, computability theory examines the limits of computability and serves as a compass in the interpretation of dynamic processes. The Church-Turing thesis contributes to theory by offering a criterion for judging computable functions. When applied to dynamic systems, this theoretical contribution becomes pivotal when axiomatizing economic agents or analyzing algorithmic choice processes. The eclectic approach of fusing computable and constructive mathematics allows researchers to handle the complexities of dynamic systems more effectively. This integration of mathematical reasoning not only strengthens the study of dynamic processes but also illuminates paths for using theories for real applications, particularly in future digitalized economic settings as highlighted in corresponding literatures. Furthermore, the theory of Abstract State Machines (ASMs) is a fundamental element in the mathematical model of dynamic systems because it provides a precise method for describing changes of states between different circumstances. ASMs support both discrete and continuous state transitions to be represented, enabling them to serve as general-purpose tools for the investigation of dynamic behavior. This flexibility is particularly useful in applications where the study of system performance and stability over time is involved. Understanding the state transitions offered by ASMs needs a sound background in mathematical logic and the mathematics that governs computational systems. Moreover, the comparison of the annotated bibliography of research that utilizes ASMs indicates their extensive applications, and thus corroborating the importance of mathematical basics for dynamic systems. The investigation of these building blocks is thus crucial to make theory and applied breakthroughs in the analysis of dynamic phenomena in various scholarly and professional areas, such as those highlighted in Boerger et al. (1998) and Velupillai et al. (2004) (Figure 2.2).

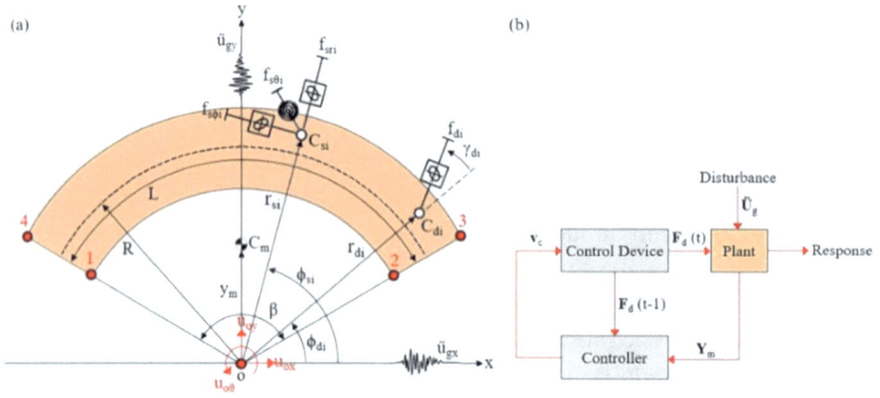


**Figure 2.2**. *Mathematical modeling of the control system.*

*Source: https://www.researchgate.net/profile/Mohsen-Amjadian/publication/326058796/figure/fig1/AS:642746490568705@1530254113581/Mathematical-modeling-of-the-control-system-a-3DOF-dynamic-model-of-a-horizontally.png.*

## 2.1.1. Differential Equations

Differential equations serve as essential tools for understanding dynamic systems across various scientific fields, from ecology to engineering. They model the relationships between variables that change over time, capturing the intricacies of these systems. For instance, in ecological models, differential equations can describe the interactions between species, such as predator-prey dynamics. A two-species model, where one species feeds on another, illustrates the application of constant coefficient differential equations to represent population dynamics. The model can help researchers predict changes in species populations over time, demonstrating how differential equations can yield insights not just on individual species, but on the health and stability of entire ecosystems (Brennan et al., 1970). The versatility of these equations enables researchers to analyze complex behaviors, offering a structured approach to understand how systems evolve and respond to external influences. The integration of differential equations into computational frameworks amplifies their utility in dynamic modeling. Advances in software, such as the R package deSolve, provide sophisticated methods for solving ordinary differential equations (ODE), as well as differential algebraic equations (DAE) and partial differential equations (PDE) (Soetaert et al., 2025). This framework allows users to initiate and simulate dynamic systems with increasing complexity, enhancing both the speed and efficiency of computations.

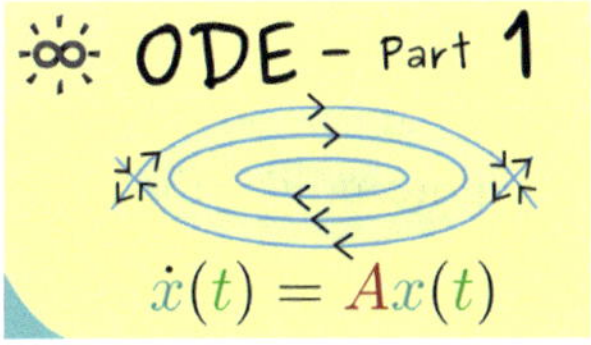


**Figure 2.3**. *Ordinary differential equations (ODE).*

*Source: https://thebrightsideofmathematics.com/images/thumbs/ode01.png.*

For instance, deSolve employs powerful algorithms and integrators to handle various types of differential equations, offering flexibility to accommodate stiff and non-stiff systems. By translating mathematical models into code, researchers can seamlessly leverage these solutions to explore a wide array of dynamic phenomena, fostering innovation in fields such as biostatistics, environmental science, and engineering. Moreover, the application of differential equations extends beyond theoretical modeling, facilitating practical decision-making in dynamic systems. For example, in ecological management, models that incorporate differential equations can inform strategies for wildlife conservation and resource management. The interplay of population dynamics, such as gender and age composition in wildlife populations, can be modeled to optimize outcomes for species conservation, thereby directly influencing management practices (Brennan et al., 1970). With advanced computational tools, researchers can simulate various scenarios and predict the consequences of different management strategies, underpinning the need for sound scientific approaches in environmental decision-making. Consequently, the mastery of differential equations not only enhances theoretical understanding but also translates into actionable insights, aligning scientific research with real-world applications in managing dynamic ecosystems (Figure 2.3).

## 2.1.2. Linear Versus Nonlinear Systems

The distinction between nonlinear and linear systems is one of the core concepts in dynamic systems analysis, particularly how they respond to inputs and their own stability characteristics. Linear systems under the law of proportionality are predictable and stable, and they accommodate the law of superposition of effects. For instance, if inputs are provided to a linear system in multiples, the aggregate effect is simply the aggregate of the separate effects. This is simpler to analyze and regulate, and linear systems are hence amenable to methods such as the Kalman Filter. Nonlinear systems have richer dynamics that cannot be superposed linearly, and generically give rise to phenomena such as bifurcations and chaos. These complex responses require advanced mathematical structures for investigation and prediction, and make the analysis of linear and nonlinear dynamics important in many applications. Nonlinear systems are often processed with special methods that are quite unlike those used when processing linear counterparts.

One very useful one is global and local coordinate systems to avoid constraints in nonlinear behavior. For instance, methods like the Implicit Extended Kalman Filter (IEKF) allow recursive identification of nonlinear models in a way that may be able to handle complexity within nonlinear measurements. This is specifically dominant in scientific studies on motion estimation of rigid bodies under various conditions that can be framed as identifying dynamic models over a topological manifold. These studies exhibit the need for specific analytical tools in order to successfully address the problems that are brought about by nonlinear systems, where generic linear analysis tools may not provide significant information. Therefore, it is essential to comprehend these distinctive methods in order to successfully model and control systems.

Nonlinear dynamic systems also pose interesting challenges in real-world engineering problems, for example, in the dynamics of structures such as suspended cables. In these instances, nonlinearities are caused by geometric arrangements and external loads, resulting in complex vibrational properties that are heavily influenced by various resonances. The experiment with a second-order multiple scales approach is a case in point, showing the complex response of the system to forced dynamics. Analytical predictions of performance and stability are supported by numerical simulations, confirming theoretical solutions and refining our knowledge of the dynamic behaviors of nonlinear systems. By comparing analytical and numerical methods, researchers can build stronger models that account for the coexistence of rival dynamics, thereby extending practical uses in nonlinear dynamics in fields such as robotics and structural engineering (Frezza et al., 1994; Rega et al., 2008) (Figure 2.4).

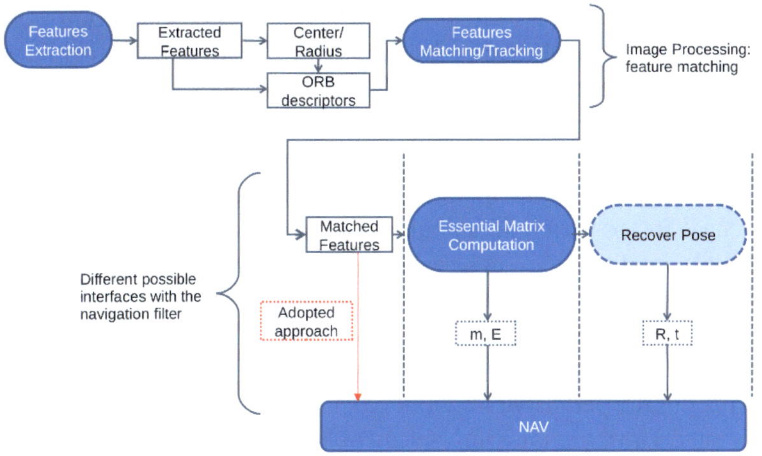


**Figure 2.4**. *Implicit extended kalman filter (IEKF).*

*Source: https://www.mdpi.com/aerospace/aerospace-09-00503/article_deploy/html/images/aerospace-09-00503-g001.png.*

## 2.1.3. Stability Analysis

Stability analysis is an underlying component of studying dynamic systems and is essential in terms of gaining crucial understanding about how a system will react to disturbances as time passes. System stability can be assessed to see if a system will be resistant to outside disturbances and internal variances. The study of such properties as input-to-state stability (ISS) is of particular significance here, as it offers the potential to characterize systems in terms of how well they can reject disturbances. The work of late has utilized the application of dynamic programming techniques to derive computable conditions that are essential for ISS evaluation, hence filling gaps in existing theory. This development in stability analysis not only enhances the understanding of nonlinear dynamics but also involves optimization theory principles, offering a single framework of tools for system analysis (Dower et al., 2005). Along with theoretical advances, the application of numerical simulations has been crucial in experimental studies of stability in dynamic systems. Techniques such as bifurcation analysis and Lyapunov exponent calculations provide visualization tools that explain system behavior for various parameter values. Commercial simulation packages such as E&F Chaos have rendered these advanced simulation tools universally available, allowing non-experts to become members of the complexity in nonlinear systems. Researchers can analyze the stability of fixed points and ascertain conditions for periodic or chaotic behavior using these simulations. This practical aspect of stability analysis serves to validate theoretical prediction and aid in examining the effects of model changes, thereby deepening discussion of dynamic systems with empirical evidence (Diks et al., 2025). Ultimately, the crossroads between theoretical stability analysis and practical simulation techniques underscores the mixed nature of dynamic systems. The ability to marry rigorous analytical methods with computing technology allows for a deeper probing of stability phenomena that cross disciplinary boundaries such as control theory, economics, and environmental science. As researchers continue to develop stability criteria and innovate unprecedented simulation techniques, the impacts of these studies extend far from the academic community. Applications range from improving the reliability of engineering systems to improved economic prediction models, all in respect of the pervasive importance of understanding stability in dynamic environments. Thus, ongoing improvement in theory and practice will continue to further understand the complexities involved in dynamic systems and inspire innovations solving real-world problems.

## 2.2. APPLICATIONS IN ENGINEERING

The versatility of dynamic systems finds its broad application in engineering, indiscriminately boosting efficiency as well as efficacy. Possibly the most striking context is flexible distributed applications' designing, where static systems just don't cut it. Engineers working in this area are interested in creating structures that do not just respond to operational needs but also evolve over time along with them. This versatility is necessary as it assists systems in adapting to fluctuating workloads and user demands without needing overhaul. The blending of adaptive and adaptable middleware policies assists designers in creating solutions that are not only resilient but also intelligent in their response to live data, ultimately providing a more seamless operation experience in engineering applications (Bigwood et al., 2010). Furthermore, dynamic systems interaction with engineering enhances innovation within the creation of complicated applications in various fields, including aerospace, civil engineering, and information technology. Dynamic modeling and control techniques are applied by engineers to create systems that can foresee and adapt to change, thereby optimizing performance. For example, in the field of aerospace engineering, dynamic systems are employed to enhance flight control devices so that real-time adjustments are made in line with atmospheric conditions. These designed systems generally involve the latest sensing technologies and data analytics to create a responsive system that may be molded to react to unexpected challenges. These new uses demonstrate how dynamic systems are key to revolutionary advancements in engineering, as they represent their pivotal role within the field (N/A, 2017). In addition to flexibility and innovation, the use of dynamic systems in engineering emphasizes thoroughly the policy environments that govern the interaction between systems. This enables uniform communication and functioning among varied components of an engineering project. A new language-independent adaptive policy framework is developed, which is a milestone that allows engineers to successfully integrate various middleware systems. The framework provides dynamic policy specification and update, which enables developers to tailor applications according to aimed execution environments. This type of flexibility is particularly important in distributed systems, where the ability to adapt policy according to changing situations can be a major factor in system reliability and performance. Ultimately, conscientious consideration of dynamic systems in engineering projects not only leads to more durable designs but also lays the foundation for future innovations (Bigwood et al., 2010).

### 2.2.1. Control Systems

Control systems are the dynamic systems' essential component responsible for regulating and controlling processes to perform optimally and be stable. Control systems are a system of interconnected components that collectively influence dynamic systems' behavior. Feedback lies at the center of control systems, which is a mechanism whereby inputs can be controlled in relation to output performance. This dynamic action can lead to greater accuracy and productivity in a variety of applications, such as industrial automation and robotics. Increasingly complex mathematical frameworks may be necessary for analysis of these control systems to enable engineers to model and predict system behavior for a variety of conditions. Using techniques such as stability analysis and robust control synthesis, practitioners can design systems that not only respond to disturbances but also optimize performance under various operating conditions. The rich ground of control systems is also to be seen in advanced applications in iterative learning control, where process dynamics are refined with many interactions. Treating complex processes, such as those in industrial applications, requires a structured understanding of differential repetitive processes, which reflect how output from one process is fed into the next. These systems require an equal weighting of constraints and performance requirements, particularly where both linear and nonlinear behavior is prevalent. Researchers increasingly stress unveiling new results in optimal control for these repetitive processes to further assist in applications in robotic manipulators. These innovations are based on extrapolating proven principles that capture the dynamics of these systems, pushing the boundaries of conventional control techniques in addressing real-world issues (Dymkov et al., 2008). Control systems also provide the foundation for enhancing the reliability and performance of aerospace systems, which are significantly dynamic in nature. The union of aeroelasticity, vibration control, and enhanced computational methodologies is essential in achieving effective control of the numerous dependent dynamics involved in aero propulsion systems. Active and passive control methods of vibration, which involve employing new devices such as electromagnetic dampers and piezoelectric crystals, are some of the technological advancements to enhance system responses. Also, the dynamic systems programs focus on research and verification of the interaction between complex systems reflects the ongoing need for systematic approaches to address the challenges posed by high-temperature and nonlinear processes. With ongoing research in this field, the interplays between structural dynamics and control design reflect the need to create resilient control strategies for future aerospace missions (Brown et al., 2025) (Figure 2.5).

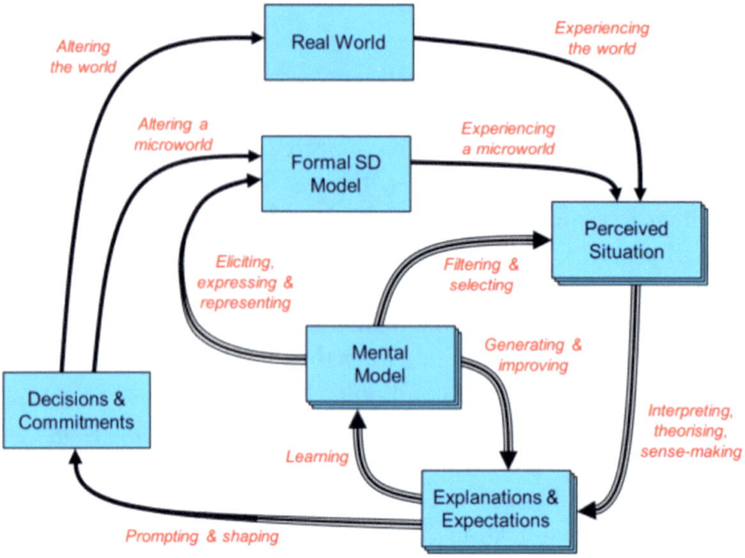


**Figure 2.5**. *Behavioural system dynamics.*

*Source: https://ars.els-cdn.com/content/image/1-s2.0-S0377221722006609-gr3.jpg.*

## 2.2.2. Structural Dynamics

The field of structural dynamics plays a prime role in dynamic systems, particularly in terms of knowing how structures behave when subjected to various forces and interactions. It includes the study of the natural frequency and mode of structures, without which they can never be assured stability and performance when subjected to dynamic loading. In aerospace use, the field is crucial since flutter and forced vibrations can lead to catastrophic failure. As pointed out in the research study conducted at LeRC, analytical and experimental methods are designed to alleviate such issues, employing aeroelasticity and vibration control principles to enhance the durability and strength of aerospace propulsion systems. The application of advanced computational methods also makes it easier to solve complex structural problems, allowing accurate parametric identification and analysis of nonlinear, damped systems. Such ventures put a premium on the necessity of proper structural dynamics in dynamic system designs that are safe and efficient (Brown et al., 2025).

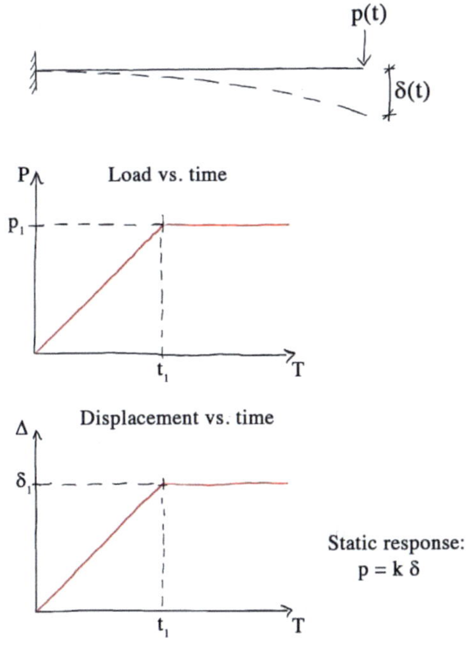


**Figure 2.6**. *Structural dynamics and inertia.*

*Source: https://www.engineeringskills.com/_next/image?url=%2Fimages%2Fposts%2Funderstanding-structural-dynamics-and-inertia%2FStructural-Dynamics-1.jpg&w=1080&q=75.*

Moreover, dynamic systems often introduce problems related to stability, especially when the applications involve human interfaces, that is, haptic devices. Stability of such systems can be disturbed by parameters like time delays and structural flexibility and can manifest itself through unpredictable behaviors. For instance, in impedance-type haptic devices' design and operation, these parameters result in a reduced stable operating range. Examination of these dynamics in a systematic manner is significant to derive stability conditions that are appropriate for the operator's position effectively. Through extensive modeling and laboratory testing, researchers have established how operators can be employed for stabilization under various parameters to enhance user experience and safety to the fullest. That kind of awareness shows the intricate interaction between the user and structural dynamics, and the importance of such a study in dynamic system designs (Budai et al., 2014). In the end, structural dynamics is of key importance to dynamic system advancements and dependability in most engineering fields. It is not exclusive to aerospace; it affects civil engineering, automotive engineering, and even robotic systems. Increased understanding of structural behavior to dynamic loads can lead to more efficient designs that are capable of sustaining dynamic loads, thereby minimizing failure during periods of extreme stresses. Further, with the continually evolving nature of technology, the procedures applied in structural dynamics are being developed to be more sophisticated. Advances in computations and analytical techniques make it possible for hitherto infeasible solutions to be tested in establishing structural performance under dynamic load. Together with the development of the field, as other fields' results, such as vibration control and aeroelasticity, increasingly impact best practice and innovation within dynamic systems design, it will lead to safer and more efficient engineered solutions (Brown et al., 2025) (Figure 2.6).

### 2.2.3. Robotics and Automation

The combination of automation and robotics in dynamic systems has been revolutionary, with far-reaching effects across several industries through enhanced efficiency and precision. At the center of this revolution has been the possibility of the creation of digital twins or virtual models of actual systems for real-time monitoring and analysis. A software-based method with the use of Kinect V1, Robot Operating System (ROS), and Gazebo has demonstrated the viability of incorporating physical data into virtual models. By adding depth information for point cloud construction, it is now possible to create accurate three-dimensional models that represent current orientations and positions of real-time updates, which are required for the majority of automation applications. This level of precision and accuracy allows industries to optimize their operations with reduced opportunity for human error, showing an impressive evolution in the science of robotics and automation as parts of broader dynamic systems platforms (João et al., 2025). With advancements in technology of robotic automation, the intersection of artificial intelligence (AI) and machine learning propels future working functions in all sectors. Industries like manufacturing, agriculture, and defense undergo revolutionary transformations through these technologies with machines now able to perform tasks once reserved for human beings. The collaborative robots merit special mention as they operate alongside human workers, boosting productivity through virtue of unproblematic interaction. But with the rapid uptake of such technologies are some accompanying disadvantages, including technical constraints, data security risks, and economic barriers. Despite all these disadvantages, the potential to considerably improve the effectiveness of operations means that the benefits of integrating robotics and automation into systems are greater than the harm that is currently preventing large-scale adoption (Kakolu et al., 2024). In the future, robotics and automation trends mean higher complexity and interconnectedness of systems. The development of multi-bot systems, for instance, shows the potential of more collaboration and capacity in complicated settings. More advances in energy control and real-time connectivity should propel field robotics' capabilities and make them more flexible and effective. As such technologies become more advanced, they also exhibit high promise in tackling challenges in hostile environments where human engagement is limited or perilous. Ultimately, the combination of automation and robotics in dynamic systems not only enhances productivity but also redefines the boundaries of problem-solving across various industrial processes, opening the door to even greater innovation and growth in the industry (Kakolu et al., 2024).

## 2.3. DYNAMIC SYSTEMS IN PHYSICS

Dynamic systems in physics are a complex interaction of numerous factors governing motion and behavior in the long run. Static systems, which remain constant unless affected, are contrary to dynamic systems, which change constantly because of forces and initial conditions. The study of these systems is central to comprehending phenomena ranging from simple mechanical oscillations to intricate fluid dynamics. Such systems are typically described in the form of differential equations, for which researchers can predict behavior from initial given parameters. As the sophistication of a system increases, so does the sophistication both in modeling and predicting with accuracy, emphasizing the necessity for advanced analysis and simulation methods. The analysis of dynamic systems thus requires both applied application and theoretical comprehension to best mimic their behavior in actual situations. A key feature of dynamic systems is the process of transition between various operational modes, as in the aerodynamics phenomenon of dynamic stall. The full dynamic stall process comprises a number of developmental stages that reveal the complexity of flow separation and vortex-vortex interaction. Static stall is different in that dynamic stall is hysteresis-prone, meaning the response of the systems is path-dependent. To bypass these complexities, the Goman–Khrabrov (G-K) dynamic stall model simplifies this process to a lower-parameter state-space representation with focus on time delays relevant to system dynamics. However, as noted in recent updates, this traditional model can render nonphysical prediction errors. Hence, the introduction of a new G-K model based on a time-varying state space gives a more suitable framework to simulate such transitions, focusing on model development advancements in dynamic systems (Zheng et al., 2025). Dynamic system utilizations in engineering also show their relevance in real-world applications, bridging theoretical physics and actual functionality. In this discipline, dynamic systems become imperative in the structures' design that must withstand dynamic loads and require precise system response estimations. Traditional analytical models fall short of providing exact solutions, especially for complex systems. With this issue, energy-based methods have emerged as possible alternatives, reducing calculations under uncertainties in engineering applications. These methods capture inconsistencies and discrepancies in dynamic systems, stepping beyond the limitations of classical models by creating a better understanding of dynamic response. As evidenced through a number of engineering examples, these newer methods highlight the relevance of dynamic systems not just for theory, but for enabling practical design and safety in a lot of fields of engineering (Andrade, 2024) (Figure 2.7).

**Figure 2.7**. *Dynamic systems in physics.*

*Source: https://upload.wikimedia.org/wikipedia/commons/thumb/5/5b/Lorenz_attractor_yb.svg/800px-Lorenz_attractor_yb.svg.png.*

## 2.3.1. Classical Mechanics

Classical mechanics serves as a foundational pillar within the study of dynamic systems, providing a comprehensive framework for analyzing motion and forces at macroscopic scales. This branch of physics, primarily formulated through the works of Newton, revolves around key principles such as inertia, force, and momentum, which elucidate the behavior of objects under various conditions. The efficacy of classical mechanics becomes evidently clear when exploring systems governed by deterministic laws, where initial conditions yield predictable outcomes. The ability to model real-world phenomena—ranging from planetary motion to collision dynamics—demonstrates not only its practicality but also its enduring relevance in both theoretical and applied physics contexts. In the realm of dynamic systems, computational mechanics arises as a valuable tool for advancing the analysis of complex interactions governed by classical mechanics.

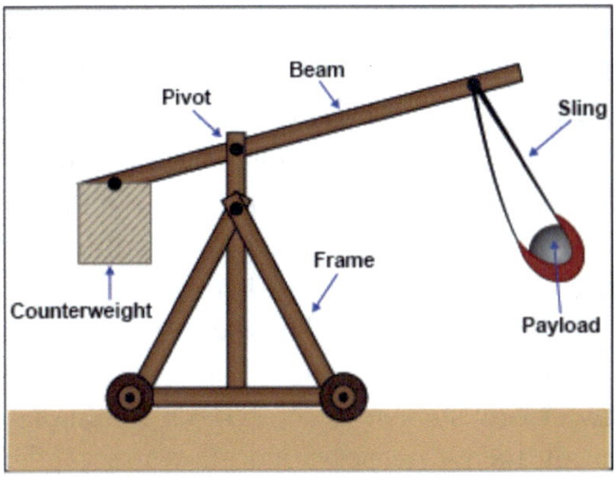


**Figure 2.8**. *Classical mechanics.*

*Source: https://ocw.mit.edu/courses/8-223-classical-mechanics-ii-january-iap-2017/10f76c13caec5bf5cc2610c53e94d6fc_8-223iap17.jpg.*

Its evolution from the study of strange attractors in turbulent flows highlights the intricate relationship between mathematics and physical phenomena. By leveraging array-based information from time series, computational mechanics provides insight into the underlying structures of dynamic systems, integrating aspects of stochastic processes and nonlinear dynamics. The rigorous methodologies established in computational mechanics have extended beyond their original scope, allowing researchers to explore chaotic systems and their implications across disciplines, thus illustrating the rich interconnectivity present within classical mechanics and modern applications (Crutchfield et al., 2017). Additionally, the incorporation of weakly nonlocal thermodynamic theories into the study of dynamic systems exemplifies the sophisticated approach needed to address complex interactions within classical mechanics. Drawing on modified forms of entropy and internal variables, these theories facilitate a deeper understanding of nonlocal phenomena and dynamic thermodynamic variables. The derivation of equations such as the Guyer-Krumhansl and Cahn-Hilliard equations underscores the adaptability of classical mechanics in different contexts, providing valuable insights into nonlocal interactions which were previously difficult to quantify. Such advancements enrich the field, allowing for a nuanced exploration of emergent behaviors in complex systems, and reinforcing the significance of classical mechanics within the broader framework of dynamic systems (Van et al., 2003) (Figure 2.8).

## 2.3.2. Thermodynamics

Thermodynamics, being a basic principle in the science of dynamic systems, closely weaves together the topics of energy, work, and heat transfer. It is a set of laws that describe the dynamics of energy in physical systems, emphasizing energy conservation and the irreversibility of energy conversions. The first law of thermodynamics—energy is neither created nor destroyed—calls attention to the constant interaction among the various forms of energy within dynamic systems. This law is particularly significant in describing phenomena such as phase transitions and chemical reactions, where energy changes dictate system behavior. The application of this law allows scientists to explore the manner in which energy constraints govern system behavior and hence allow scientists to predict the behavior of a system under various conditions. Establishing the overriding function of energy conservation allows researchers to study more effectively the complexities surrounding dynamic systems. Aside from the first law, the second law of thermodynamics also introduces the concept of entropy, a quantity of disorder or randomness in a system. The second law declares that natural processes will always take the direction of maximizing entropy and adds that there will be an unavoidable dissipation of energy in closed systems. In dynamic systems, this notion has profound implications, especially in cycle energy transformation, such as heat engines or biochemistry. The tendency for growing entropy indicates the boundaries of efficiency in transferring and converting energy and necessitates innovative ways to energy use. Thus, entropy not only lends itself to conceptual paradigms but also triggers pragmatic inventions within energy systems that guide the manufacture of technologies striving to maximize energy saving and efficiency. The interplay between entropy and energy consumption remains a topic of interest for scientists who aim to optimize dynamic system performance. Computational models combined with thermodynamics have significantly enhanced knowledge of dynamic systems. Modern computational techniques allow for detailed studies of the character of such phenomena as the solid electrolyte interphase (SEI) in electrochemical systems. SEI growth models, discussed in the context of battery technology, yield data on the thermodynamics and kinetics of the reduction reactions of the electrolyte, vital in designing energy storage devices optimally. By applying multi-scale simulations, researchers can predictively model the fundamental features of SEI, including electron tunneling and ionic conduction, thereby revealing the interrelation between microscopic processes and macroscopic effects. These advances demonstrate the transformative effect of computational approaches on thermodynamics, enabling a more profound insight into energy processes in devices and systems at both nanoscale and engineering scale (Wang et al., 2018; Streltsov et al., 2017) (Figure 2.9).

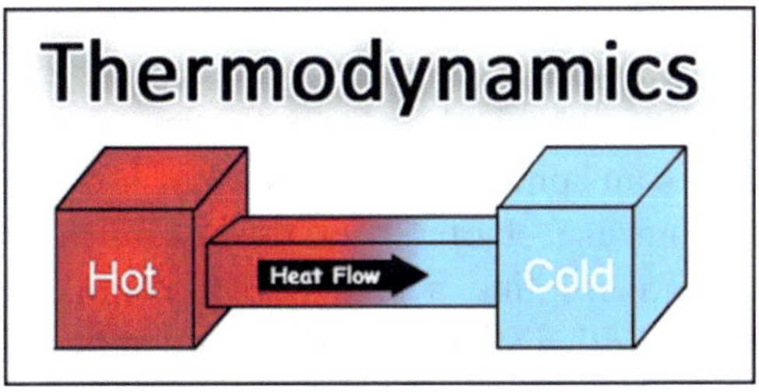


**Figure 2.9**. *Thermodynamics.*

*Source: https://images.ctfassets.net/cbk8m389sggz/zfuMiL3Ns19CbqantgvR2/0c782d5b60ac69ea68df437b9be00e89/unnamed.png?w=512&h=254&q=50&fm=png.*

## 2.3.3. Quantum Dynamics

Within dynamic systems, quantum dynamics is important in the study of how isolated many-body systems behave as they evolve, particularly when considering phenomena distant from equilibrium. Quantum systems, as guided by superposition and entanglement principles, can display behavior far divergent from classical expectations. For example, studies involving spin spiral states in the three-dimensional isotropic Heisenberg model showcase the intricacies of quantum dynamics. Theoretical framework provided via field-theoretical formulation rigorously improves on mean-field characterizations of real-time quantum evolution. Through a mapping of spins onto Majorana fermions and the use of 1/N expansion of the ensuing two-particle irreducible effective action, scientists uncover the fluctuation-induced relaxation phenomena that define such spins and provide insight into dynamic system behavior that is critical for full understanding of quantum mechanics (Babadi et al., 2015). The further away systems drift from equilibrium, the more intricate the dynamics become, recalcitrant to easy interpretation. Giamarchis analysis highlights that states of equilibrium are the exception rather than the norm in nature. When physical systems are perturbed, their restoration to equilibrium can be thwarted by either their complex interactions or the long time required for stabilization. The challenge of describing non-equilibrium states is the lack of universal laws like thermodynamic principles that would dictate their behaviors. Instead, scientists are required to study every situation carefully and identify some kind of commonality between a wide range of systems. This study is essential for achieving a collective understanding about out-of-equilibrium dynamics, but once more reinforcing the need to explore quantum behaviors that have the propensity to demonstrate unusual attributes smearing the differences between different dynamical regimes (Cugliandolo et al., 2013). Additional to this, studies on quantum dynamics also border on the presence of prethermalized states and how they arise in physical systems. In certain quantum systems, such as the aforementioned spin spirals, the formation of long-lived prethermalized plateaus is an intriguing aspect of relaxation dynamics. Through parameter adjustment, such as spiral winding, systems can experience divergent lifetimes and thereby create states with distinct effective temperatures in different modes. This hierarchical relaxation can simulate phenomena in glassy systems, where usual equilibrium rules fail. The information gained from these studies not only enriches theoretical models but also provides opportunities for experimental validation, particularly in the case of ultracold atoms manipulated in quantum gas microscopes. Such a discovery brings into focus the synergy between experiment and theory in the study of quantum dynamics (Babadi et al., 2015).

## 2.4. BIOLOGICAL SYSTEMS

Biological systems' operation and response to external disturbances are strongly controlled by their dynamic behavior. Biological systems are founded on the principles of biomechanics, a cross-disciplinary field that unites engineering mechanics with biological and physiological processes. The union enables greater insight into organisms' movement, growth, and environmental adaptation. These discoveries are at the heart of the development of medical applications for everything from methods of rehabilitation to enhancing sporting performance. Techniques such as inverse dynamics and finite element analysis modeling, for instance, allow scientists to simulate intricate human movement and calculate the impact of injury. The continuous evolution of biomechanics gives credence to its role in the discovery of the complexity and adaptability of biological systems, as it relies on feedback from a variety of sources that range from engineering, biology, and medicine, ultimately enhancing human performance and health (Harris et al., 2012). In addition, biological systems are not only mechanical but also exhibit complex interactions and networks that determine their behavior. Epidemic process studies provide a framework for deconstructing such interactions, in this instance, how disease can travel in biological and social networks. Recent studies have also compared the dynamics of two rival viruses in such networks and documented how different pathogens can interact within populations.

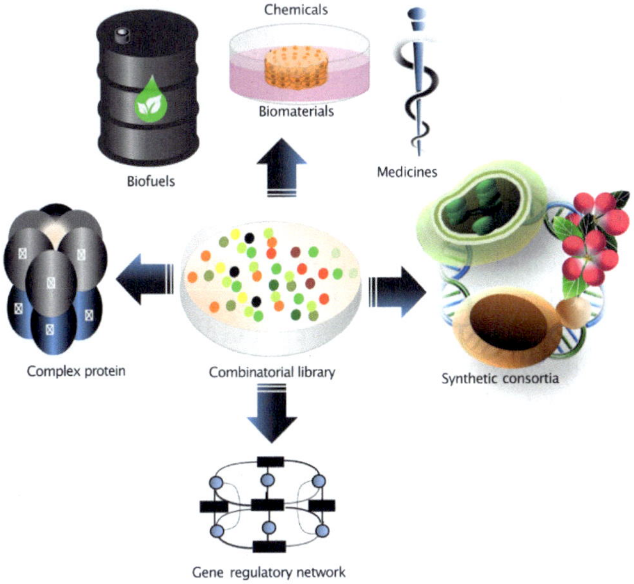


**Figure 2.10**. *Biological systems engineering.*

*Source: https://upload.wikimedia.org/wikipedia/commons/thumb/7/7b/Applications_of_combinatorial_gene_circuit_optimization_strategies.svg/1200px-Applications_of_combinatorial_gene_circuit_optimization_strategies.svg.png.*

Such a framework is central in developing disease-control measures because it addresses the necessity of understanding how viruses compete both in static and dynamic environments. With the use of mathematical modeling and analysis of stability, researchers can predict the outcomes of epidemic phenomena as well as come up with effective response measures. Hence, biological system dynamics are more than physical motion to incorporate the intricate web of interactions that characterize life's processes (Başar et al., 2017). The synthesis of biomechanical modeling and epidemic theory is an illustration of the complexity of biological systems. In the dynamic systems perspective, biomechanics provides crucial information on individual movement and action, while epidemic models offer significant information on collective health dynamics. As the science advances, scientists are confronted with the additional challenges of understanding both the mechanical and networked aspects of biological systems in studies. This integrative thinking permits a full description of organism responses to extrinsic stresses and interconnectedness of health at the individual and at the community level. Further, interdisciplinary collaboration remains unavoidable in addressing the challenges of the day in biological systems due to its utilization of the strengths of different scientific disciplines. Through such collaborative endeavors, the future of research in biological systems appears rosy, paving the way for innovative solutions to complex health-related issues (Harris et al., 2012; Başar et al., 2017) (Figure 2.10).

## 2.4.1. Population Dynamics

Population dynamics, a central concept in dynamic systems thinking, examines population size and composition variation over time caused by environmental and biological factors. The dynamics are influenced by birth rate, death rate, immigration, and emigration, which create complex relationships within ecosystems. Understanding these factors is significant since population dynamics not only reflect the health of a species but also have broader economic and ecological roles to play in establishing biodiversity and the stability of ecosystems. By mathematically modeling these factors, scientists can predict future population dynamics and establish the probable impacts of human activities, such as habitat destruction and climate change, on ecological integrity. The intricate feedback loops between population species play to highlight the importance of holistic management practices that prioritize sustainability and ecological resilience, a point underscored in the call for change in water resource management practices to incorporate dynamic ecological paradigms. Biotic homogenization was found to be an excellent focal point for exploration of population dynamics, illustrating the manner in which anthropogenically driven changes are reshaping indigenous biodiversity. With the movement of populations to different locations based on globalization and environmental factors, there is an emerging process whereby various ecosystems are being homogenized and lose their local species uniqueness. This process poses a significant risk to ecosystem services, which are crucial for human survival. As practioners understand the interdependence of growth and underinvestment in ecological systems, they are able to better comprehend the work of inadequate investments in conservation towards the decline of species diversity. Moreover, population dynamics models need to be re-calibrated to include these concerns. The integration of concern for ecosystem health into resource management paradigms not only permits an improved understanding of population change that is better informed, but it highlights the need for adaptive policy frameworks that recognize ecological integrity's interconnectedness with socioeconomic circumstances. Mathematical models are powerful tools in explaining population dynamics principles and inform policymaking. Such models, particularly those based on replicator dynamics, are helpful in describing how selection processes move population frequencies from one generation to the next. Through differential equation analysis, scientists may uncover complex relationships inherent in population genetics, ecology, and even economics. One can see the necessity of developing models that are founded on dynamic interactions when using them to address issues of conservation of biodiversity and distribution of resources. By incorporating feedback loops and thresholds for regime changes in ecosystems, policymakers can arrive at better judgments regarding long-term impacts. This is particularly true in the management of water systems, where the processes of species interaction must be understood in order to resist biotic homogenization and facilitate ecosystem-based management programs that are friendly to resilience and sustainability (Hjorth et al., 2014; Sigmund et al., 1984) (Figure 2.11).

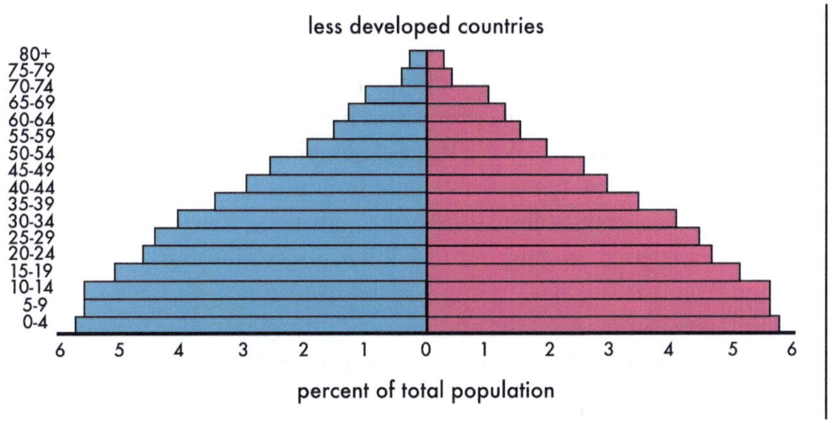


**Figure 2.11**. *Population dynamics.*

*Source: https://i.ytimg.com/vi/XFpM5e_6rtM/maxresdefault.jpg.*

## 2.4.2. Ecosystem Modeling

Ecosystem modeling is a vital tool in understanding the complexity of ecological interactions among organisms within dynamic systems. The models enable the exploration of diverse species interactions and their environments to shed light on ecosystem stability and resilience. For instance, the use of P systems to model the Pyrenees' Bearded Vulture ecosystem illustrates how computational models can capture the structural and dynamic characteristics of an ecosystem. By using rewriting rules on multisets of objects, such models capture the discreteness of population dynamics as well as the stochasticity of ecological systems. This modeling approach not only enhances our knowledge of species interactions but also provides practical applications in conservation, particularly for endangered species such as the Bearded Vulture, thereby reflecting the practicality of ecosystem modeling in real environments (Margalida et al., 2009). The addition of fuzzy cognitive mapping (FCM) to ecosystem modeling further enhances our analysis capacity, particularly in complex environments such as shallow lake ecosystems. The application of FCM facilitates researchers to consolidate the tacit knowledge of many experts and generate a robust collective cognitive map that encompasses the intricate interdependencies of the ecosystem. For example, the development of a shallow lake model with FCM revealed internal cycles and non-linear dynamics that could otherwise be concealed in simple models. The ability of such models to model ecological processes democratically highlights the requirement for capturing the highly complex character of ecosystems independently of top-down hierarchies. By demonstrating that only specific disturbances possess the ability to transition ecosystems between stable states, FCM-based models highlight the pivotal role of dynamic changes in the environment towards maintaining ecological balance (Chao et al., 2005). As we look ahead to the implications of ecosystem modeling, we note that the models not only provide a framework for the examination of ecological interactions but also serve important functions as aids to environmental management and policy-making. The conclusions drawn through these models inform practice from species conservation projects to habitat restoration initiatives. Also, the simulations from these models permit scenario testing, and hence researchers and practitioners can forecast the outcomes of different management strategies. For instance, biomanipulation treatment through fish removal and nutrient reduction has proven to be effective in converting a turbid shallow lake ecosystem to a clearwater state, thus establishing the applicability of dynamic systems modeling to ecological management. Lastly, the integration of various modeling paradigms such as P systems and FCM increases the understanding of ecosystem dynamics and ultimately contributes to better and more effective environmental management.

### 2.4.3. Human Physiology

The complicated landscape of human physiology is a dynamic system where diverse biological elements repeatedly interact to provide homeostasis. This continuous interaction among organ systems—most dramatically with physical activity—gives weight to viewing body functions as not isolated events, but rather an interactive system. Recent advances in network physiology emphasize this interrelatedness as it explains how cardiovascular, respiratory, muscular, and neural systems interact to handle physical stressors. For instance, during exercise, these kinds of systems synchronize activity so that sportsmen can cope more effectively with adverse situations. The stability and adaptability of these physiological networks, especially in the case of trained subjects, may be attributed to the complex regulating mechanisms involved, which ensure effective communication and corrective action for the maintenance of overall performance and minimizing risk of injury. Those perspectives highlight the crucial function of dynamic systems to investigate human physiology and enhance sports abilities, as identified in research on biomarkers reflecting physiological reaction to training stress (James, 2025). Over the past decade, advancements in organ-on-a-chip (OOC) technology provide an important breakthrough in simulating human physiological reactions for research. These micro physiological models are designed to mimic the unique characteristics of intact living human tissues through dynamic flow conditions, yielding improved accuracy compared to traditional animal models. The versatility of OOC devices enables researchers to model various physiological and pathophysiological conditions, ultimately leading to innovation in biomedical exploration and drug discovery. OOC systems can be categorized based on flow types, operating mechanisms, and configurations, making them amenable to applications that address the needs of a particular experiment. These developments not only bridge the gap between *in vitro* experiments and whole-organism studies but also provide a more insightful view into human physiology as it reacts to environmental and drug perturbations. With the developments of these technologies, they can increase our insight into complex physiological interactions among dynamic systems (Samantasinghar et al., 2025). The bringing in of the concept of understanding human physiology based on dynamic systems has several implications, particularly on therapeutic intervention and training procedures. Conceptualizations of the brain as a central node within the physiological system determine responses of various systems, including the muscular, cardiovascular, and respiratory, under various conditions. Disruptions to this network have a serious effect on recovery and performance, particularly when stress- or fatigue-related. As such, the application of network analysis appears to be priceless in optimizing performance because it allows trainers and clinicians to design personalized training protocols and monitor recovery more accurately. Furthermore, the use of predictive modeling via machine learning procedures holds potential for early identification of overtraining syndromes and autonomic nervous system imbalance. Ongoing research into these multifaceted physiological interactions will continue to advance our understanding of health and resilience, ultimately maximizing athletic training and therapeutic interventions in various populations (James, 2025; Samantasinghar et al., 2025) (Figure 2.12).

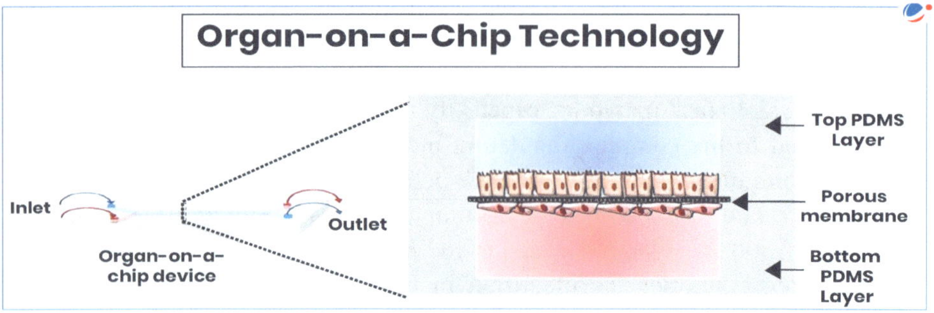


**Figure 2.12**. *Organ-on-chip (OOC) technology.*

*Source: https://d2av8kbir6lh9m.cloudfront.net/uploads/i9qfTovaxosPSHEsijHWS-bzJEwwkfoyMOkLI3ARo.jpg.*

## 2.5. ECONOMIC MODELS

The study of economic models is a critical component of learning dynamic systems because economic models are vital analytical tools used to explore complex interactions in economies. By modeling the behavior of economic agents, the models allow researchers to explore the way various factors—such as competition, market forces, and consumer behavior—are interrelated in the long run. Micro-dynamic models, for example, have worked as effective tools to illustrate the complexity of market behavior. These models capture empirical features of competition networks arising naturally from a minimal set of general mechanisms governing the dynamics of competition, as highlighted by research considering interfirm organizational networks (Bar-Yam et al., 2025). Economists can develop strategies that adhere to both competitive dynamics and social behaviors governing market interactions as economies develop and transform due to reliance on such models. A fundamental characteristic of economic models is that they may incorporate elements from physical systems to further render them effective in various situations. Sophisticated control techniques can significantly improve the performance of dynamic economic systems by emulating a variety of interactions in market structures. A vapor compression cycle system is an illustration of how emerging modeling structures are more efficient and perform better, and it can be a model for future economic models (Alleyne et al., 2006). Through the application of analytical techniques that combine empirical approaches with theory, researchers are better able to generate approximations of economic reality. This is significant to policymakers and corporations because it allows them to anticipate changes in the economic landscape and react by adjusting strategies, indicative of the nonlinear, interdependent character of economies. At the intersection of economic modeling and dynamic systems lies the potential for increased insight into regional economic development. Understanding how competitors behave in relation to one another in a spatial environment can enlighten regional differences and the mechanisms of economic growth. Inclusion of geographic information into economic models adds a greater complexity with the consideration of distance and density impacts in competitive interactions. For example, while corporations expand and form networks, the locational distribution of corporate headquarters has the ability to produce varying competitive advantages in local markets (Bar-Yam et al., 2025). With a comprehension of these phenomena, economists can better inform urban planning and policymaking that leads to enhanced regional development plans that promote economic resilience and development in an ever-accelerating environment. Thus, a holistic approach to economic models in dynamic systems helps better understand both local and overall economic environments (Figure 2.13).

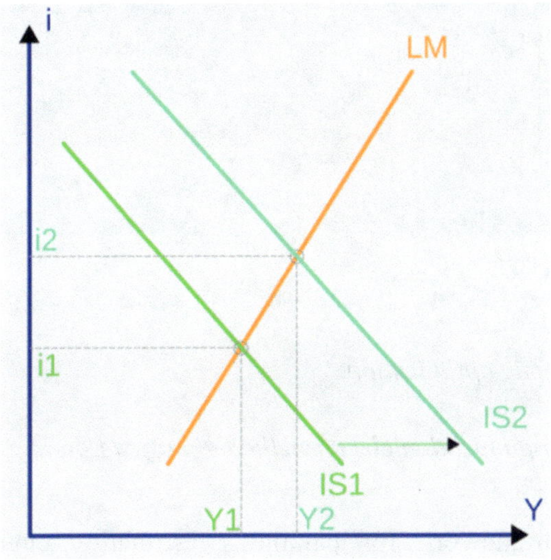


**Figure 2.13**. *Economic model.*

*Source: https://upload.wikimedia.org/wikipedia/commons/thumb/b/b9/Islm.svg/1200px-Islm.svg.png.*

## 2.5.1. Dynamic Optimization

Dynamic optimization is at the core of dynamic systems analysis as a way to deliver the tools necessary to make decisions at their optimum level through time under uncertainty and variability. With a step-by-step approach, dynamic optimization seeks to maximise or minimize a particular objective function subject to constraints by the system's dynamics. This involves developing the problem as a mathematical representation of the complexity of dynamic interactions, enabling one to infer strategies that not only take into account the immediate situation but also impacts in the future. The fundamental principles can be further clarified by techniques such as dynamic programming, which is an approximation of complex multistage problems into manageable subproblems, optimizing decision at each stage while keeping the optimality of the entire strategy. At the center of dynamic optimization is parameter estimation for dynamic systems, an exercise pursued with aggressive methods such as maximum likelihood estimation. Parameter estimation is critical to good modeling because it actually specifies the predictions and inferences drawn from the system.

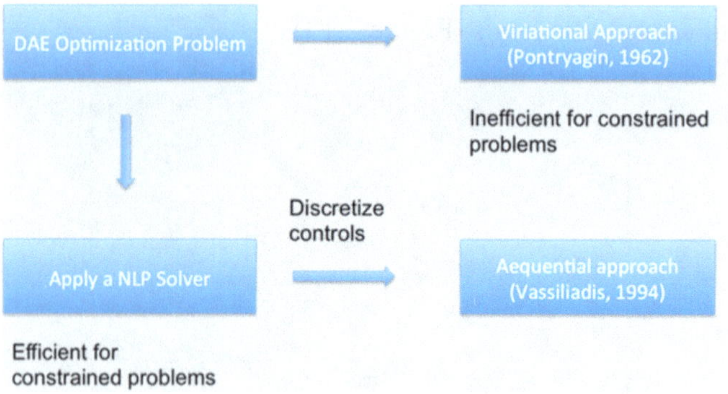


**Figure 2.14**. *Dynamic optimization.*

*Source: https://optimization.cbe.cornell.edu/images/4/46/Shy_345_wiki_fig_02.png.*

Theoretical framework for parameter estimation cuts across various mathematical disciplines like statistics and linear algebra, which improve the accuracy and reliability of the optimization methods applied. The estimators' performance and convergence are extremely important, particularly where stochasticity in dynamic systems models is involved, with reliable parameter estimates dictating adaptive responses and strategies for facilitating optimal decision-making under uncertainty and variability that exist in real-world problems (Iliff et al., 2025). Moreover, the use of strong algorithms, such as those in Hilbert structures, enhances the performance of dynamic optimization. The structures allow one to handle structured monotone inclusions, which form a key basis in optimal control problem formulation. By employing methods such as regularized Newton dynamics and gradient projection dynamics, researchers are able to attain convergence behaviors that enable the solution to converge to the desired optimal state in an efficient manner. This synergy of theoretical and algorithmic design supports dynamic optimization in solving complex and multi-criteria decision-making issues, primarily in finance, engineering, and environmental management. Improvements in these techniques continually open up new avenues for applying dynamic optimization to solve multi-criteria decision-making problems, thereby solidifying its application to the broader context of dynamic systems (Abbas et al., 2014) (Figure 2.14).

## 2.5.2. Game Theory Applications

Game theory, a mathematical methodology of strategic interaction among rational players, has profound applications in most fields, particularly in dynamic systems. Negotiation of resources allocation is a key issue where the parties must strike a balance on conflicting interests to achieve best deals. The application of non-cooperative bargaining theory to water issues is a classic case, where game theory is employed to analyze conditions under which cooperative contracts are likely to occur. Through bargaining process modeling, stakeholders are able to construct frameworks that will enhance cooperative action, ultimately minimizing conflict over scarce resources. The knowledge gained from these applications not only enhances the efficiency of negotiations but also provides policymakers with pivotal tools for shaping the rules of the game governing resource management, and thus ensures the pragmatic relevance of game theory in addressing community issues (Sgobbi et al., 2025). Moreover, game theory expands its range of application to economic analysis of highly complicated systems, beyond traditional models that are likely to oversimplify interactions. One example is the development of a comprehensive system for measuring the sustainability of agri-food systems through the multi-dimensional Governance Assessment Matrix Exercise (GAME). The new system uses game-theory principles to synthesize qualitative analysis of various capitals—social, human, economic, and natural—into quantitative values that reflect their interdependencies. This holistic vision contrasts sharply with conventional economic models that do not consider the richness of externalities and therefore provide a more advanced understanding of resource interaction. These conclusions form a secure basis for adaptive management protocols grounded in the complexity of dynamic system interactions, suggesting the likelihood of game theory being a useful factor in guiding sustainable agriculture practice (Lemon et al., 2018). The application of game theory in negotiations and resource management not only enriches our understanding of dynamic systems but also addresses topics of conflict resolution and sustainability in the widest possible sense. Environmental stakeholders can leverage game-theoretical models to enhance the complexity of environmental negotiations, particularly in trans-boundary environments where cooperation is needed but typically difficult. By employing non-cooperative game theory, negotiators are able to simulate various possibilities and ascertain the resulting impacts of varying strategies, hence perfecting their decision-making processes. This preventive measure is highly essential in facilitating agreements that not only end up being justifiable but also ensure long-term sustainability, hence highlighting the intersection of game theory and environmental policy. With evolving game theory, its application in dynamic systems will become even more vital in influencing the development of new solutions to new global problems (Sgobbi et al., 2025).

### 2.5.3. Market Equilibrium

The concept of market equilibrium is a foundation for study in dynamic economic systems, in which demand equals supply at a price level and, as a result, there exists stability in an economy. In a state of equilibrium, individuals are willing to buy an equal amount of goods that producers are ready to sell, and there is equilibrium in a market. Shocks to consumer tastes, technology, and production costs are among the forces that can perturb this equilibrium and produce changes in prices that ripple through the economy. Consequently, an equilibrium analysis must incorporate dynamic forces that embody up-to-the-minute adjustments and the reactions of market participants. By examining these interactions, economists can anticipate market reactions and develop policies that offset the detrimental forces of disequilibrium, like persistent unemployment or inflation. Dynamic systems have their own complexities, and these are particularly significant in the case of labor markets where socioeconomic situations are influenced by equilibrium. Research has shown that the introduction of basic income schemes can actively reduce levels of equilibrium unemployment in unionized economies. These schemes, instituted to counteract the inefficiencies of traditional unemployment insurance mechanisms, create a setting through which a more preferable labor market equilibrium can be achieved. The normative analysis points out that only people who are participating in the labor market should be entitled to such income, as this would encourage participation and work. By providing an active citizen's income, the economies can potentially achieve a Pareto improvement such that no one is worse off and at least one party is better off, which ultimately leads to a more stable market equilibrium as well as complementing the dynamic systems of labor economics (Van der Linden et al., 2025). Furthermore, market equilibrium cannot be isolated from the overall financial environment, especially with increased global financial openness. Empirical evidence displays that countries whose financial systems are highly developed tend to have less business cycle volatility, and this means more stable market equilibrium. However, the interactive dynamics of market frictions and financial openness show that an increase in openness does not necessarily result in a drop in volatility. On the other hand, the dynamics of financial market nature and frictions could hide the expected gains from integration. With dynamic general equilibrium models, researchers can explore the nuanced impacts of financial market conditions on economic stability. Policymaking requires policymakers to understand such interactions in order to establish circumstances under which equilibrium can be achieved and maintained even in the face of external shocks and market forces (Buch et al., 2003).

## 2.6. COMPUTATIONAL APPROACHES

Computational progress has significantly influenced the study and examination of dynamic systems, particularly in modeling biology and economics. Stochastic simulation approaches, representative of the randomness that characterizes biological processes, are crucial for simulating the complexity of systems like the yeast cell cycle. Such systems are inadequate with conventional continuous models, as they fail to account for the discrete effects of small molecule populations. Computationally intensive are methods like executing thousands of simulation runs in parallel. Because task durations for individual tasks are uncertain, efficient load balancing is essential to take full advantage of computational resources. The stochastic model proposed in recent work highlights the necessity for developing dynamic load balancing programs tailored to handle such stochastic biochemical simulations, such that efficiency is not only achieved but also precision in modeling biological processes, validating the role of advanced computation methods in dynamic systems analysis (Ahn et al., 2010). In economic models, computational techniques, including Agent-based Computational Economics (ACE) as a prominent example, supply a robust system for the analysis of interactions between different decision-makers. Human subjects' uniqueness whose decisions are made on the basis of past and future potential actions need models that describe local decision routines under changing conditions. This viewpoint puts actual economies within the framework of locally-constructive sequential games where the local actions contribute significantly to future states. With agent-based models, complex economic interactions can be modeled such that one can potentially derive insights from macroeconomic events using the micro perspective. This not only makes it possible to study emergent market behaviors but also counteracts the uncertain nature of the economic system, thereby proving to use innovative computational methods in exploring dynamic systems (Tesfatsion et al., 2017). The marriage of economic and biological modeling through the application of complex computational techniques demonstrates the applicability and strength of these methods in explaining dynamic systems. The application of stochastic simulations to biological systems allows for detailed study of biochemical pathways that would otherwise be indistinguishable in deterministic frameworks. At the same time, the agent-based modeling used in economic systems allows for complex examination of the behaviors of individual participants and how those behaviors collectively influence overarching market trends. This cross-disciplinary perspective emphasizes that strong computational methods must be developed to be resilient under changing conditions so that scientists can successfully investigate the complexities of dynamic systems. With increasing computational power, the ability to simulate, analyze, and predict complex interactions in biological and economic systems will become more efficient, further cementing the middle ground position of computational approaches in contemporary dynamic systems studies (Figure 2.15).

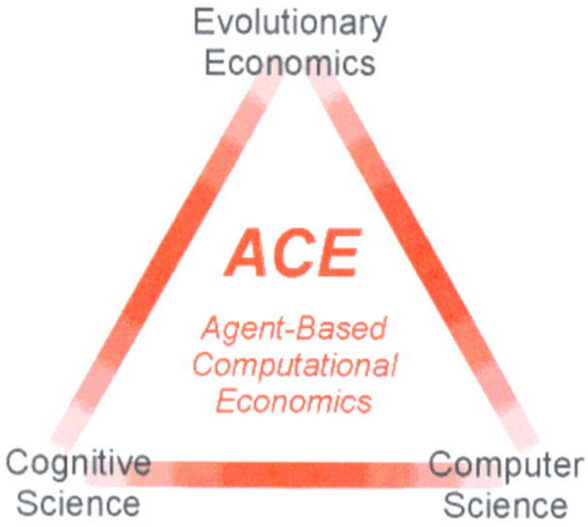


**Figure 2.15**. *Agent-based modeling approach.*

*Source: https://faculty.sites.iastate.edu/tesfatsi/archive/tesfatsi/images/acetri5.gif.*

## 2.6.1. Simulation Techniques

Simulation techniques are valuable tools in the analysis and investigation of dynamic systems in fields including engineering, biology, and aerospace. Simulation techniques enable scholars and professionals to capture complex circumstances that would be impossible or difficult to replicate in real situations. Through computer-based simulations, analysts can examine system performance under varying conditions, making it possible to gain insights into performance, stability, and potential failures. This ability to experiment in a controlled virtual environment allows the process to be perfected before implementation, reducing costs and risks of physical experimentation. Ultimately, simulation techniques promote innovation, enhance decision-making, and assist in developing dynamic systems across a range of applications. The integration of knowledge-based programming and conventional simulation techniques is a key advancement in the topic. As highlighted in new studies, the merger of these methodologies allows dynamic stochastic simulation models that can modify and learn from data inputs to be developed. The knowledge-based simulation model development environment provides a platform that takes advantage of the capabilities of artificial intelligence, allowing researchers to construct systems that mimic complex human decision-making. This intersection of technologies not only expands the scope of application of simulation techniques but also renders them more effective in simulating dynamic systems, evidenced in the development towards the Advanced Launch System Model (ALSYM) (Shiva et al., 2025). This new method is the epitome of dynamic systems analysis being enhanced by combining various technological advancements. Moreover, applications of simulation methodologies in such areas as aerospace and biotechnology prove their diversity and effectiveness. Programs such as the NASA/American Society for Engineering Education Summer Faculty Fellowship prove how simulations transform space exploration and life sciences research, resolving extremely challenging problems presented by microgravity and dynamic structural analysis (Bannerot et al., 2025). These simulations are not just intellectual exercises; they generate significant understanding of systems behavior that informs design and operation of high-risk activities. By replicating conditions—low-gravity environments, bioreactor operation—scientists can predict difficulties and build more robust solutions, enhancing our understanding of dynamic systems on earth and elsewhere. The continuous refinement of these techniques promises to further unlock the potential of simulation in addressing complex real-world problems.

### 2.6.2. Numerical Methods

Numerical techniques are powerful tools in the analysis and prediction of the behavior of complicated systems in the area of dynamic systems to a high degree of accuracy. The techniques facilitate the approximation of solutions of differential equations, which in the majority of cases are not solvable analytically. Applications include the utilization of numerical techniques such as the finite difference and finite element methods in engineering problems for modeling physical processes ranging from structural dynamics to fluid flow. However, the validity of such numerical techniques does largely depend on the choice of discretization techniques and step sizes, needing deep understandings of the methods and systems to be simulated. As complex systems possess non-linear dynamics and interaction, successful implementation of numerical techniques has the potential to provide insights into complicated dynamics that inform design and operating decisions. Surrogate modeling has been recognized as one of the possible ways to enhance the effectiveness of numerical methods to nonlinear dynamic systems. By using surrogate models that produce simplified versions of complex simulations, scientists can achieve computational needs effectively while ensuring the accuracy is very high. A number of surrogate modeling techniques, such as Gaussian process (GP) regression and Long Short-Term Memory (LSTM) networks, have been investigated for their potential in long-time-scale system response prediction. The recent studies show that the use of these methods is capable of exploiting training data for prediction that reduces the burden of explicit numerical integration. Specifically, the GP-NARX model was found to be very stable and accurate compared to deep learning approaches, marking its use in predictive analytics as well as in designing systems for dynamic settings (Zhao et al., 2022). Further, optimization of numerical integration schemes adds robustness to simulations in dynamic systems. A new methodology based on matrix pencil methods has been developed to quantify the accuracy and reliability of such approaches to modeling power systems. By comparing differences between dynamic modes of actual systems and those computed numerically, this methodology facilitates comparison of different numerical approaches. This not only enhances understanding of performance of individual methods but also enables the uniform application of numerical methods to many applications. Outcomes of its uses, for example, simulation on the WSCC 9-bus system, demonstrate how integration schemes can significantly affect the accuracy of simulations, informing best practices for engineers and researchers (Tzounas et al., 2022) (Figure 2.16).

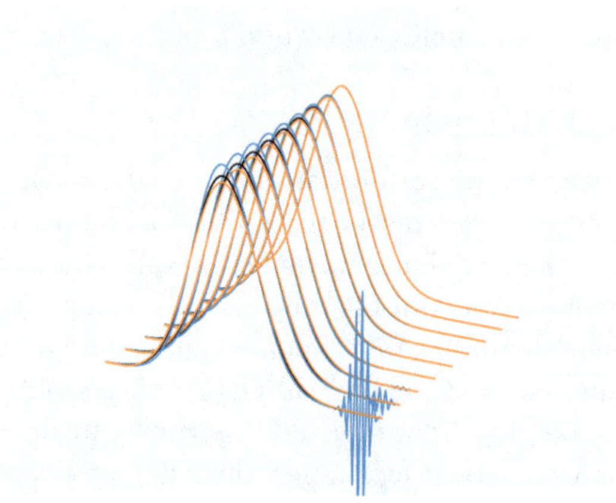

**Figure 2.16**. *Numerical methods.*

*Source: https://numericalmethodssullivan.github.io/images/bookcoverimage.png.*

### 2.6.3. Software Tools for Analysis

In dynamic system fields, selection and use of analysis tools in software are significant to enhancing efficiency and accuracy. Among these, whole-building energy simulation software is an important tool to evaluate the performance of new systems such as passive cooling techniques. For instance, simulations can analyze the viability of using phase change materials (PCM) in building enclosures, whereby they can reduce cooling energy loads considerably under specific climatic conditions. This capacity gives a precise insight into how various design options might influence thermal performance, which ultimately leads to more energy-efficient architectural designs. The study of such dynamic interactions highlights the importance of advanced computational models that can optimize building performance, as incorrect use of PCM can produce outcomes that drastically deviate from initial expectations, highlighting the importance of thorough software analysis in modern building design (Cabeza et al., 2017). Besides their use in architecture, software utilities are invaluable in the application of urban drainage systems (UDS), and more so in managing real-time control (RTC) strategies. The MatSWMM toolbox offers a simulation of how software may be employed to unite control-based models with traditional storm water management methods. By applying a combination of programming languages such as Matlab and Python, the toolbox allows engineers to model and design UDS more precisely than typical methods. The integration of population-dynamics-based controllers in MatSWMM facilitates extensive system behavior exploration that can support short-term as well as long-term urban water management dynamics. Through the marriage of theoretical principles and practical application, software tools are really helping in the understanding of complicated dynamic systems, gaining new approaches for the long-established urban issues (Barreiro et al., 2016). The creation of analysis tools through software also not only progresses innovation within the design process but also promotes the collective methodology within the research of dynamic systems. With these technologies developing further, they improve the capability of professionals from any field to collaborate with each other, utilizing the simulation data for cross-disciplinary uses. The use of data analytics and modeling capabilities provokes the convergence of expertise from engineering, environmental studies, and architecture, leading to integrated solutions that tackle multi-faceted issues. Here, the ongoing development of software platforms will be paramount to future application and research in dynamic systems. By improving analytical capacity with the help of software tools, industrial practitioners can develop more sustainable and resilient solutions that are critical in solving urbanization and climate change challenges, thereby enhancing the contribution of software innovations in contemporary dynamic system analysis (Barreiro et al., 2016; Cabeza et al., 2017).

# SUMMARY

A study of dynamic systems, it is necessary to recognize the pluralism and the complexity of such systems as well as the complexity involved in analyzing them. Dynamic systems are a variety of phenomena from every walk of life, especially in the social sciences where time-varying data is typical. Dynamic series compositional analysis review displays significant methodological advances, such as autoregressive integrated moving average (ARIMA) analysis and compositional vector autoregression systems (CVAR), that have emerged via Bayesian approaches. The area remains, however, open for new research, as methodologies to date have not fully exploited the unique challenges presented by compositional data structures. That social scientists remain aware of these developments will not only advance theory, but it will also strengthen research procedures empirically as new methods of analysis are created (Larrosa, 2025). The application of quantitative computational models is also helpful in explaining the intricacy of biological dynamic systems. The models facilitate close examination of complex pathways such that investigators can test scenarios and ascertain controlling parameters. For instance, the narrative modelling language developed for the gp130/JAK/STAT signaling pathway is a classic case in point, enabling biological observations to be mapped directly to executable representations. In fact, through simulation and sensitivity analysis, influential features like compartmentalization and phosphorylation state were demonstrated to be the controlling determinants in the pathways' behaviour. This approach not only facilitates the modeling process but also enhances the use of advanced biological data analysis, vindicating the effectiveness of narrative models to combine theoretical constructs and actual applications of dynamic systems research (Herrmann et al., 2008). To conclude these reflections, the future of dynamic systems analysis holds increased integration of advanced methodologies with traditional approaches. With scientists making the most of the capability of multiple analytical methods, there is a vast potential to refine our knowledge of both social and biological systems. Continual research into the interactive dynamics among all the various components will provide additional insights and more potent predictive abilities. Thus, embracing a multidisciplinary approach will be essential to facilitating enhancements in this field. It is crucial that biologists and social scientists remain up-to-date with the newest trends and approaches, ensuring that their studies reflect the richness and diversity in dynamic systems. By doing this, not only do they contribute to the research-based body of knowledge, but they also create space for practical real-world applications (Larrosa, 2025; Herrmann et al., 2008).

One of the most crucial things learned under dynamic systems is the study of information flow within software environments, particularly that of security and confidentiality of data. The core concern issue at hand is the threat posed by sensitive information influencing computational outcomes in a negative way, leading to information leaks. As a solution for this issue, dynamic information flow analysis entails runtime monitoring in which instrumentation code is inserted into the subject software. This strategy is designed to actively scan data flows and trigger warnings in cases of breached flow policies. Such measures not only enhance data security but also overall system dependability by stopping data from unintentionally seeping due to unforeseen interactions or vulnerabilities (Barany et al., 2016). Constant innovation and refinement of such dynamic analysis techniques reinforce their utility in contemporary software engineering. Another important characteristic of dynamic systems is media technology improvement in processing, particularly through user-oriented approaches in summarization. This can be seen in contemporary work that proposes a formal video summarization model focusing on collaborative work between several experts in the generation of a summary. The algorithm progresses in three stages: initially, video frames are rated by a diverse set of operators according to their relevance; then, scores are aggregated to create a single value per frame; and lastly, the highest-scoring frames together with associated audio and text content are retrieved for the final summary. This revolutionary approach not only enhances the quality of video summaries but also demonstrates the superiority of collaborative scoring mechanisms versus older automatic summarization algorithms (Adesnik et al., 2014). This is the significance of dynamic systems for syntheses and multimedia content processing on the rise. The concurrent research of the convergence of information flow analysis and adaptive summarization methods illustrates the dynamism as well as complexity in modern systems. Since these two fields are developing further, their effects remain persistent on data security, software integrity, and media processing. Through the integration of novel analytical techniques and collaborative methods, practitioners and researchers can develop strong systems that better handle dynamic requirements. While the volume of data and content increases exponentially, the application of these dynamic methods will be crucial in sustaining efficient processing, security, and end-user satisfaction. Ongoing research in these areas has the potential to yield more effective, more robust methodologies, and new applications in a wide variety of areas, further validating the relevance and timeliness of dynamic systems in an increasingly changing technology world.

# REFERENCES

1. Abbas, B., & Attouch, H. (2014). Dynamical systems and forward-backward algorithms associated with the sum of a convex subdifferential and a monotone cocoercive operator. http://arxiv.org/abs/1403.6312 (accessed on 19th September 2025).
2. Adesnik, H., Cabrini, S., Dhuey, S., et al. (2014). Video summarization by group scoring. https://core.ac.uk/download/29203942.pdf (accessed on 19th September 2025).
3. Ahn, T. H., Baumann, W. T., Cao, Y., et al. (2010). Parallel load balancing strategies for ensembles of stochastic biochemical simulations. https://core.ac.uk/download/10676251.pdf (accessed on 19th September 2025).
4. Ale, B., Kramer, E. H., Moorkamp, M., et al. (2014). Safety management theory and the military expeditionary organization: A critical theoretical reflection. https://core.ac.uk/download/30731391.pdf (accessed on 19th September 2025).
5. Alleyne, A. G., & Eldredge, B. D. (2006). Improving the accuracy and scope of control-oriented vapor compression cycle system models. https://core.ac.uk/download/4821554.pdf (accessed on 19th September 2025).
6. Alves de Almeida, J. P., El Youssef, E. S., & Rincon, L. M. (2025). Leveraging Kinect and ROS for object representation and real-time environment interaction in dynamic systems. 2025 Brazilian Conference on Robotics (CROS), 1, 1–6. https://www.semanticscholar.org/paper/a466351ad250e2c27fb52d893d385cb92b2b8fc6 (accessed on 19th September 2025).
7. Andrade, L. (2024). A survey on energy methods for vibroacoustic analysis of random dynamic systems and applications. Journal of Physics: Conference Series, 2796. https://www.semanticscholar.org/paper/8624b0fcd9a744cbecf2eab164c0e9cc91e47fd8 (accessed on 19th September 2025).
8. Anonymous, Bagiev, M., Brockett, R., Buchholz, J. J., Camacho, E. F., Murray-Smith, D. J., et al. (2011). Feedback methods for inverse simulation of dynamic models for engineering systems applications. https://core.ac.uk/download/9399528.pdf (accessed on 19th September 2025).
9. Babadi, M., Demler, E., & Knap, M. (2015). Far-from-equilibrium field theory of many-body quantum spin systems: Prethermalization and relaxation of spin spiral states in three dimensions. https://core.ac.uk/download/216214303.pdf (accessed on 19th September 2025).

10. Baianu, I. C., & Glazebrook, J. F. (2010). Categorical ontology of complex systems, meta-systems and theory of levels: The emergence of life, human consciousness and society. https://core.ac.uk/download/pdf/332869.pdf (accessed on 19th September 2025).
11. Bannerot, R. B., & Goldstein, S. H. (2025). *NASA/ASEE Summer Faculty Fellowship Program, 1990, Volume 1.* https://core.ac.uk/download/pdf/42817325.pdf (accessed on 19th September 2025).
12. Bar-Yam, Y., Braha, D., & Stacey, B. (2025). Corporate competition: A self-organized network. https://core.ac.uk/download/pdf/6849874.pdf (accessed on 19th September 2025).
13. Barany, G. (2016). Hybrid information flow analysis for programs with arrays. http://arxiv.org/abs/1607.02225 (accessed on 19th September 2025).
14. Barreiro Gómez, J., Ocampo-Martínez, C., Quijano Silva, N., Ramírez Jaime, et al. (2016). MatSWMM–An open-source toolbox for designing real-time control of urban drainage systems. https://core.ac.uk/download/81571172.pdf (accessed on 19th September 2025).
15. Başar, T., Beck, C. L., Liu, J., et al. (2017). Multi-competitive viruses over static and time–varying networks. http://arxiv.org/abs/1702.07754 (accessed on 19th September 2025).
16. Bigwood, G., Dearle, A., Kirby, G., et al. (2010). Towards adaptable and adaptive policy-free middleware. https://core.ac.uk/download/9821360.pdf (accessed on 19th September 2025).
17. Boerger, E., & Huggins, J. K. (1998). Abstract state machines 1988–1998: Commented ASM bibliography. https://core.ac.uk/download/304691565.pdf (accessed on 19th September 2025).
18. Brandenburger, A., Masten, A. S., Jennings, B. J., Folke, C., Holling, C., Kephart, J. O., Boulding, K., et al. (2015). On resilient behaviors in computational systems and environments. http://arxiv.org/abs/1503.08421 (accessed on 19th September 2025).
19. Brennan, R. D., Quanttrin, E. V., Williams, W. A., et al. (1970). The utility of a digital simulation language for ecological modeling. https://core.ac.uk/download/pdf/29390590.pdf (accessed on 19th September 2025).
20. Brown, G., Dirusso, E., Ernst, M. A., et al. (2025). Determining structural performance. https://core.ac.uk/download/pdf/42813200.pdf (accessed on 19th September 2025).
21. Buch, C. M., & Pierdzioch, C. (2003). The integration of imperfect financial markets: Implications for business cycle volatility. https://www.

econstor.eu/dspace/bitstream/10419/17903/1/kap1161.pdf (accessed on 19th September 2025).

22. Budai, C., Kovács, L., & Kövecses, J. (2014). Haptic rendering and human stabilization in presence of structural flexibility. https://core.ac.uk/download/42932641.pdf (accessed on 19th September 2025).
23. Cabeza, L. F., de Gracia Cuesta, Á., Saffari Tabalvandani, M., et al. (2017). Passive cooling of buildings with phase change materials using whole-building energy simulation tools: A review. https://core.ac.uk/download/304371814.pdf (accessed on 19th September 2025).
24. Chao, A., Boström, B., Hobbs, B. F., Kosko, B., Kosko, B., Kosko, B., Moss, B., et al. (2005). A cognitive model of an epistemic community: Mapping the dynamics of shallow lake ecosystems. http://arxiv.org/abs/q-bio/0509022 (accessed on 19th September 2025).
25. Crutchfield, J. P. (2017). The origins of computational mechanics: A brief intellectual history and several clarifications. https://core.ac.uk/download/324162687.pdf (accessed on 19th September 2025).
26. Cugliandolo, L. F. (2013). Out of equilibrium dynamics of classical and quantum complex systems. https://core.ac.uk/download/580036119.pdf (accessed on 19th September 2025).
27. Diks, C. G. H., Hommes, C. H., Panchenko, V., et al. (2025). E&F Chaos: A user-friendly software package for nonlinear economic dynamics. https://core.ac.uk/download/pdf/7178452.pdf (accessed on 19th September 2025).
28. Dower, P. M., Huang, S., James, M. R., et al. (2005). Analysis of input-to-state stability for discrete time nonlinear systems via dynamic programming. https://opus.lib.uts.edu.au/bitstream/10453/4040/1/2005000840.pdf (accessed on 19th September 2025).
29. Dymkov, M. P., Rogers, E., Galkowski, K., Kostyukova, O. I., Dymkov, M., & Dymkou, S. (2008). Constrained optimal control theory for differential linear repetitive processes.
30. Frezza, R., Perona, P., & Soatto, S. (1994). Dynamic estimation of rigid motion from perspective views via recursive identification of exterior differential systems with parameters on a topological manifold. https://core.ac.uk/download/4892715.pdf (accessed on 19th September 2025).
31. Harris, G. F., & Slavens, B. A. (2012). Biomechanics. https://core.ac.uk/download/213080657.pdf (accessed on 19th September 2025).
32. Herrmann, A., Mahdavi, A., Regev, A., Regev, A., Singh, A., Verma, A., Sharrocks, A. D., et al. (2008). Narrative-based computational modelling of the Gp130/JAK/STAT signaling pathway. https://core.ac.uk/download/

pdf/1631079.pdf (accessed on 19th September 2025).

33. Hjorth, P., Huckins, C. J., Madani, K., et al. (2014). Water resources management in a homogenizing world: Averting the growth and underinvestment trajectory. https://core.ac.uk/download/77003894.pdf (accessed on 19th September 2025).
34. Iliff, K. W., & Maine, R. E. (2025). Identification of dynamic systems, theory and formulation. https://core.ac.uk/download/pdf/42846216.pdf (accessed on 19th September 2025).
35. James, A. (2025). Network physiology in exercise: A novel framework for integrative analysis of multisystem responses to physical training. *Journal of Sports and Rehabilitation Sciences*. https://www.semanticscholar.org/paper/ad5bc5b5e310989c2885dd1ec5de45ee7ed38048 (accessed on 19th September 2025).
36. Kakolu, S., & Faheem, M. A. (2024). Advanced robotics automation for field operations: Leveraging AI and machine learning for dynamic, real-world applications. International Journal of Science and Research Archive. https://www.semanticscholar.org/paper/7847a004e8f3afcc583c42d0e58e657c426afbf5 (accessed on 19th September 2025).
37. Kirilyuk, A., & Ulieru, M. (2009). IT complexity revolution: Intelligent tools for the globalized world development. http://arxiv.org/abs/0910.5495 (accessed on 19th September 2025).
38. Larrosa, J. M. C. (2025). Compositional time series: Past and present. https://core.ac.uk/download/pdf/9310215.pdf (accessed on 19th September 2025).
39. Lemon, M., Mitchell, A., & Sajeva, M. (2018). Reconfiguring household management in times of discontinuity as an open system: The case of agro-food chains. https://core.ac.uk/download/228194672.pdf (accessed on 19th September 2025).
40. Margalida, A., Margalida, A., Margalida, A., Margalida, A., Margalida, A., Margalida, A., Watson, A., et al. (2009). Modeling ecosystems using P systems: The bearded vulture, a case study. https://core.ac.uk/download/157760182.pdf (accessed on 19th September 2025).
41. Rega, G., & Srinil, N. (2008). Space-time numerical simulation and validation of analytical predictions for nonlinear forced dynamics of suspended cables. https://core.ac.uk/download/9026797.pdf (accessed on 19th September 2025).
42. Samantasinghar, A., Sunildutt, N., Ahmed, F., Memon, F. H., Kang, C., & Choi, K. H. (2025). Revolutionizing biomedical research: Unveiling the

power of micro physiological systems with advanced assays, integrated sensor technologies, and real-time monitoring. *ACS Omega, 10*, 9869–9889. https://www.semanticscholar.org/paper/85d2e80847ee619d7684edad1330bdb3a887c37f (accessed on 19th September 2025).

43. Sgobbi, A., Carraro, C., & Marchiori, C. (2025). Applications of negotiation theory to water issues. https://core.ac.uk/download/pdf/6264228.pdf (accessed on 19th September 2025).
44. Shimizu, R., & Tanaka, H. (2015). A novel coarsening mechanism of droplets in immiscible fluid mixtures. http://arxiv.org/abs/1509.03385 (accessed on 19th September 2025).
45. Shiva, S. G., & Williams, M. D. (2025). Construction of dynamic stochastic simulation models using knowledge-based techniques. https://core.ac.uk/download/pdf/42822506.pdf (accessed on 19th September 2025).
46. Sigmund, K. (1984). A survey of replicator equations. https://core.ac.uk/download/33893932.pdf (accessed on 19th September 2025).
47. Soetaert, K., Setzer, R. W., & Petzoldt, T. (2025). Solving differential equations in R: Package deSolve. https://core.ac.uk/download/pdf/6287976.pdf (accessed on 19th September 2025).
48. Streltsov, A., Adesso, G., & Plenio, M. B. (2017). *Colloquium*: Quantum coherence as a resource. *Reviews of Modern Physics, 89*. https://doi.org/10.1103/revmodphys.89.041003 (accessed on 19th September 2025).
49. Tesfatsion, L. (2017). Modeling economic systems as locally-constructive sequential games. https://core.ac.uk/download/131212401.pdf (accessed on 19th September 2025).
50. Tzounas, G., Dassios, I., & Milano, F. (2022). Small-signal stability analysis of numerical integration methods. IEEE Transactions on Power Systems, 37, 4796–4806. https://www.semanticscholar.org/paper/b9a0e01eee62f17c2cf1f3536a1c510eefa79bb8 (accessed on 19th September 2025).
51. Van der Linden, B. (2025). Active citizen's income, unconditional income and participation under imperfect competition: A normative analysis. https://core.ac.uk/download/pdf/6270767.pdf (accessed on 19th September 2025).
52. Van, P. (2003). Weakly nonlocal irreversible thermodynamics. http://arxiv.org/abs/cond-mat/0112214 (accessed on 19th September 2025).
53. Velupillai, K. V. (2004). A primer on the tools and concepts of computable economics. https://core.ac.uk/download/11829256.pdf (accessed on 19th September 2025).

54. Wang, A., Kadam, S., Li, H., Shi, S., & Qi, Y. (2018). Review on modeling of the anode solid electrolyte interphase (SEI) for lithium-ion batteries. *npj Computational Materials, 4.* https://doi.org/10.1038/s41524-018-0064-0 (accessed on 19th September 2025).

55. Zhao, Y., Jiang, C., Vega, M. A., Todd, M. D., & Hu, Z. (2022). Surrogate modeling of nonlinear dynamic systems: A comparative study. J. Comput. Inf. Sci. Eng., 23. https://www.semanticscholar.org/paper/4a3b2002eeb6edb3b5ac470bf60ae7c4fcf57b6b (accessed on 19th September 2025).

56. Zheng, B., Yao, W., & Xu, M. (2025). Enhancing the Goman–Khrabrov dynamic stall model through flow delay analysis. Physics of Fluids. https://www.semanticscholar.org/paper/96883142a193eb94a5e1413661a682c480c3c4ff (accessed on 19th September 2025).

# CHAPTER 3

# TIME DOMAIN ANALYSIS

## Contents



## INTRODUCTION

Time domain analysis is a critical tool in dynamic systems research and modeling, particularly for systems modeled via partial differential equations. When scientists examine the factor of time, they can deduce critical information regarding system behavior under various time scales and conditions. This is most relevant in fields such as physics, engineering, and applied mathematics where the dynamics of systems must be properly modeled to make predictions about how they evolve with time. The relevance of this analysis rests on its ability to provide a high level of insight regarding the behaviors of time, which is paramount in creating systems that are efficient and resilient. Thus, the time domain analysis parameters embrace both theoretical and applied research, hence making it a crucial topic to study in ongoing research. One of the key areas of concentration under time domain analysis is the applications which facilitate computation on large-scale systems, particularly when handling parabolic equations. Recent studies have indicated how temporal approximation techniques, founded on overlapping domain decompositions, can be employed with efficiency to minimize computation on parallel and distributed platforms. By formulating such techniques in a variational framework with weighted Sobolev spaces, researchers have made significant developments in deriving optimal temporal error analyses for standard integrators. Above all, the application of alternating direction implicit schemes and additive splitting schemes has been theoretically and numerically confirmed and proven to exhibit stability when tackling complex temporal dynamics while maintaining stability and accuracy (Hansen et al., 2016).

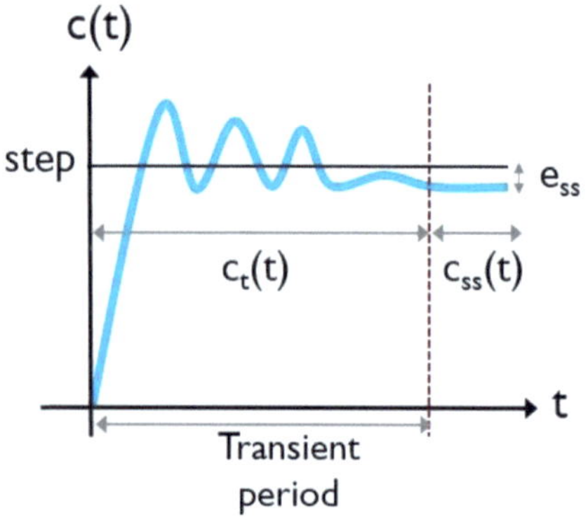


**Figure 3.1**. *Time domain analysis.*

*Source: https://electronicscoach.com/wp-content/uploads/2020/02/time-domain-analysis-of-signal.jpg.*

Also, new techniques in activity recognition have posed new frameworks that apply time domain analysis to solve data labeling problems in various domains. The use of stratified transfer learning is a technique that surmounts the lack of labeled data, enabling cross-domain action recognition through intra-class knowledge transferring. By applying majority voting techniques to pseudo labeling and domain shifts into common subspaces, this method significantly enhances classification accuracy. These developments highlight the overall importance of time domain analysis in a wide range of applications, such as ubiquitous computing, where temporal behavior needs to be understood to achieve improved performance and guide future research endeavors. As more development is made in time domain analysis, its incorporation with emerging techniques will undoubtedly introduce more sophisticated models able to solve the complex real-world issues (Chen et al., 2017) (Figure 3.1).

## *Definition of Time Domain Analysis*

Time Domain Analysis is a basic signal processing technique that establishes how signals vary with time. It is primarily interested in the examination of a signal's amplitude variation as a function of time and forms the backbone of dynamic system behavior and properties. Through the explanation of signals in terms of time, the analysts can study transient and steady-state behavior, where patterns might not be evident from the frequency domain. This approach is particularly effective in handling non-stationary signals, where time-varying characteristics are of significance. The time-domain approach allows engineers and researchers to develop methods for system design, monitoring, and diagnostic checking, thus enhancing the efficiency and reliability of complex systems as a whole. Time Domain Analysis is a critical role in Requirements Engineering (RE) to examine system design specifications under development. As RE deals with capturing and modeling system requirements, dynamic system behavior needs to be comprehended. Time Domain Analysis facilitates visualizing how such measures of performance as response time and stability vary under varying conditions. This analytical approach supports in-depth decision-making processes, particularly when dealing with ill-defined and new problem space problem-solving tasks. Since systems typically get conceived with different environmental contingencies, the ability to analyze them over time ensures final specifications' robustness and flexibility (Jureta et al., 2015). Therefore, time domain solutions are not speculative but the essence of generating pragmatic engineering results. Besides that, Time Domain Analysis intersects with theoretical frameworks for abstract ideas in security and safety, say. As per research, our understanding of abstract concepts is highly reliant on experience, and therefore context takes center stage in the analysis (Borghi et al., 2016). In the context of system evaluation, it becomes essential to specify time-based parameters that influence abstract concepts such as risk and reliability. By observing systems in time domain perspectives, the task of modeling abstractions becomes an easier endeavor, hence the simpler development of simpler definitions and representations essential to cognitive processing. This fusion between time domain data and abstract concept formation focuses on the multidimensional nature of analysis by integrating empirical data with interpretative frameworks responsible for the comprehension and optimization of systems over time.

## *Importance in Various Fields*

Time domain analysis finds crucial applications in a broad spectrum of areas, particularly physics and engineering, where the temporal dynamics will have essential impacts on system behaviors. In wave equations, such as scalar field dynamics, effective potential comprehension close to singularities in space-time illustrates the significant implications of time domain analysis. For example, experiments demonstrate that the effective potential possesses a universal leading x^{-2} inverse square law when viewed in scalar waves surrounding singularities that satisfy the Dominant Energy Condition (DEC) (Giveon et al., 2006). This information not only contributes to theoretical models within gravitational physics but also gives back to applications in practice, such as the building of stable particle trajectories in high-energy environments. Consequently, the scope of time domain analysis extends to engineering disciplines whose design issues are founded on the predictability of time-dependent behavior. Building on disciplines such as fluid dynamics, time domain analysis provides a good basis for explaining turbulent-like complex behavior. By employing artificially synthesized velocity fields, researchers have been able to develop an integral method of analyzing turbulent flow and its structures (Chernyshenko et al., 2006). This decomposition means that traditional assumptions about the relation between streak and eddy structures need to be challenged now, leading to more complex models of fluid dynamics. Such observations are especially significant to many engineering problems, like aerodynamics optimization for transportation systems as well as enhancing mixing processes in chemical reactions. With the emergence of complexity of time-dependent behaviors through this window of analysis, engineers are able to better forecast problems and develop better solutions that honor the underlying principles governing fluid motion. Furthermore, the implications of time domain analysis translate over into telecommunications and signal processing applications, defining the way in which data is sent and interpreted over time. With the use of time domain methods, engineers are able to isolate and analyze signal behaviors, which provide clearer integrity for communication systems. By converting signals from their time domain representation into the frequency domain, they can be able to detect noise and interference so that filters and equalizers can be effectively designed and implemented to maintain signal fidelity. With the advancement of telecommunication technology, efficient time domain analysis becomes increasingly evident, providing useful insights into temporal signal properties. In summary, the numerous applications of time domain analysis illustrate its standing in many areas, reaffirming its status as a pillar of scientific investigation and applied use.

## 3.1. HISTORICAL BACKGROUND

The time domain analysis is rooted in the history of the development of signal processing techniques, which has had a wide impact in the field of engineering, physics, and information technology. The field traces its early beginnings to the creation of basic mathematical models and physical models of signal evolution as a function of time. As technology evolved, particularly with the advent of digital computing, researchers found it possible to study more complex phenomena, leading to more complex time domain analysis. This laid the groundwork for modern applications across a variety of domains such as audio indexing and seismic data interpretation, all based on the concepts of time domain to process and analyze vast amounts of data with ease. With the progress of the discipline, the role of audio indexing itself was significantly significant as far as protecting cultural heritage through the effective use of temporal data processing was concerned. Studies like that documenting the application of audio indexing tools illustrate how a series of historical audiovisual content can be indexed and re-purposed for use nowadays. These technologies use language models and acoustic analysis for cross-media linking and explicitly reveal a critical linkage between archiving historical data and existing data processing techniques. This research work points out how indexing tools need to be optimized in order to meet disparate user needs for further enhancing the usability and availability of past cultural heritage and ultimately developing a bridge between technology progress and history (Heeren et al., 2006). Moreover, time domain analysis has made significant progress in geophysics, notably with the integration of contextual neural networks or machine learning techniques into seismic processing. The recent developments include technologies that synergize traditional time domain analysis with advanced computational techniques to facilitate more accurate interpretations of geological structures and behaviors. For example, the method of incorporating 4D seismic data into history matching through the use of machine learning demonstrates a sophisticated approach to back-reconstructing petrophysical properties in subsurface reservoirs. This is done through the combination of techniques such as the Ensemble Smoother with Multiple Data Assimilation, which addresses inversion of historical data challenges (Etienam et al., 2019). These advances testify not only to the adaptability of time domain analysis but also to its use in meeting challenges facing it today, demonstrating the enduring presence of classical methodological traditions in modern scientific practice.

### 3.1.1. Development of Time Domain Analysis Techniques

The development of methods of time domain analysis has been influenced significantly by the advances in other fields of science, such as quantum physics and engineering. When researchers explored dynamics of quantum systems, new techniques emerged that enhance our ability to monitor and regulate the systems in real time. The building of quantum feedback control theoretical models is a testament to the necessity of adapting existing time domain approaches to incorporate quantum processes' subtleties. This union of mature control theory and quantum mechanics not only illuminates entanglement and measurement back-action complexities but also guarantees enhanced understanding of the non-equilibrium dynamics implicated in complex quantum systems, therefore making it worthy to emphasize the central place of sophisticated analytical methods in addressing new experimental issues (Doherty et al., 2000). In addition, applications of time domain analysis techniques cross basic physics into various areas of engineering, particularly for intelligent transportation systems (ITS).

The integration of innovative sensing and communication technologies into vehicle networks is a cross-disciplinary initiative driven by the challenges of modern transport. As researchers and practitioners join hands to optimize such networks, time domain analysis assumes a role in ensuring real-time data processing as well as traffic control. These advances usher in opportunities for greater safety and efficiency of ground travel, addressing the urgency of demands in metropolitan areas. In addition, recognizing the potential of IoT technology for creating intelligent infrastructure further highlights the significance of time domain analysis in realizing future innovation in the transport sector (Mao et al., 2020). As time domain analysis techniques evolve, they provide unprecedented access to many complicated systems. The availability of newer-fangled technologies, such as IoT and quantum feedback control, highlights the significance of interdisciplinarity for pushing analytical methodologies forward. In pushing these techniques forward, researchers are able to tackle significant issues, ranging from optimizing vehicle networks to exploring theoretical questions in quantum mechanics. This expansion not only contributes to the power of time domain analysis but also underscores its central role in encouraging innovation across a broad variety of scientific and engineering fields. Ongoing investment in research and the incorporation of analytic structure will undoubtedly contribute to future advances, emphasizing the long-term relevance of time domain analysis to dynamic system complex behavior (Figure 3.2).

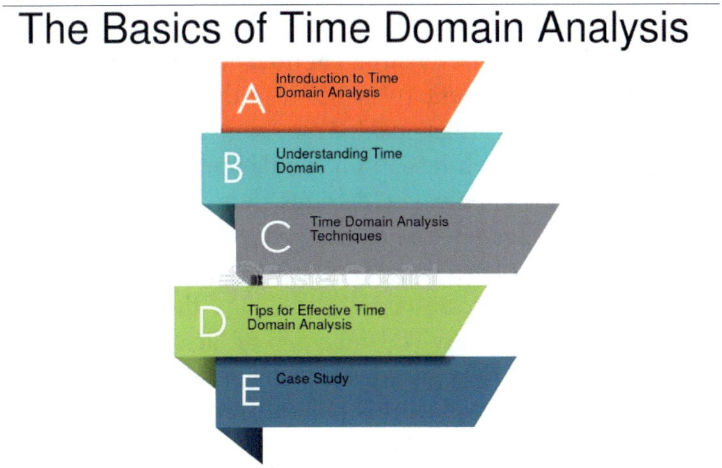


**Figure 3.2**. *Development of time domain analysis techniques.*

*Source: https://fastercapital.co/i/From-Time-to-Frequency--FFCS-for-Comprehensive-Insights-update--The-Basics-of-Time-Domain-Analysis.webp.*

## 3.1.2. Key Figures in the Evolution of the Field

The history of time domain analysis has been greatly influenced by numerous notable figures whose efforts have set the method and uses of the field. Among all the influential characters, mathematical physicists' contributions have played the most significant roles, particularly in the understanding of complex phenomena such as phase separation and flow dynamics. For instance, the application of the Navier-Stokes-Cahn-Hilliard model equations gives a comprehensive framework to simulate phase separation processes under flow, specifying the intricate coarsening phenomenon that happens under established parameter regimes. This modeling offers insight into the importance of flow dynamics as well as diffusion in the phase transition scenario, specifying the complexity of time-dependent systems and their research in the current science, as illustrated in Naso et al. (2017). Moreover, advances in time domain analysis have also been stimulated by feedback from researchers at the boundary between fluid dynamics and magnetohydrodynamics. These researchers have demonstrated how external magnetic fields affect fluid behavior and circulation, particularly in low magnetic Prandtl number fluids. Their work proves the breakdown of current circulation theorems under certain conditions and therefore challenges previous assumptions made in fluid dynamics. Via examination of various vortex solutions, such as Gaussian and elliptical vortices, these studies provide crucial information on circulation loss mechanisms, with a emphasized dependency on applied forces. Such studies emphasize the interdisciplinary nature of flows and the overall need for rigorously stringent time domain analysis, highlighted in Diamond et al. (2018). The interplay between theoretical development and numerical computation has further enabled the comprehensive understanding of time domain phenomena. High-resolution simulations have been employed by investigators to validate theoretical predictions along with the detection of new regimes that do not obey standard scaling laws. By comprehensive examinations of flow-pattern maps, interdependencies of viscosity, diffusion, and coarsening behaviors have been clarified, with an illustration of the use of these analyses in various physical conditions. The meeting of theory and computational techniques has offered a sound platform to deal with intricate systems, and this has been subject to further studies that are still broadening the field of time domain analysis. In brief, the contributions of these pioneering scientists enable us to gain a thorough understanding of dynamic systems and push the horizons of traditional methods and investigations, resulting in enormous breakthroughs in scientific research.

### 3.1.3. Milestones in Time Domain Analysis Research

Among many milestones on the path of time domain analysis research development, the advent of Virtual Research Environments (VREs) is a break-through. VREs have revolutionized collaborative research practice by facilitating interaction between researchers globally. Growing use of VREs reflects growing realization of the intricacies involved in time domain analysis, particularly in interdisciplinary research. As indicated by recent studies, researchers must leverage such environments in order to adequately address the intricate issues caused by temporal data. While the concept of VRE is still in its infancy, user perception measurement is crucial towards defining its success and consequential potential for innovations in this field with an aim to enhance overall efficiency in collaborative research in time domain analysis (Amaratunga et al., 2009). Secondly, the use of agile software development methods in time domain analysis is also a major milestone. Agile methods, widely utilized in the software world, are used more and more in cross-disciplinary projects where timely data analysis is crucial. The SOFIE project represents a good example, showing how agile methods can be used for effective development of systems with high levels of complexity, such as the stability test bench designed to improve bicycle safety. Against this background, efficient application of agile methods provides potential for timely reaction to research procedures and thus enhances the time domain analysis flexibility conducted in collaborative environments. Secondly, the comparisons made between the complexity of the SOFIE project and traditional large-scale projects emphasize the importance of these methods in the innovation of time domain analysis research (Bonnema et al., 2012). Finally, the convergence of technological advance and cooperative paradigms has ushered time domain analysis into a new era of scientific inquiry. In its ongoing evolution, researchers are making increasingly greater use of new methodologies that harness the power of advanced computational tools and high-performance data analysis. This transition is crucial, as it will allow for deeper understanding of temporal dynamics and improved simulation of complex phenomena. The ongoing confluence of various disciplines and approaches also works to highlight the growing relevance of time domain analysis to a broad range of applications, from engineering to social science. Furthermore, ongoing engagement with new tools and paradigms will help ensure that advancements in time domain analysis research in the future are informed, building the discipline into a dynamic field marked by continuous evolution and refinement in response to changing realities of research (Figure 3.3).

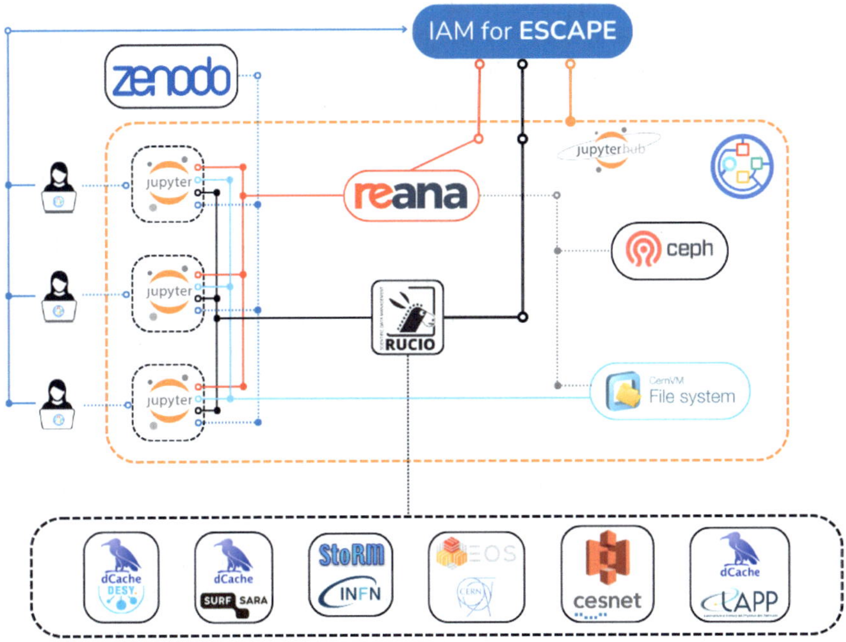


**Figure 3.3**. *Virtual research environments (VREs).*

*Source: https://vre-hub.github.io/assets/images/VRE-diagram-82148ece12c-1fa3d116c4d0b38ef4af0.png.*

## 3.2. FUNDAMENTAL CONCEPTS

The examination of time as a fundamental concept in various disciplines is at the core of exploring complex systems and their dynamics. In physics, time is not merely measured; it is the causal connection between events and how these events are in causal succession. The philosophical foundation of time suggests that our understanding of reality is temporal in nature, and this influences how we construct knowledge. In time domain analysis, time manipulation of signals unwinds the hidden patterns, and that is why the method is a useful tool for scientists and engineers. By examining how systems respond to change over time, we can uncover the richness of their behavior so that we can model and predict effectively to inform technological advances and theoretical models. Additionally, time domain analysis has methodological foundations which converge with central ideas in social science and psychology. By employing methods such as the second-best theorem, researchers are able to ask questions about how variations in timing and context influence decision-making. Wiens et al. (2020) highlights the application of such mathematical models beyond those traditionally charted, emphasizing the need to identify temporal dynamics for effective policy application. Such convergence of theoretical frameworks and practical applications inspires out-of-the-box thinking to overcome challenges from various domains. Finally, inclusion of time in these topics is a way through which one may understand highly complicated things more profoundly, affirming the thesis that time is not just the platform on which things occur but rather a constituent element of analytic constructs. In cognitive psychology, representational concept in the temporal realm is another instance of the reasons why time domain analysis is significant. The conceptual spaces theory illustrates how concepts operate in a setting that is heavily tied to time, pointing out that our understanding of natural concepts is based on their temporal aspects. Douven et al. (2019) argues that natural concepts are identified based on their place within an ordered similarity space that is temporal dynamics-dependent. The combination of time and conceptual representation provides evidence of human thinking, as it points to the importance of timing in shaping our perception and classification of information. Thus, the incorporation of time as a central concept enhances our ability to analyze, propelling inter-disciplinary discourse that increases our understanding of theoretical frameworks as well as practical implementations in various disciplines.

### 3.2.1. Time Domain Versus Frequency Domain Analysis

Frequency and time domain analysis difference is essential to establishing signal behavior and design of various electronic systems. Time domain analysis focuses on how signals develop over time, measuring their instantaneous amplitude variation. Time domain analysis is particularly helpful for observing the transient effects, such as the response of a system to a step input, and thus is vital in measurements involving timing and describing performance. For instance, time-to-digital converters (TDCs) are examples of time domain analysis that provide highly precise digitization of timing information important in mixed-signal circuits like digital phase-locked loops (PLLs) and analog-to-digital converters (ADCs). These circuits detect intervals of pulse signals or clock signal edges with very good resolution, as depicted in recent work where TDC architectures were reported with 0.25 ps resolution, reaffirming the relevance of time domain analysis for future high-performance electronics (Katragadda et al., 2018). Frequency domain analysis provides a complete picture of frequency components that constitute a signal so that sinusoidal patterns in complex signals can be detected. This approach is beneficial for simplifying the analysis of linear time-invariant systems, where signals may be decomposed into their frequency components through methods such as the Fourier Transform. With an understanding of how different frequencies impact a signal's response, engineers are able to build filters and systems specific to specific responses at specific frequencies.

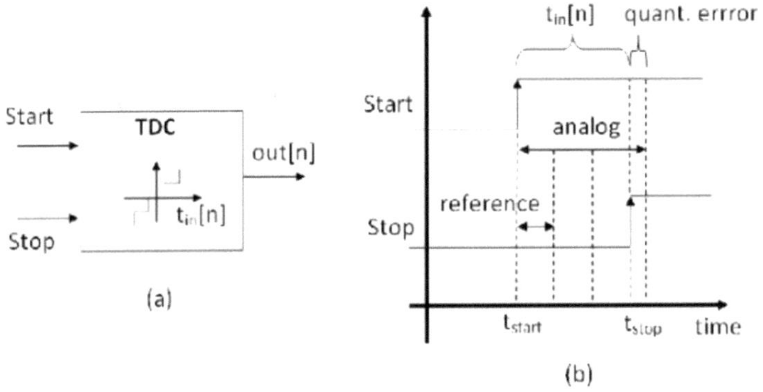


**Figure 3.4**. *Concept of a time-to-digital converter.*

*Source: https://www.researchgate.net/profile/Michael-Kennedy-28/publication/252022855/figure/fig1/AS:669022429392914@1536518785619/Concept-of-a-Time-to-Digital-Converter.png.*

Frequency domain techniques also play a critical role in signal processing applications such as image and video processing, where graph signal processing (GSP) is a critical paradigm. GSP allows frequency domain concepts to be applied to irregular data structures, and it vastly enhances such operations as image restoration and compression, thus broadening the analytical tools that engineers have available to them in many different fields (Cheung et al., 2018). It is difficult to compare time domain and frequency domain approaches without discussing their complementary function and function within an integrated analytical approach. Time domain is better suited for applications involving real-time where important timing measurements must be made to assess system performance. Frequency domain analysis can assist with optimization and signal manipulation where sophisticated filtering techniques can be performed based upon spectral characteristics. Engineers frequently employ both methods collectively, as each contains certain information that can be utilized in design. For example, the time response and frequency performance can be predicted to minimize circuit design, which can enhance noise performance and system efficiency. This collective insight not only adds to enhanced knowledge but also brings forth new avenues for utilizing the best of both in applications of contemporary technology (Figure 3.4).

### 3.2.2. Key Parameters in Time Domain Analysis

An important aspect of time domain analysis is to discover key parameters influencing the efficacy of signal processing algorithms. Latency and accuracy are some such parameters that form the core of determining the effectiveness of real-time applications, especially in cases like live vocal pitch detection. The issue is one of balancing high pitch estimation accuracy with minimizing latency so as not to cause perceivable delays in audio rendering. Consequently, scientists have been keen on exploring the effects of signal processing techniques on these indicators, with the evidence of a literature gap about the implications of latency for real-time audio processing. These analyses are significant in that they reveal the basic trade-offs practitioners must deal with when designing stable systems for use by the ears, where small delays can impair user performance and experience, making it even more imperative to continue research in adjusting these parameters (Firth et al., 2015). Further along within the category of techniques used under time domain analysis, one can observe the significance of choosing appropriate algorithms. For instance, different pitch detection algorithms yield various levels of performance based on the input parameters. Recent research has shown that IACF outperforms other methods in terms of balancing latency and precision. Knowledge such as this is helpful for practitioners and engineers because it provides a benchmark for selecting the best signal processing techniques based on specified applications. Furthermore, the continued refinement of algorithmic precision highlights the necessity for rigorous testing to establish best parameter settings for maximum effect, so systems can remain adaptive and sensitive to instant demands. In identifying these algorithmic dynamics, researchers assist in developing more sophisticated and efficient acoustic programs that require refined adjustments of multiple parameters (Firth et al., 2015). In fault diagnosis of equipment, time domain analysis also turns out to be a valuable tool by availing dominant parameters in order to interpret the signal better. Techniques such as stochastic resonance (SR) have turned out to be promising as far as the enabling of weak impulsive signals goes for the detection of faults in rotating machines. Correct application of such methods relies to a great extent on the management of both non-linear distortions and spectral characteristics of signals. With a moving least squares fitting method, researchers have demonstrated how time domain analysis is able to reconstruct signals effectively with a focus on preserving indispensable features needed for accurate diagnostics. This highlights the intersection of time domain analysis with real life, where it epitomizes the need for thorough parameter assessments to optimize signal integrity in the presence of interference. These advancements not only enhance diagnostic precision but also set the course for future research in signal processing technology (Figure 3.5).

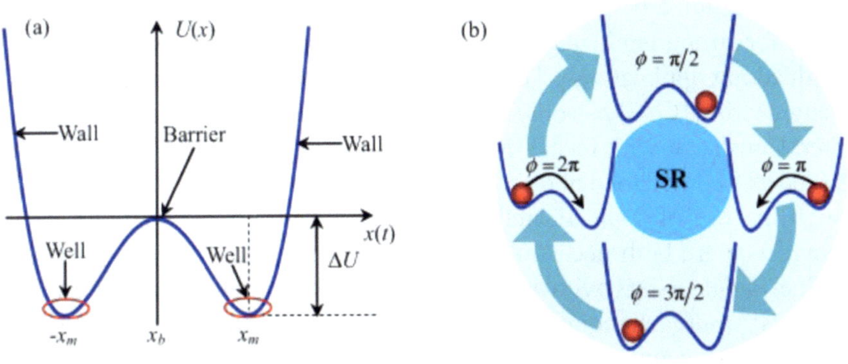


**Figure 3.5**. *Stochastic resonance.*

*Source: https://ars.els-cdn.com/content/image/1-s2.0-S0888327018308094-gr1.jpg.*

## 3.2.3. Mathematical Foundations of Time Domain Analysis

The mathematical foundations of time domain analysis are of utmost relevance across many fields, most importantly signal processing and systems dynamics. Since describing the response of systems to inputs in terms of time involves writing differential equations that define such behaviors, the laws of calculus apply because they help us analyze discrete as well as continuous signals. For instance, an ordinary differential equation is utilized to describe the output-input relation of a linear time-invariant system in order to gain insights into its transient response and stability. Thanks to development in computational algorithms, their role has increased even more for solving complex time-domain equations. This progress highlights the conjunction of mathematical theory and computational techniques to bridge the gap between theoretical mathematical ideas and practical technology and engineering applications. On this assumption, time domain analysis uses some mathematical tools for analyzing and questioning the time-dependent system behavior. They are the Laplace and Fourier transforms, which allow the transforming of time-domain signals into frequency-domain representations. This transformation allows engineers to take advantage of the properties of linear systems more easily, which are simpler to analyze. Further, the ability to transform these frequency-domain solutions into the time domain simplifies the interpretation of system responses in terms of their original context. The combination of these mathematical techniques and computational science enhances our capability to model and simulate dynamic systems significantly, allowing us to learn more and optimize better. This interconnectedness is a measure of the social demand for strong mathematical underpinnings in the study of time dynamics in engineering applications. The rapid advancement of computational science and engineering (CSE) has only broadened the horizon of time domain analysis further. Consistent with recent emphasis in the literature, CSE has come to be regarded as a cornerstone of discovery, generating innovation across disciplines through the use of sophisticated algorithms and numerical methods. Theory and evidence are both necessary as researchers confront tough questions that combine big data sets with sophisticated mathematical models. This cross-disciplinary approach requires a good understanding of mathematical logic and its application, as noted in the discussion of the future direction of this research area. The mathematical underpinnings are not only a collection of instruments to address current problems but also are credited with provoking future advances in science and technology, and hence are an indicator of the necessity for continuous development in mathematical concepts and computational techniques (Biros et al., 2016; Alexander et al., 2001).

## 3.3. APPLICATIONS IN ENGINEERING

Applications of time domain analysis in engineering span across several disciplines like electrical, mechanical, and civil engineering. Important information regarding system performance, stability, and response under varying conditions can be obtained by observing signals in the time domain by engineers. Specifically, in electrical engineering, time domain analysis enables one to analyze transient response in circuits and the way the systems respond to sudden changes, for example, voltage surges. This is achieved via computer software that allows engineers to visualize and model the dynamics. For instance, software like Win-eLab offers real-time Fourier analysis of audio-frequency signals, improved comprehension of complex systems, and improved learning for students in both theory and practice (Leis et al., 2009). The ability to analyze time-domain signals provides foundational knowledge that informs ensuing frequency-domain considerations that are crucial for advanced engineering applications. Time domain analysis finds application beyond electrical engineering into mechanical and aerospace applications where dynamic responses and vibrations must be comprehended. For example, in mechanical systems, time domain analysis is used by engineers to study the behavior of structures subjected to vibrations loads, which is of critical significance for structural performance and integrity. It can prevent deadly failure by predicting impending resonances that can destabilize systems.

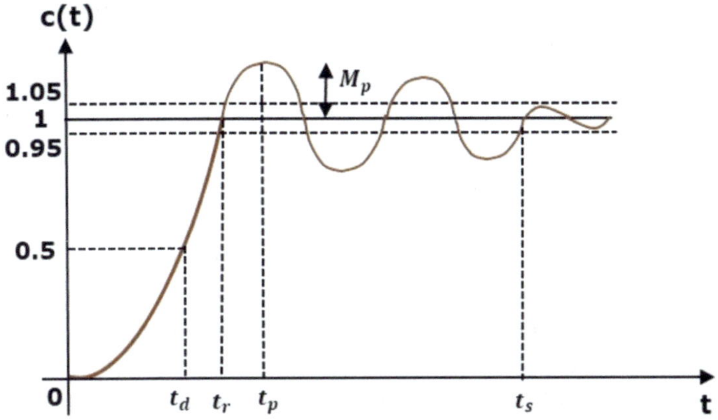


**Figure 3.6**. *Time domain specifications.*

*Source: https://www.tutorialspoint.com/control_systems/images/time_domain.jpg.*

In addition, uncertainty analysis software like TMsim provides depth to engineering designs. By using Taylor models, TMsim helps engineers perform parametric analysis and worst-case calculations so that designs are maintained within acceptable safety margins regardless of variable parameters (Manfredi et al., 2017). This two-fold depth of time domain analysis unites a comprehensive outlook towards engineering problems by effectively bridging theoretical analysis with real-world applications. Time domain analysis is an important area of study for civil engineering, whereby the dynamic response of structures when subjected to various loads, for example, seismic activity or wind, is determined. By observing how structures react over time, engineers can design strong structures that can withstand dynamic stress and continue to function and be safe. Moreover, real-time monitoring systems based on time domain analysis can provide vital information on structural integrity, and timely interventions and maintenance can be carried out. This crossing of technological and engineering practice emphasizes the requirement for time domain analysis in today's applications. Since engineers are becoming more reliant on high-end modeling and simulation tools, the crossing of time-domain and frequency-domain analyses becomes pivotal in ensuring the reliability and safety of engineering solutions across all disciplines. The evolution of analysis methods thus continues to improve the potential and effectiveness of engineering practices towards addressing real-world issues (Figure 3.6).

### 3.3.1. Signal Processing Applications

In the realm of signal processing applications, understanding time domain analysis is fundamental to a variety of fields, including communications, audio engineering, and medical diagnostics. Time domain analysis focuses on the behavior of signals as they vary over time, making it essential for the interpretation and manipulation of signal data. For instance, audio signals may be analyzed to determine pitch, amplitude, and timbre, all of which are significant in telecommunication and sound production. Computer software, including Win-eLab, reveals the potential for time and frequency domain displays to improve comprehension of complex signals. The software may be employed to manipulate signals in real time, making learning more engaging and improving comprehension of signal characteristics. Therefore, the time domain provides a basis for real-world applications that guide our technology world, highlighting its importance in educational environments as well as in working environments (Leis et al., 2009). Furthermore, development of advanced hardware platforms enabling signal processing, such as Field Programmable Gate Arrays (FPGAs), has significantly changed the application of time domain analysis in real-world applications. FPGAs offer reconfigurable architectures that are adapted to sophisticated signal processing applications, from image processing to network communications. Traditional FPGAs are limited by their characteristics, including large power dissipation and silicon use compared to custom ASIC implementations. Emerging technologies are aimed at bridging this gap with the design of reconfigurable fabrics that offer the power efficiency of ASICs and the programmability of FPGAs. Through the application of techniques like simulated annealing and local search, these advances offer improved design space exploration, which optimizes performance for DSP applications. This confluence of time domain analysis and adaptable hardware signals a trend toward signal processing solutions that are more efficacious and efficient (Mehta et al., 2009). The uses of employing sophisticated time domain analysis techniques extend beyond traditional uses and into cutting-edge technologies such as Internet of Things (IoT) devices and autonomous systems. Because such technologies rely extensively on real-time information for their operation, the ability to analyze and process time domain signals takes on paramount importance. Efficient algorithms based on time-domain and frequency-domain transformations can lead to better signal fidelity, reduced response times, and reduced latency—parameters that are significantly influential in the decision-making process of autonomous systems. With industries gravitating increasingly towards automated and intelligent solutions, the role of time domain analysis in signal processing applications is going to grow. Here, the improvement of a better understanding of time-dependent signal variations will remain the cornerstone to pushing technology forward and the optimal utilization of signal processing methods (Leis et al., 2009; Mehta et al., 2009) (Figure 3.7).

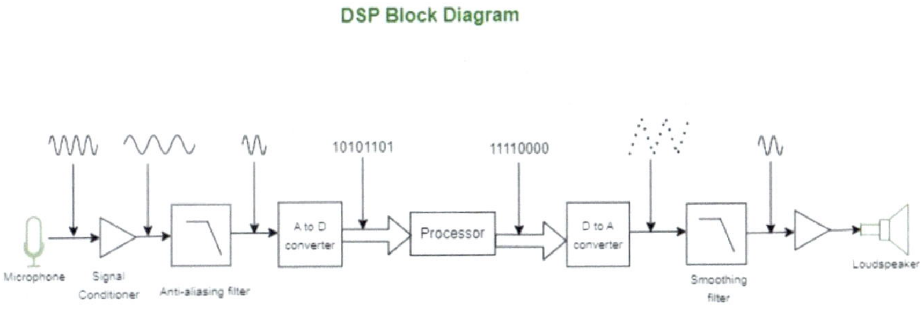


**Figure 3.7**. *Digital signal processing (DSP).*

*Source: https://media.geeksforgeeks.org/wp-content/uploads/20240403130126/Untitled-Diagram---2024-04-03T130110286.webp.*

## 3.3.2. Control Systems Analysis

Analysis of control systems is pivotal in comprehending and improving dynamic system performance via systematic measures and compensations. One of the primary techniques, time domain analysis, concerns itself with how systems respond with respect to varying inputs over time. Time domain analysis allows engineers to measure key performance measures like response time, stability, and steady-state error, which are vital toward ensuring system reliability and functioning. Along these lines, methods such as discrete time domain methods enhance the degree of granularity of analysis, enabling engineers to model the behavior of systems under varying conditions of operation and to adjust control mechanisms accordingly. The inclusion of powerful analytical methods enables more insight into complex interactions of systems, thus creating pathways towards significant enhancements in system design and optimization as evidenced in numerous applications in most engineering fields (Lee et al., 2025). Central to the analysis of control systems is the development of simulation tools that mimic real-world usage so that control methods can be suitably tested and tuned before they are applied. Time domain simulations expose useful information regarding the transient characteristics of systems and provide a controlled environment in which to experiment with control laws and system topologies. The advances in modeling techniques have allowed practitioners and researchers to examine many dynamic conditions, revealing the system behavior weaknesses. By means of accurate simulation of small and large signal behaviors, engineers can make systems such as switching regulators and dc-dc converters more optimal. These models not only facilitate theoretical research but are also central to practical implementation in methodology design so that the systems can attain high performance levels (Lee et al., 2025). Besides, contemporary control system analysis employs various methodologies that address the specific complexities of different applications, particularly in aerospace. Rotorcraft flight dynamics is a prime example of the requirement for sophisticated modeling due to their sophisticated nature and high-performance needs. The marriage of time domain methods with frequency domain analysis presents a highly integrated approach to modeling the difficulties in the case of high-g maneuvering tasks. Nonlinear models and parameter identification techniques have been found that cater specifically to these special demands, enhancing the ability to conduct flying qualities investigations efficiently. Thus, the complementary interaction of time domain and frequency domain approaches gives the engineer a toolbox needed for designing efficient control laws in the end, increasing the system's performance level in challenging environments.

### 3.3.3. Structural Analysis in Civil Engineering

In civil engineering, the analysis of structures is of utmost significance to create safe, long-lasting structures, particularly when loaded dynamically, such as earthquakes. It is important to understand how structures react under these external loads, and time domain analysis is one of the effective methods which gives adequate information. This technique allows engineers to emulate the time-varying behavior of structures under various conditions, mimicking the dynamic responses that are very important in ensuring structural integrity. By taking into account temporal loading conditions, structural engineers achieve better predictability of failure points, resulting in a more effective design process. Moreover, the employment of advanced computational techniques contributes to creating more realistic models simulating real conditions in a situation, which becomes simpler to verify potential weaknesses in structural systems under adverse environmental conditions. Time domain analysis is particularly helpful in seismically active areas, as standard static analysis methods may not be efficient. For seismic movement, structures are subjected to rapid and varying loads that standard methods may not be able to replicate efficiently. With the application of time domain techniques, engineers can explore the entire spectrum of dynamic responses, including resonance behavior and nonlinear material response (Burrus et al., 2007). This is done through a variety of analytical tools ranging from accelerated response spectra to nonlinear dynamic analysis that contribute towards the overall picture of a structure's response to seismic loading.

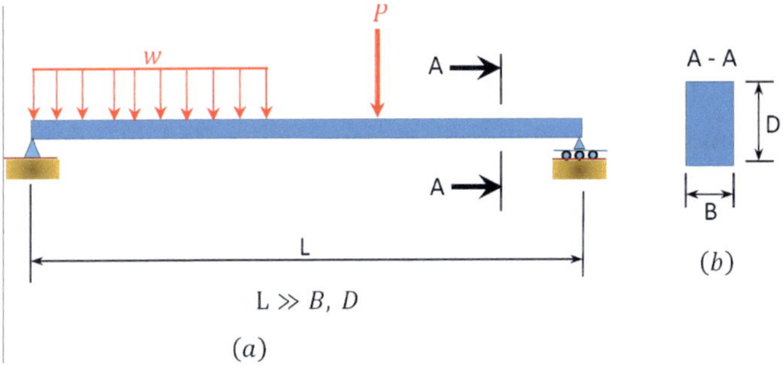


**Figure 3.8**. *Structural analysis in civil engineering.*

*Source: https://temple.manifoldapp.org/api/proxy/ingestion_sources/1dbcd297-2c1e-456b-8488-f181a996b96b.*

Moreover, techniques such as the harmonic wavelet transform introduced in Burrus et al. (2007), enable civil engineers to capture very delicate aspects of ground motion and structural response. The accurate analysis not only enhances design safety but also enables engineers to incorporate adaptive interventions for retrofitting structures, thus promoting longevity. In the past couple of decades, integration with advanced technologies such as Artificial Intelligence (AI) has changed structural analysis. AI methods such as machine learning and artificial neural networks are increasingly being utilized to enhance Finite Element Analysis (FEA), which is a pillar of structural engineering design. Such intelligent systems can easily handle large data sets to optimize designs and predict structural behavior under different conditions. They enable the formation of fuzzy cognitive mappings that describe complex relationships between various structural parameters. By employing such complex frameworks, engineers are able to enhance decision-making regarding safety and reliability issues in structural designs. With the increasing advancements in AI capabilities, the future of structural analysis within the field of civil engineering promises greater accuracy and efficiency, eventually contributing to safer and more resilient infrastructure as buttressed by time domain analysis methods (Figure 3.8).

## 3.4. APPLICATIONS IN PHYSICS

Physics' dynamic systems need to be probed with a deep understanding of time domain analysis, particularly its use in any branch of physics. Time domain analysis is the fundamental tool through which quantum systems can be controlled and observed in quantum physics. Scientists can now observe individual quantum states in real-time using sophisticated experimental techniques, providing invaluable insights into quantum dynamics. This leap of capability not only speeds up the development of quantum feedback control techniques but also makes us understand better quantum phenomena such as entanglement and measurement back-action. By detailed time domain analysis, physicists are able to sharpen control theories required to handle the complexity of quantum systems, which can revolutionize experimental physics in the quantum regime, as highlighted in Doherty et al. (2000). Another significant field of time domain analysis is in the case of dielectric mixtures, which are extremely significant in materials science and electrical engineering. The study of dielectric properties using time domain methods makes it possible for scientists to determine the material's response toward various electric fields and frequencies. Particularly, polarization at the interface which occurs in dielectric mixtures could produce various dielectric responses according to the mixtures' constituents and concentration. This aspect is especially significant for developing new materials for capacitors and insulators uses. Application of numerical simulations along with time domain analysis makes it possible for scientists to abstract a better picture of behavior of these mixtures under different operating conditions, which in turn indicates means of enhancing material properties in actual applications as mentioned in Gubanski et al. (2001). Moreover, time domain analysis goes beyond traditional boundaries by providing a framework for studying complex quantum systems in statistical mechanics and quantum information theory. As theorists continue to bridge traditional control theory approaches with quantum mechanics, understanding acquired from time domain analysis is important to comprehend non-equilibrium quantum system dynamics. The unique properties of quantum feedback and the ability to analyze large quantum systems by an input-output approach can provide useful insights into the information processing in the quantum regime.

Basically, the capability to use time domain analysis on these new fields of research not only strengthens theoretical debate but also provides entry points into new technologies and computational methods, pointing to the extensive reach of time domain techniques in modern physics (Figure 3.9).

**Figure 3.9**. *Electrical engineering.*

*Source: https://engineeringonline.ucr.edu/blog/media_15239f713a03fb0ffdb7b8b9692095c3863bca339.jpg?width=750&format=jpg&optimize=medium.*

### 3.4.1. Time Domain Analysis in Wave Mechanics

Time domain analysis of wave mechanics is a valuable tool in the understanding and prediction of wave motion when waves propagate through various media. Time domain analysis allows scientists to track changes in the waveform as a function of time, displaying all types of interference, reflection, and diffraction effects. Unlike frequency domain analysis, in which data are smoothed over a broad spectrum of frequencies, time domain analysis offers a closer view of the manner in which particular wave events develop. With numerical models such as the three-dimensional boundary element-finite element (BEM-FEM) method, researchers can simulate complex interactions between buildings and the soil around them. It encourages an overall perception of soil-structure interaction, more so in terms of the impact of out-of-core wave fields on structure health (Galvín et al., 2014). One of the major characteristics of time domain analysis is its capability to plot transient responses to dynamic excitations, which is particularly important in engineering applications. By simulating wave propagation and interaction in real time, specialists can develop realistic means of minimizing adverse impacts on structures exposed to earthquake activity, underwater explosions, or other transient loadings. As an illustration, the approach in the time domain enables one to investigate how Rayleigh waves influence foundations, a crucial aspect of building design for seismically active areas. Moreover, the incorporation of internal soil damping as a feature enhances the precision of such analyses and thus predictions in structural responses. The in-depth modeling like this not only enables profound theoretical insight but also assures practical application leads to safer and more precise outcomes (Colonius et al., 2020). Time domain analysis, in examining how waves develop, illustrates the fundamental role of nonlinearity in wave mechanics. Whereas linear models create basic approximations, nonlinear analysis describes the subtle wave interactions among waves of various amplitudes and frequencies. This is particularly important when there is turbulent flow, and disturbances can change from laminar to chaotic regimes, depending on higher harmonics and triadic interactions. The incorporation of these nonlinear parameters into time domain models makes the forecasting of drag forces and stability limits in fluid dynamics more robust. Ultimately, an integrated system of both linear and nonlinear components in the time domain leads to better comprehension of wave mechanics, which paves the way for innovations in fields as disparate as civil engineering through to aerodynamics and beyond (Colonius et al., 2020) (Figure 3.10).

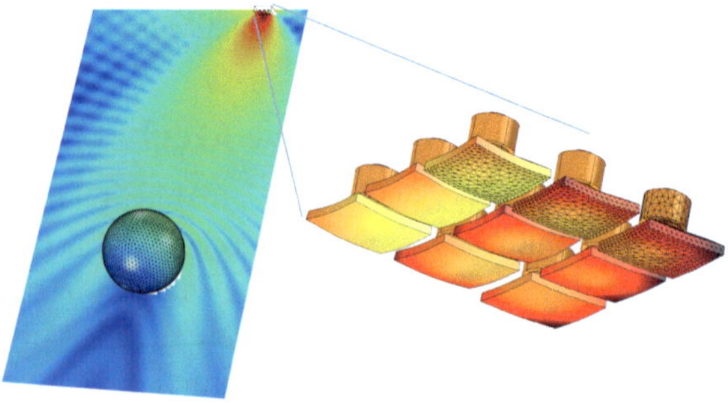

**Figure 3.10**. *Boundary element method.*

*Source: https://cdn.comsol.com/wordpress/2018/03/boundary-element-method-comsol-simulation.png.*

## 3.4.2. Role in Quantum Mechanics

Quantum mechanical involvement in time domain analysis is the most important since it provides a framework within which temporal processes can be explained and described. For quantum mechanics, the temporal behavior of a system is governed by its wave function, which determines the probability of a system being in various states at any given time. This intrinsic property allows physicists to study how quantum states evolve, providing an understanding of quantum phenomena such as tunneling, superposition, and entanglement. Time domain analysis allows the study of transient effects, which are necessary to understand dynamics in quantum systems. Such analyses can define core information about system performance and stability and ultimately inform technology developments and design innovative solutions to practical problems. In addition, the synergy between quantum mechanics and time domain analysis has bearings on multiple theoretical and experimental frameworks. For example, the AdS/CFT duality has a crucial function in unifying alternate physical theories by connecting quantum field theories to gravitational theories, especially in high-energy physics. By extending this duality to time-dependent settings, scientists can explore intricate systems such as quark-gluon plasmas. The investigation of such systems holds profound implications for the comprehension of the very essence of matter under extreme conditions, as a bridge between theoretical constructs and experimental reality. Even if empirical verification is lacking here, the role that gauge/gravity duality plays in capturing physical reality in high-energy regimes cannot be overstated (Dardashti et al., 2018). In addition, the use of quantum mechanics in time domain analysis brings supersymmetry into the picture, which further widens our understanding of particle interactions. An example is the study of scattering off domain walls for fermions, which reflects the complexity of structures enabled by supersymmetric quantum mechanics. This study demonstrates the need to create higher-dimensional representations and plot the impact of external factors, for example, magnetic fields, on quantum systems (Kleidis et al., 2014). The study of behaviors such as the Witten index also reiterates how these quantum interactions are stable over time and subjected to external perturbations. Together, the interaction between quantum mechanics and analysis of time not only enhances our knowledge of basic physics but also presents doors for real-world applications in new technologies.

### 3.4.3. Applications in Thermodynamics

The field of thermodynamics is highly related to many applications for a better understanding of physical systems. The area dealing with mixtures, particularly ideal gases, is extremely significant. The understanding of such mixtures can significantly enhance our understanding of thermodynamic behavior, mainly when placed in the context of an action principle. This method allows researchers to get more accurate results, such as the Saha equation, which elegantly describe chemical reactions like hydrogen dissociation. In addition, the gravity corrected Gibbs-Dalton hypothesis makes predictions in agreement with experiment more closely than the usual hypothesis. Ultimately, such systematic treatments thus highlight the importance of thermodynamics to produce insights into dynamics of mixtures towards facilitating increased applicability in theoretical as well as experimental studies in time domain analysis (Fronsdal et al., 2010). In addition to ideal gas mixture study, recent developments have also stressed the necessity for inhomogeneous temperature gas-liquid two-phase flow models. These models employ the Peng-Robinson equation of state and provide superior accuracy compared to the classic van der Waals model. This is an improvement that allows for a more accurate description of complex fluids, especially hydrocarbons, and contributes to our understanding of their thermodynamic behavior. One significant achievement of this study is having found a relationship between pressure gradients and temperature and chemical potential gradients, which implies that these gradients are forces that dominate fluid motion. By unifying the first and second laws of thermodynamics, such sophisticated models not only enhance our knowledge of fluid dynamics but also provide sound foundations for time domain analysis in thermodynamic systems (Kou et al., 2017). The consequences of these outcomes extend beyond mere theoretical results; they pave the way for designing efficient numerical schemes that honor the laws of thermodynamics. With rigorous proofs, researchers have established that the Helmholtz free energy density is a concave function under some physical constraints. This information can be utilized to develop novel, thermodynamically consistent numerical schemes that respect the basic laws of thermodynamics. The ability to simulate gas-liquid interactions appropriately not only enjoys daily uses in various sectors, such as petroleum engineering and environmental science, but also assists in the exploration of long-standing problems in thermodynamics. Such convergence of theoretical models and computational techniques is a feature of the forward progress of the discipline and its suitability for time domain analysis, ultimately leading to a better appreciation of dynamic systems (Kou et al., 2017).

## 3.5. TOOLS AND TECHNIQUES

In time domain analysis, the selection of the right tools and techniques is of great importance to process the signals efficiently and extract valuable information. The conventional digital signal processing (DSP) techniques provide the foundation in which many advanced tools can be constructed. One of the significant characteristics of such tools is that they can perform sampling and filtering, something that is extremely crucial in managing noise and ensuring that the time-dependent signal's proper characteristics are properly represented. Advances in graph signal processing (GSP), for instance, shed new light on how data defined on irregular graph domains can be high-efficiency processed. By the core principles of GSP, researchers are able to extend conventional DSP methods so that not only do they continue to serve previous applications but also adapt in ways of keeping up with newer scenarios of data complexity expansion (Frossard et al., 2018). Other than sampling and filtering techniques, the presence of distributed diagnostic software further broadens the toolset for time domain analysis. One of the best examples of this advancement is the introduction of high-performance grid-based signal search software, particularly in fields such as aerospace and medicine. The development of the Signal Data Explorer is a case in point, demonstrating the power of distributed processing of data through real-time analysis of vast data sets produced by advanced systems. This equipment makes possible the diagnosis and prognosis of many conditions with the ability to analyze tremendous volumes of electrocardiograph and sensor data with extremely high rates and accuracy. These machines not only enhance the effectiveness of time domain analysis but also provide a canopy under which to examine time-critical data in various industries with a resultant innovation in diagnostics and predictive maintenance (Austin et al., 2006). Moreover, application of advanced techniques in time domain analysis promotes multidisciplinary collaborations that further enrich the tools. The exploration of GSP across different application fields—ranging from sensor networks to bioinformatics data analysis—indicates the flexibility and adaptability of existing methodologies. These techniques continue to improve with an increasing volume of research that combines concepts from engineering and computer science. As new challenges arise, particularly with increasing complexity of data environments, technologies that facilitate efficient analysis and visualization will remain the backbone of effective time domain analysis. Apart from directing optimal practice in specific fields, these breakthroughs drive the overall level of insight necessary to counter future innovations in data handling (Frossard et al., 2018) (Figure 3.11).

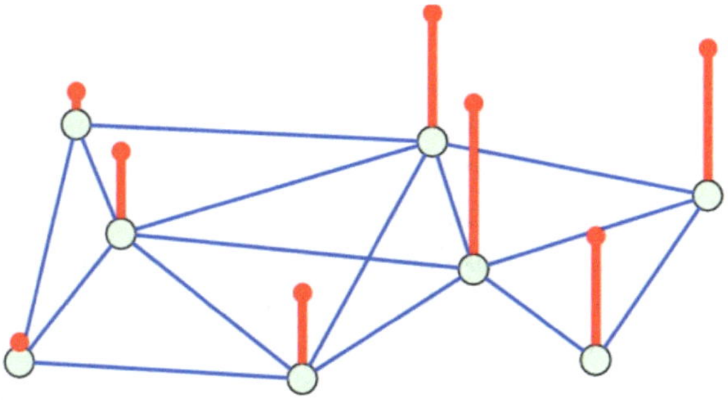

**Figure 3.11**. *Graph signal processing.*

*Source: https://miro.medium.com/v2/resize:fit:1400/1*CeErZYwR-N3N-lqcB3Yk1UQ.jpeg.*

### 3.5.1. Software Used for Time Domain Analysis

Time domain analysis relies on sophisticated computer software packages that handle and analyze time series data. Most time domain analysis software packages include the application of nonlinear modeling methods such as GAMs to deal with nonlinear relationships in data. Through the application of Bayesian methods, these packages enable a better examination of complex phenomena through Markov Chain Monte Carlo (MCMC) simulation techniques of inference. One example especially worth mentioning is the BayesX package, which allows for users to describe different response distributions efficiently. Such versatility in coping with complex datasets is incredibly useful in a range of applications, from ecological forecasting to health insurance analysis, ultimately allowing for researchers to extract meaningful information from their time-dependent data. With evolving analytical needs, ongoing development of such software ensures analysts continue to be able to address problems of contemporary time domain analysis in a timely fashion. In addition to specialized statistical programming tools, understanding and interpretation of time domain data also are made easier through integrated software environments that enhance program understanding. A traditional example is Conclave, designed to enable software maintainers to navigate and modify massive codebases that incorporate sophisticated time domain analysis functions. With its innovative blend of natural language processing and programming language compatibility, Conclave enables programmers to locate and understand critical features with ease. With the ability to create mappings from natural language problem descriptions and the corresponding code, Conclave streamlines the process of identifying features and functionality that are relevant to time domain analysis. This support is crucial for developers with the daunting task of maintaining or extending applications for time series analysis, eventually resulting in a better workflow and confidence in what changes are being made. Moreover, the use of several analysis tools simultaneously in such environments constitutes an interactive analysis methodology for examining time domains. The diverse facilities offered by software tools allow scientists to unleash the potential of ontologies that interlink code, problem domains, and program execution results. By supporting feature locating and code searching, computing environments like Conclave not only help to understand existing codes but also encourage innovation in developing new analytical algorithms. With the growing area of work, communication among various software tools will be crucial to enhancing the accuracy and efficiency of time domain analysis. Their ongoing advancement is proof of increasing recognition of the importance of sophisticated software for unlocking the full power of data analysis and improving decision-making across every discipline (Figure 3.12).

**Figure 3.12**. *Markov chain Monte Carlo (MCMC) methods.*

*Source: https://www.statlect.com/images/MCMC-vs-Monte-Carlo.png.*

## 3.5.2. Experimental Methods for Data Collection

In time domain analysis, experimental methods of data collection are essential in establishing the validity and reliability of results. For instance, observational studies, controlled experiments, and field studies are employed in collecting data that reflects actual events that take place in the world. One successful experimental method is manipulating independent variables to observe dependent variable changes, thus establishing causal relationships. For example, in time-domain analysis of visual information retrieval, local feature analysis techniques such as Scale Invariant Feature Transform (SIFT) can be employed to enhance the classification of visual data using robust data gathering mechanisms. Not only does this ensure the authenticity of the results but also permits meaningful interpretations of data in the context of the research, promoting a comprehensive understanding of temporal dynamics in areas of interest (Kesorn et al., 2010). Another critical characteristic of experimental approaches is an integration of qualitative and quantitative data for developing an even analysis. Triangulation, the incorporation of multiple sources or methodologies to study the same phenomenon, brings richness and breadth to the research findings. In time domain analysis, qualitative data gathered through interviewing or focus groups can be merged with quantitative findings from controlled experiments to present a more nuanced description of how temporal factors influence user behavior or system response. In addition, analysis and retrieval of sets of advanced data can be supported by ontology-based knowledge structures for better interpretation of the results. Such hybrid approaches enhance not only the quality of data collection but also bridge gaps between heterogeneous types of data, enabling a broader exploration of time-dependent variables (Nogueira et al., 2001). To further establish conclusions drawn from experimental methods, scientists must ensure that sound data handling and analysis protocols are in place. The application of advanced statistical techniques with traditional analysis methods is needed to interpret correctly time domain data. For instance, through the application of time series analysis, scientists can reconcile trends within stated times, delivering insights into patterns that would otherwise be invisible using single data points. Besides, the incorporation of machine learning algorithms in data processing can further enhance model predictive power for time domain analysis. By fusing rigorous data collection processes with sophisticated analytical methods, researchers can effectively make logical inferences about temporal dynamics to produce powerful findings that advance broad expertise in many areas. Though the quantity and depth of information are growing, experimental techniques for data collection will remain a priority in order to achieve insightful outcomes in time domain analysis (Kesorn et al., 2010).

### 3.5.3. Data Interpretation Methods

A critical aspect of time domain analysis is effective interpretation of data, which depends on the implementation of appropriate methods that yield useful information from complex sets of data. Traditional methods such as Frequency Response Analysis enable analysts to examine system behavior through a comparison of the response of an input signal within a given period of time. These techniques utilize metrics like correlation and spectral skewness to identify variations in system responses, increasing the precision of outcomes even further. Leveraging advanced graphical user interfaces, like those depicted in related works, users can effectively visualize and compare time-domain data, further simplifying analysis procedures (Ramírez et al., 2006). Such interpretative strictness is very important in use cases like from engineering diagnosis to seismology, where accurate interpretation of the data leads to fault detection and system evaluation success. Additionally, machine learning-enabled development, especially with Artificial Neural Networks (ANNs), has revolutionized the methods of data interpretation in a myriad of applications. As opposed to traditional algorithms that are bogged down by the amount and intricacy of input data, ANNs are designed to mimic human reasoning capacity and thereby improve the precision of event identification and categorization processes. In seismic analysis, for example, such networks can mechanize seismic trace interpretation by accelerating time-domain data processing of large amounts of data so that researchers can conduct more detailed analyses within shorter time frames (Alvarado et al., 2019). This transition to automated systems not only alleviates the burden on human analysts but also decreases inconsistencies typically found in conventional systems, establishing new standards of data interpretation reliability and effectiveness. Moreover, an integration of time domain analysis with multimodal data interpretation techniques enhances the overall robustness of analytical practice. By combining time-domain data and frequency response, as is practiced in the case of transformer lightning impulse testing, the possibility exists to better characterize system dynamics (Ramírez et al., 2006). This combination allows for quantitative indicators that can result in significantly improved accuracy in failure diagnostics and system diagnosis. As sensors become increasingly accurate and data accumulation accelerates, use of these combined approaches will be critical in boosting the interpretive power of data. Thus, acquiring this blend of approaches will not only increase the range of analyses but will also improve innovative solutions in predictive modeling and risk assessment across many applications.

## 3.6. CHALLENGES AND LIMITATIONS

In time domain analysis, among the key issues is the complexity of collection, processing, and interpretation of data. With scientists attempting to perform real-time monitoring with various sensor technologies, several data streams often end up causing inconsistencies and inaccuracy. For instance, computer-based methods for monitoring human activity and well-being, while innovative, tend to produce massive amounts of data that need to be analyzed efficiently through intricate algorithms. The diversity of sensors and the structures used exacerbates this, as it is difficult to harmonize results across different uses, as recent research uncovers. Thus, although monitoring progress is heartening, it is at the cost of gargantuan technical hurdles which need sustained improvement in data handling methodology for reliability and precision of results (Achumba et al., 2013). A second significant drawback of time domain analysis is the inherent nature of constraints exerted by the capabilities of cognitive radio networks, particularly in multiband settings. While multiband cognitive radio networks (MB-CRN) have the potential to excel over single-band systems—for example, increased throughput and improved channel maintenance—they also complicate spectrum access and management issues. The operating requirements of wideband front-ends deemphasize the intricacies of effectively sensing and leveraging the spectrum. Besides, recent research identifies that while considerable advances have been experienced in multiband spectrum sensing techniques, various design compromises remain that come in the way of optimal performance (Hattab et al., 2014). As such, the limitations inherent in spectrum management directly affect the performance of time domain analysis in communication systems with changing environments, opening up opportunities for future research. Finally, time domain analysis applied in real-world contexts is challenged for scalability and flexibility. In the majority of real-world contexts, particularly those involving monitoring based on sensors, being capable of scaling without compromising performance is crucial. As these types of systems are deployed to progressively diverse environments—spanning clinical settings through to assisted living settings—their functionality has to be capable of handling differing variations in input data and user behavior. However, the limitations in available methods do pose problems in the capacity to ensure analysis frameworks are adequately flexible and resilient to address such diversity effectively. This is also combined with the necessity for real-time monitoring and assessment, whereby available solutions tend to be slow in delivering timely insights. Solving these issues will involve creative methods that focus on flexibility while at the same time improving the analytical processes involved in time domain research (Achumba et al., 2013; Hattab et al., 2014) (Figure 3.13).

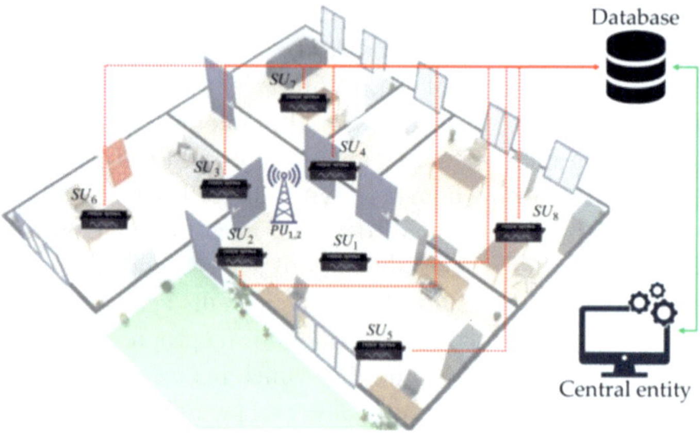


**Figure 3.13**. *Cooperative multiband spectrum sensing using radio environment maps and neural networks.*

*Source: https://www.mdpi.com/sensors/sensors-23-05209/article_deploy/html/images/sensors-23-05209-g011-550.jpg.*

### 3.6.1. Limitations of Time Domain Analysis Methods

The time domain analysis techniques, though central to many disciplines, have great limitations that provide difficulties in real-world usage. The use of assumptions that are not necessarily true in actual data situations is one main issue. Many of the techniques assume that the signals to be analyzed are stationary or have constant statistical properties over time. However, such an assumption is often violated, particularly in complex systems with shifting environmental conditions or non-linear behavior. Consequently, such assumptions' validity will detract from the credibility and precision of analyses. The issue of high-dimensional data is another challenge, where multi-dimensional time-series signals' complexities can be employed to deceive, discounting potentially crucial insights that are generated in noisy environments, which in turn depicts the need for improved methodologies to honor such complexities. Also, time domain analysis methods might suffer from computational intensity, particularly when they work with massive data sets characteristic of modern scientific research. Utilizing sturdy statistical procedures, like those used in the development of Kernel Additive Modelling (KAM), demonstrates this problem. Although KAM enhances separation performance by employing robust statistics over time-frequency bins, computational complexities involved with such methods would restrict their usage in practice in real-time applications or high-dimensional data sets. Such a constraint becomes increasingly important for areas like pharmaco-EEG, where recommendations found in updated protocols emphasize the need for standard procedures that facilitate high-quality data. However, inherent complexity in time domain analysis can retard efficient processing of pharmaco-EEG data, thereby affecting accuracy of neurological outcomes and drug evaluation, as explained by Babiloni et al. (2012). Further, the lack of standardization across time domain analysis methods may generate variability that affects reproducibility and validity of research findings. For instance, variations in analytical techniques between studies can lead to opposing findings, posing difficulties in making comparisons and impacting meta-analyses to aggregate data from different sources. This inconsistency renders it impossible to have benchmarks applicable universally, thereby impacting the stability of outcomes and the development of effective algorithms. Therefore, this case shows the critical need for transcending such limitations to advance the integrity of time domain approaches. Researchers will need to strive for increased standardization and optimizing analytical techniques, hence better placing them to make sound inferences and supporting solid methodologies, particularly for new fields that are based on advanced interpretations of time data, as noted in Babiloni et al. (2012).

### 3.6.2. Challenges Faced in Practical Applications

In time domain analysis, the professionals are usually confronted with a series of complex hurdles that form an obstacle to proper implementation and utilization. Integration of information from various sources is one of the main obstacles, particularly for multimodal devices. For instance, in the case of lifelogging, several devices such as smartphones and biometric wearables collect various types of information but are independent of one another in such a manner that it is difficult to create an integrated dataset. The absence of interoperability between such systems adds complexity because individuals cannot leverage the maximum potential of the accessible information. Solution to such integration issues relies not only on technological advancement but also new data management and user interaction design. Thus, effective use of time domain analysis in practical contexts largely relies on the solution to the issues raised by data interoperability and use (Byrne et al., 2012). Dynamic data analysis over time poses another challenge of significance when dealing with geospatial characteristics. The intricacies of how time variability affects geospatial data demand sophisticated modeling techniques that can capture these dynamics accurately. Conventional Geographic Information Systems (GIS) are prone to falter in handling real-time updates or integrating qualitative data sources, thus inhibiting their applicability in changing settings. The supporting technology must improve to facilitate high-level semantic and qualitative analytical capabilities as well as solve ontological and computational challenges that accompany dynamic data conditions.

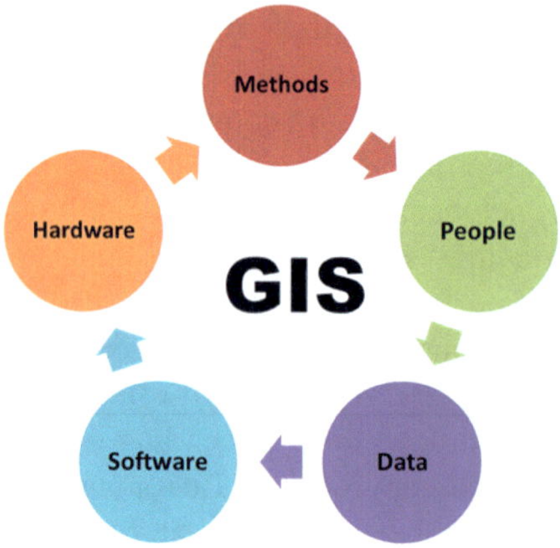


**Figure 3.14**. *Geographic information systems (GIS).*

*Source: https://www.arcweb.com/sites/default/files/2021-08/Geographic%2520Information%2520Systems%2520%28GIS%29%2520Components%5B1%5D.PNG.*

Therefore, creating time domain analysis in the GIS outlook grows more vital for the purpose of creating tools able to effectively model and visualize change with time, leading to more reliable findings and better decision-making processes (Bhatt et al., 2013). Finally, practitioners must contend with the limitations of current analytical tools, which may bewilder meaningful interpretations of results. For time domain analysis, suitable powerful analytical frameworks that can accommodate natural variability inherent in temporal data are required. Even so, most statistical and computational models suffer from issues such as noise, missing data, and real-time processing demands, which, by their very nature, tend to skew results and lead to potentially wrong conclusions. And when those models are combined with more advanced AI and machine learning techniques, complexity is again introduced, as clear and explainable models are now a flat requirement. Hence, a sound knowledge of the following mechanisms and analytical method limitations is of vital significance to practitioners looking to successfully address the overall challenges of practical applications of time domain analysis (Byrne et al., 2012; Bhatt et al., 2013) (Figure 3.14).

### 3.6.3. Future Directions for Overcoming These Challenges

In time domain analysis, the future is rich in directions for resolving longstanding issues of data interpretation and processing. One of them is the integration of multimedia summarization techniques, which can more easily allow us to understand complex time-dependent data. The conceptual paradigm of video summarization is most noteworthy as an example of why it is important to integrate internal content analysis with external information from the user to bridge the semantic gap in understanding visual streams of information. By means of such frameworks, practitioners can create more context-driven and meaningful representations of time data, leading to better decision-making processes in the end. Such a hybrid approach not only addresses the problem of relevance but even presents tailored insights that are in line with the user's needs in a big way, a crucial step towards comprehensive time analysis methodologies in industries like telecommunications and multimedia. Furthermore, technology advancement in massive MIMO technology has a revolutionary potential for the processing of time-domain signals transmitted across wireless networks. With the widespread use of full-digital transceiver arrays becoming increasingly viable, technological innovations in the spatial capability to process signals are now imminent. The transition to massive MIMO, which is achieving industrial traction at an unprecedented rate, is a source of stimulus to improve signal processing techniques that understand time-varying channels. By addressing legacy challenges like pilot contamination, researchers can adopt prospective new fields in time domain analysis for prospective applications like ultra-reliable communication and low-power machine-type communications. The invention of Massive MIMO will help in bringing the knowledge that advanced time-domain methods provide, enabling the networks to support more spectrum efficiency and longer service capacity, according to the increasingly growing needs of today's communications systems. Also, exploration of new research areas related to massive MIMO, for example, very large aperture arrays and six-dimensional positioning, is needed to enhance time domain analysis. These developments are likely to offer unprecedented spatial and temporal resolution, and therefore it is conceivable that researchers can capture and analyze data with much higher fidelity. Huge-scale MIMO radar systems represent the trend of this trend since these emphasize communication and sensing capability's synergy. By leveraging these new technologies, researchers and experts can respond to challenging issues like signal interference and noise reduction in real-time environments. The integration of these new methods into existing methods in time domain analysis not only creates room for innovative advancements in data interpretation but also aligns with the essential requirement of adaptive and responsive systems in the fast-evolving digital world.

## SUMMARY

Time domain analysis proves to demonstrate enormous insight into the inherent temporal dynamics of phenomena, as the necessity for a multi-dimensional approach to examining these contexts becomes imperative. The data points gathered over the period of the study illustrate that, if one were focusing on frequency or amplitude alone, an incomplete understanding of behavior or response of the system can be gleaned. By making time a significant variable, researchers can obtain a more evolved understanding of how different parameters interact with one another over time, thereby enhancing predictive value. This points towards the necessity of including time domain analysis in more integrated analytical frameworks so that future studies can derive a better and integrated understanding of complex systems and their dynamics. In addition, the consequences of research results in time domain analysis extend beyond theoretical frameworks to practical actions, as seen by recent discoveries in oncology and psychology. For instance, new developments in ultra mutated colorectal cancers reveal that mutations in the DNA polymerase ε have clinical consequences that require early management and individualized therapeutic strategies (Bläker et al., 2014). The emergent recognition of these dynamics demands that medical interventions, provided in a timely manner, have the ability to alter patient outcomes, thereby strengthening the assertion that time is not simply background but an operating force within systems analysis and intervention strategies. Beyond this, the complexities identified through time domain analysis have application to broader therapeutic issues, most notably on recovery processes for mental health. Research shows that agency and social support are significant in overcoming obstacles to recovery from disorders such as Anorexia Nervosa, exemplifying how timing dictates therapeutic effect and decision-making (Kuss et al., 2016). By considering timely factors within treatment models, practitioners can better target interventions at times most suited to the patient's needs, leading to more successful pathways to recovery. The exploration of these dimensions thus itself reflects the necessity for ongoing development within research practices which embrace and access the potential of time domain analysis in fields.

In time domain analysis pursuit, several fundamental points emerge revealing its significance across various fields, particularly in computational and physical sciences. Among the fundamental breakthroughs is the development of a new hybrid memory analysis method called Structural Analysis. This method integrates an expressive form analysis style with efficient points-to style transfer functions. According to empirical research benchmarks on runtime heap structures of modern object-oriented languages, this analysis demonstrates the capability of capturing complex structural and shape features that are typically overlooked by conventional points-to analyses. This greater expressiveness is particularly crucial in optimization and error detection applications, and it presents dramatic improvement over traditional analytical methods (Marron et al., 2012). Another significant aspect of time domain analysis appears in the field of gravitational wave research, such as in the simulation of test masses orbiting Kerr black holes and their resulting radiation process. Researchers have introduced a way to approximate the associated Dirac delta function with unprecedented accuracy using a minimum number of grid points to make it capture its essential features. This advancement makes it possible to include the source term within numerical codes that solve the Teukolsky equation in the context of time domain. Surprisingly, the outcomes exhibit remarkable advances in performance with a significant reduction in numerical errors within different parameter spaces. These developments not only enhance the precision of gravitational wave modeling but also indicate the increased sophistication of time domain analysis methods in solving complex physical problems (Hopman et al., 2007). Convergence of these developments illustrates the multifaceted applications of time domain analysis, highlighting its innovative potential in computational theory and physical science applications. The integration of robust analysis frameworks facilitates the better understanding of complex structures and behaviors, and thus outcomes are more accurate. Moreover, the driving force behind hybrid memory analysis and enhanced gravity wave modeling is a reflection of ongoing progress in advancing analytical techniques with future advancements awaiting. With further research developments of techniques, the potential for optimization, detection of errors, and theoretical physics will keep increasing, leading to time domain analysis becoming a vital pillar of analytical work in the coming years. Awareness and use of these basic building blocks of time domain analysis thus become more significant for theoretical investigation and practical application.

# REFERENCES


1. Achumba, I. E., Azzi, D., Bersch, S. D., et al. (2013). Real-time human ambulation, activity, and physiological monitoring: Taxonomy of issues, techniques, applications, challenges and limitations. https://core.ac.uk/download/29586954.pdf (accessed on 19th September 2025).
2. Aiken, E. W., Chen, R. T. N., Lebacqz, J. V., et al. (2025). Helicopter mathematical models and control law development for handling qualities research. https://core.ac.uk/download/pdf/42833592.pdf (accessed on 19th September 2025).
3. Almeida, J. J., Carvalho, N. A. R., Henriques, P. R., et al. (2014). Conclave: Writing programs to understand programs. https://core.ac.uk/download/154277166.pdf (accessed on 19th September 2025).
4. Alvarado, L., Mus, S., Otero Calviño, B., et al. (2019). Artificial neural networks as emerging tools for earthquake detection. https://core.ac.uk/download/231704922.pdf (accessed on 19th September 2025).
5. Amaratunga, R. D. G., Haigh, R. P., & Keraminiyage, K. P. (2009). A HCI principles-based framework to assess the user perception of web based virtual research environments. *Special Issue on Capacity Building for Post Disaster Infrastructure Development and Management*. https://core.ac.uk/download/1661631.pdf (accessed on 19th September 2025).
6. Austin, J., Fletcher, M., Jackson, T., Jessop, et al. (2006). The Signal Data Explorer: A high-performance Grid based signal search tool for use in distributed diagnostic applications. https://core.ac.uk/download/56504.pdf (accessed on 19th September 2025).
7. Babiloni, C., Boeijinga, P. H., Brunovsky, M., Drinkenburg, W. H., Ffytche, D. H., Freeman, J., et al. (2012). Guidelines for the recording and evaluation of pharmaco-EEG data in man: The International Pharmaco-EEG Society (IPEG). https://core.ac.uk/download/54522901.pdf (accessed on 19th September 2025).
8. Bhatt, M., & Wallgruen, J. O. (2013). Geospatial narratives and their spatio-temporal dynamics: Commonsense reasoning for high-level analyses in geographic information systems. http://arxiv.org/abs/1307.2541 (accessed on 19th September 2025).
9. Biros, G., Bungartz, H., Corones, J., Cramer, et al. (2016). Research and education in computational science and engineering. https://core.ac.uk/download/148025463.pdf (accessed on 19th September 2025).
10. Björnson, E., Hoydis, J., Marzetta, T. L., et al. (2019). Massive MIMO is a reality–What is next? Five promising research directions for antenna

arrays. http://arxiv.org/abs/1902.07678 (accessed on 19th September 2025).

11. Bläker, H., Brenner, H., Chang-Claude, J., et al. (2014). Mutations in POLE and survival of colorectal cancer patients–Link to disease stage and treatment. https://core.ac.uk/download/199421144.pdf (accessed on 19th September 2025).
12. Bonnema, G. M., Cooke, A. G., & Poelman, W. A. (2012). Agile development for a multi-disciplinary bicycle stability test bench. https://core.ac.uk/download/pdf/11484685.pdf (accessed on 19th September 2025).
13. Borghi, A. M., Caramelli, N., & Setti, A. (2016). How abstract is risk for workers? Expertise, context and introspection in abstract concepts. https://core.ac.uk/download/80310577.pdf (accessed on 19th September 2025).
14. Brezger, A., & Lang, S. (2003). Generalized structured additive regression based on Bayesian P-splines. https://core.ac.uk/download/12162748.pdf (accessed on 19th September 2025).
15. Burrus, C. S., Cohen, L., Mallat, S., Newland, D. E., Newland, D. E., Newland, D. E., & Newland, D. E. (2007). Numerical treatment of seismic accelerograms and of inelastic seismic structural responses using harmonic wavelets. https://core.ac.uk/download/2708118.pdf (accessed on 19th September 2025).
16. Byrne, D., Jones, G. J. F., & Kelly, L. (2012). Multiple multimodal mobile devices: Lessons learned from engineering lifelog solutions. https://core.ac.uk/download/11310093.pdf (accessed on 19th September 2025).
17. Chen, Y., Hu, L., Peng, X., et al. (2017). Stratified transfer learning for cross-domain activity recognition. http://arxiv.org/abs/1801.00820 (accessed on 19th September 2025).
18. Chernyshenko, S. I., Di Cicca, G. M., Hu, Z. W., et al. (2006). Analysis of data on the relation between eddies and streaky structures in turbulent flows using the placebo method.
19. Cheung, G., Magli, E., Ng, M., et al. (2018). Graph spectral image processing. http://arxiv.org/abs/1801.04749 (accessed on 19th September 2025).
20. Colonius, T., Rigas, G., & Sipp, D. (2020). Nonlinear input/output analysis: Application to boundary layer transition. https://core.ac.uk/download/287634006.pdf (accessed on 19th September 2025).
21. Dardashti, R., Dawid, R., Gryb, S., et al. (2018). On the empirical consequences of the AdS/CFT duality. https://core.ac.uk/download/199184206.pdf (accessed on 19th September 2025).

22. Diamond, P. H., Dritschel, D. G., & Tobias, S. M. (2018). Breaking Kelvin: Circulation conservation and vortex breakup in MHD at low magnetic Prandtl number. http://arxiv.org/abs/1806.03922 (accessed on 19th September 2025).

23. Doherty, A., Doyle, J., Habib, S., et al. (2000). Robust control in the quantum domain. https://core.ac.uk/download/4876306.pdf (accessed on 19th September 2025).

24. Douven, I., & Gärdenfors, P. (2019). What are natural concepts? A design perspective. https://core.ac.uk/download/186331318.pdf (accessed on 19th September 2025).

25. Etienam, C. (2019). 4D seismic history matching incorporating unsupervised learning. http://arxiv.org/abs/1905.07469 (accessed on 19th September 2025).

26. Ewert, S., O'Hanlon, K., Sandler, M. B., et al. (2018). Shift-invariant kernel additive modelling for audio source separation. http://arxiv.org/abs/1711.00351 (accessed on 19th September 2025).

27. Firth, M. (2015). Minimizing latency of pitch detection algorithms for live vocals on low-cost hardware. https://core.ac.uk/download/30734258.pdf (accessed on 19th September 2025).

28. Fronsdal, C. (2010). Heat and gravitation. III. Mixtures. http://arxiv.org/abs/1001.2579 (accessed on 19th September 2025).

29. Frossard, P., Kovačević, J., Moura, J. M. F., et al. (2018). Graph signal processing: Overview, challenges and applications. http://arxiv.org/abs/1712.00468 (accessed on 19th September 2025).

30. Galvín, P., Maestre Torreblanca, J. M., & Romero Ordóñez, A. (2014). A 3D time domain numerical model based on half-space Green's function for soil-structure interaction analysis. https://core.ac.uk/download/232001702.pdf (accessed on 19th September 2025).

31. Giveon, A., Ishibashi, A., Ishibashi, A., van Keviczky, A., Konkowski, D. A., Frank, D., Papadopoulos, G., et al. (2006). Scalar field probes of power-law space-time singularities. http://arxiv.org/abs/hep-th/0602207 (accessed on 19th September 2025).

32. Gu, F., Xu, G., He, H., et al. (2014). The recovery of weak impulsive signals based on stochastic resonance and moving least squares fitting. https://core.ac.uk/download/20569133.pdf (accessed on 19th September 2025).

33. Gubanski, S. M., Serdyuk, Y. V., & Tuncer, E. (2001). Dielectric mixtures -- Electrical properties and modeling. http://arxiv.org/abs/cond-mat/0111254 (accessed on 19th September 2025).

34. Hansen, E., & Henningsson, E. (2016). Additive domain decomposition operator splitting–Convergence analyses in a dissipative framework. http://arxiv.org/abs/1512.05941 (accessed on 19th September 2025).
35. Hattab, G., & Ibnkahla, M. (2014). Multiband spectrum access: Great promises for future cognitive radio networks. http://arxiv.org/abs/1409.5913 (accessed on 19th September 2025).
36. Heeren, W., de Jong, F., & Ordelman, R. (2006). Exploration of audiovisual heritage using audio indexing technology. https://core.ac.uk/download/pdf/11470247.pdf (accessed on 19th September 2025).
37. Hopman, C., Sopuerta, C., Misner, C. W., Khanna, G., Arnaud, K. A., Arnaud, K. A., Sundararajan, P. A., et al. (2007). Towards adiabatic waveforms for inspiral into Kerr black holes: I. A new model of the source for the time domain perturbation equation. http://arxiv.org/abs/gr-qc/0703028 (accessed on 19th September 2025).
38. Jureta, I. (2015). Requirements problem and solution concepts for adaptive systems engineering, and their relationship to mathematical optimization, decision analysis, and expected utility theory. http://arxiv.org/abs/1507.06260 (accessed on 19th September 2025).
39. Katragadda, S. (2018). Noise shaping asynchronous SAR ADC based time to digital converter. https://core.ac.uk/download/211332437.pdf (accessed on 19th September 2025).
40. Kechris, A. S., Pillay, A., Shore, R. A., & Buss, S. R. (2001). The prospects for mathematical logic in the twenty-first century. https://core.ac.uk/download/12817608.pdf (accessed on 19th September 2025).
41. Kesorn, K. (2010). Multi modal multi-semantic image retrieval. https://core.ac.uk/download/30695203.pdf (accessed on 19th September 2025).
42. Kleidis, K., & Oikonomou, V. K. (2014). Extended supersymmetric quantum mechanics algebras in scattering states of fermions off domain walls. http://arxiv.org/abs/1408.1552 (accessed on 19th September 2025).
43. Kou, J., & Sun, S. (2017). Thermodynamically consistent simulation of nonisothermal diffuse-interface two-phase flow with Peng-Robinson equation of state. http://arxiv.org/abs/1712.03090 (accessed on 19th September 2025).
44. Kuss, D. J., & Smethurst, L. J. (2016). "Learning to live your life again": An interpretative phenomenological analysis of weblogs documenting the inside experience of recovering from anorexia nervosa. https://core.ac.uk/download/42393104.pdf (accessed on 19th September 2025).

45. Lee, F. C., Wangenheim, H., Warren, D., et al. (2025). Modeling and analysis of power processing systems (MAPPS), initial phase 2. https://core.ac.uk/download/pdf/42873263.pdf (accessed on 19th September 2025).
46. Leis, J. W., & Snook, C. (2009). A zero-cost, real-time, Windows signal laboratory. https://core.ac.uk/download/11037937.pdf (accessed on 19th September 2025).
47. Manfredi, P., Stievano, I. S., & Trinchero, R. (2017). TMsim: An algorithmic tool for the parametric and worst-case simulation of systems with uncertainties. https://core.ac.uk/download/84040953.pdf (accessed on 19th September 2025).
48. Mao, G. (2020). From vehicular networks to IoT for smart roads: How a communication engineer can help solve transportation problems (Invited Talk). https://core.ac.uk/download/296708935.pdf (accessed on 19th September 2025).
49. Marron, M. (2012). Structural analysis: Shape information via points-to computation. http://arxiv.org/abs/1201.1277 (accessed on 19th September 2025).
50. Mehta, G. (2009). A micro power hardware fabric for embedded computing. https://core.ac.uk/download/12208062.pdf (accessed on 19th September 2025).
51. Money, A. G., Babaguchi, Boyatzis, Cernekova, Chang, Chang, Crockford, et al. (2008). Video summarization: A conceptual framework and survey of the state of the art. https://core.ac.uk/download/19501132.pdf (accessed on 19th September 2025).
52. Naso, A., & O'Naraigh, L. (2017). A flow-pattern map for phase separation using the Navier-Stokes Cahn-Hilliard model. http://arxiv.org/abs/1707.06067 (accessed on 19th September 2025).
53. Nogueira, D. L., Pereira, H. B. de B., & Pérez Vidal, L. (2001). Study on behavioral impedance for route planning techniques from the pedestrian's perspective: Part I - Theoretical contextualization and taxonomy. https://core.ac.uk/download/81570741.pdf (accessed on 19th September 2025).
54. Ramírez, J., & Valecillos, B. (2006). Evaluation of lightning impulse test by frequency response analysis. https://core.ac.uk/download/29427993.pdf (accessed on 19th September 2025).
55. Sharif, A. M. (1997). The management of intelligence-assisted finite element analysis technology. https://core.ac.uk/download/336716.pdf (accessed on 19th September 2025).
56. Wiens, D. (2020). The general theory of second best is more general than you think. https://core.ac.uk/download/186331566.pdf (accessed on 19th September 2025).

CHAPTER 4

# FREQUENCY DOMAIN ANALYSIS

## Contents



## INTRODUCTION

Frequency domain analysis is a critical tool towards achieving the dynamics of systems in terms of frequency response study. Through frequency domain analysis, engineers are presented with the opportunity of examining how systems react to varying input frequencies, thus revealing stability and performance factors that might be absent in time-domain analysis. By transforming differential equations into algebraic equations, frequency domain solution not only allows for better understanding of the system's dynamics but also enables the application of methods such as Bode plots and Nyquist criteria. The frequency response concepts analysis is elementary, whereby it reveals the input and output signal relationships as a function of frequency, employed for the forecasting of system behavior under different working conditions and simplifying the complex systems' design and control. The construction of Bode plots, and how they graphically represent the magnitude and phase of the systems' transfer function, is one of the frequency domain analysis.

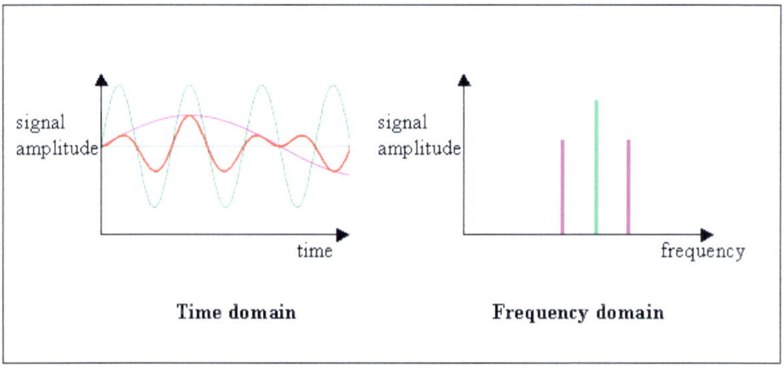


**Figure 4.1**. *Frequency domain analysis.*

*Source: https://images.deepai.org/glossary-terms/1cb28e3432394f769f8d0757bf512ae1/clip_image001_20110615223353.gif.*

These plots allow engineers to assess frequency response quickly and effectively, returning stability margins and resonant behaviors. A Bode plot is made up of two separate plots: one for the gain (magnitude) versus frequency and one for the phase shift. With this graphical tool in hand, designers can identify the important parameters, such as cut-off frequencies and bandwidth, and therefore make decisions on system tuning. In addition, Bode plots are not only theoretical tools; they have practical uses in actual applications, for example, in control system design where stability and transient response are central to optimization of system performance. The Nyquist stability criterion is a development of frequency domain analysis techniques in that it presents a method of determining stability of feedback systems. With the analysis of frequency response and Nyquist contour plotting, engineers are able to get useful information on gain and phase margins, which play a critical role in determining system stability. According to research, balanced three-phase system small-signal stability can significantly be improved using the generalized Nyquist stability criterion (GNC), an approach that emphasizes the contribution of impedance matrices to assessing stability in subsystems. With these impedance models, the analysis is able to facilitate investigations that would otherwise have been characterized as being complex, simplifying the stability consideration into a matter of scalar ratios indicative of stability margins (Wen et al., 2017). With these theoretical results, in turn, it becomes possible to design and control better distributed power systems and ensure that they are capable of satisfying stringent performance and stability requirements (Li et al., 2020) (Figure 4.1).

## *Definition and Significance of Frequency Domain Analysis*

Frequency domain analysis is a highly significant building block in the analysis of dynamic systems, particularly in signal processing and control systems. Frequency domain analysis shifts focus away from time-domain signals to the frequency components of them to enable engineers and scientists to analyze how systems react to various frequencies. By methods such as the Fourier Transform, frequency domain analysis decomposes complex signals into somewhat simpler sine and cosine components, facilitating a better understandable view of system operation. Not only does this process of analysis reveal vital information about system stability and performance, but it forms the foundation of many design strategies, such as the construction of Bode plots and application of the Nyquist stability criterion. The value of frequency domain analysis is further increased by its application to the creation of Bode plots that graphically display the phase and gain of a system as a function of frequency. These plots, founded on the principles of frequency response, are fundamental to control system designers who wish to design them. A Bode plot allows for a comprehensive analysis of system dynamics, such as resonance frequencies, stability margins, and phase shifts, that can play a decisive role in determining the performance of systems. Hence, the ability to see these responses provides engineers with concrete insight to enhance system design so that systems can sustain various operating conditions and reduce the risk of instability due to the application of feedback loops. Other than its application in the construction of Bode plots, frequency domain analysis has primary application in the examination of a system's robustness against uncertainties, as is the case with the Nyquist criterion for stability and gain/phase margins. From the examination of the frequency response of a system, engineers are able to ascertain how the deviation of gain and phase can destabilize the system, particularly with model uncertainty. The capacity to combine these parameters in one holistic analysis provides for a deeper appreciation of system reliability. For example, it has been found that the knowledge of how feedback and buffer systems limit the operation of the system greatly increases our understanding of system stability and homeostasis, showing how these ideas form the backbone of control theory and more sophisticated biological systems (Edward et al., 2017; Ge et al., 2017).

Consequently, frequency domain analysis is not purely a theoretical exercise but a valuable engineering tool for the design of systems with desired operating characteristics.

## *Comparison to Time Domain Analysis*

System analysis can be obtained by both time domain and frequency domain observations, each with certain advantages and limitations of their own. Time domain analysis is formulated from systems response in terms of time so that signals can be graphically viewed while they transform from input to output. It is generally intuitive since it follows natural phenomena that we observe every day. However, time domain analysis will struggle to portray stability and performance characteristics such as oscillations and transient responses that influence long-term system behavior. On the other hand, frequency domain analysis has a methodical procedure for examining system response to various input frequencies, which makes understanding stability margins, phase shifts, and gain characteristics possible, thus enabling easier design and tuning of control systems with precision. In frequency domain analysis, instruments such as Bode plots and Nyquist diagrams are envisioned as being pivotal in system performance determination. Bode plots, which schematically portray the amplitude and phase response of a system across a range of frequencies, enable engineers to directly estimate system stability and even compensate for design purposes. The graphical technique effectively translates complex transfer functions into graphically accessible information, making necessary visions into gain and phase margins possible. Time domain analysis can provide transient solutions, but it is less explicit in suggesting how systems are performing under various frequency conditions. Moreover, approaches like Nyquist stability criteria continue to reinforce the frequency domain analysis by giving it a solid foundation for determining stability in terms of contour plots, which are relatively more difficult to compute and interpret in time domain cases. Relative to time-domain analysis, application of the Nyquist criterion with Bode plots can significantly enhance control system design practices. For instance, a study in an adaptive phase-locked loop (PLL) reported significant performance improvement using frequency response metrics, such that a 78.6% increase in phase margin in some cases is reported, emphasizing the importance of frequency-domain stability analysis (Meng, 2020). Conversely, while time-domain simulations are able to verify output responses, they might not be as brief and streamlined as frequency domain analysis in solving potential instability issues. As complexity increases, the two methods need to be used together; frequency domain methods can make analysis and design more streamlined, with time domain simulation verifying transient behavior and applicability in the real world. Therefore, using both approaches maximizes control system design and increases knowledge of dynamic performance.

## *Applications in Engineering and Control Systems*

The use of frequency domain analysis is of primary importance in control systems and engineering, specifically in the design and analysis of stability in complex interconnected systems. The Nyquist stability criterion is one of the most important tools in the field that helps engineers determine the stability of systems described with feedback loops. As an example, for offshore wind farms connected by modular multilevel converters, stability is no longer a theoretical issue but has to be ensured for operational reliability. The interaction among the wind power inverters and converter units can create resonances that destabilize the system. By employing frequency response concept-based analytical methods using impedance, designers can create impedance models that enable phase margin prediction and potential instabilities, as outlined in our description of Lyu et al. (2015). Predictive capability is necessary in ensuring that engineering designs are sufficiently robust to withstand variations in operation without being led astray by instability. Another prominent area of application for frequency domain analysis is control system design for grid-interfacing inverters, and most especially those that incorporate LCL filters. Time delay, a universal issue in such systems, has a significant detrimental effect on stability; it is thus also important to realize its effect. Studies have shown that different feedback loops, that is, the inverter current feedback (ICF) and grid current feedback (GCF), respond differently to time delays, highlighting the need for dedicated strategies for each. With the use of systematic frequency response analysis, it is possible for engineers to compute effective delay ranges that guarantee system stability. With this caution in analyzing stability margins, procedures for the derivation of controller designs that incorporate predictive techniques for ICF while acknowledging the advantages of time delays in GCF, as quoted in Wang et al. (2015), can be deduced. Such applications are examples of how theoretical principles from frequency domain are directly applied in engineering practice. Furthermore, frequency domain analysis software spans from traditional systems to emerging technologies and integration of complex systems. With industries moving toward greater networked complexity—such as integrated renewable energy systems and smart grids—the need for stability analysis using sound techniques is more critical than ever. Techniques like Nyquist diagrams and Bode plots not only provide graphical representations of the system parameters but are also formal approaches to precisely determining gain and phase margins. These graphical tools enable engineers to condition system parameters such that different components can work in a synchronized and autonomous way without compromising the stability of the system. Integrating these frequency response concepts in existing engineering practices enables engineers to create more resilient systems capable of responding to dynamic operating conditions while at the same time offering optimum performance, thus encouraging innovation in control system technology.

## 4.1. FREQUENCY RESPONSE CONCEPTS

One of the fundamental concepts of frequency response principles in control systems is system dynamics in sinusoidal inputs. When working with linear time-invariant (LTI) systems, the frequency response defines the manner in which the system behaves to different frequencies of input signals. This characteristic is often described through the transfer function, which displays magnitude and phase over a set of frequencies. Using Bode plots, engineers are able to view gain and phase shift as functions of frequency so they can assess stability and performance at a glance. The graphical assessment made possible by Bode plots not only facilitates understanding of the frequency response, but it also allows one to design and tune control systems to optimize their performance, ultimately assuring given performance criteria within specified operational constraints. The Nyquist stability criterion is a critical technique in frequency response analysis since it provides a way of evaluating the stability of a system from the open-loop frequency response. The criterion draws on a contour method whereby the Nyquist plot, a graphical function of the complex frequency response, is utilized. Through this method, designers can tell whether feedback systems are stable or not without having to analyze the poles directly in the right half-plane. Nyquist criterion establishes stability conditions via the encirclements of the critical point in the complex plane, demonstrating the relationship between gain and phase margins. In practice, this approach is significant in the study of converter-based power systems because it allows for a general understanding of how frequency-domain performance influences stability and control performance in a variety of system configurations like ac-dc networks and grid-forming systems (Sun, 2022). Additionally, the use of gain and phase margins in frequency response analysis provides valuable information regarding system robustness. Gain margin is the increase in gain that a system can maintain until it loses stability, while phase margin is the increase in phase lag at the gain crossover frequency that still allows stability. Both margins can be derived from the Nyquist plot and are important to engineers in assessing how variations in system parameters can affect the performance of control systems. For instance, in the application of the Nyquist stability criterion, stability and instability regions can be identified from the determined gain and phase margins, thus informing compensator design such as fractional-order controllers (Bauer et al., 2020). Including these concepts, frequency domain analysis is an important methodology for ensuring the stable and sound operation of modern control systems in various engineering applications (Figure 4.2).

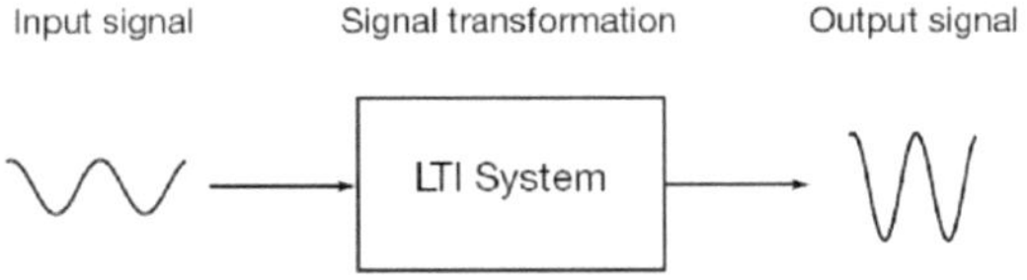


**Figure 4.2**. *Linear time-invariant (LTI) system.*

*Source: https://www.researchgate.net/profile/Mosbeh-Kaloop/publication/267469724/figure/fig5/AS:613969488248832@1523393141016/Schematic-of-a-linear-time-invariant-LTI-system-The-input-signal-is-transformed-into.png.*

## 4.1.1. Definition of Frequency Response

The frequency response is the key concept in describing systems' behavior when subjected to various frequency inputs, most notably in control systems and signal processing. Frequency response essentially defines the steady-state response of a dynamic system to sinusoidal inputs at various frequencies. This calculation simplifies to the study of how the amplitude and phase of the output signal change as functions of the input signal, providing essential information regarding system stability and performance. Plotting magnitude and phase versus frequency, the designer can visually observe how his systems will react to different operating points, a particularly useful function in the design of robust control strategies. Frequency response is typically in frequency domain analysis described by the use of Bode plots that give a graphical representation of the systems response versus frequency range. A Bode plot consists of two plots: phase and magnitude, versus a logarithmic scale of frequency. This method facilitates easy interpretation of system parameters such as bandwidth, gain margin, and phase margin, all of which are critical to ensure system stability. Moreover, frequency response is easily optimized through feedback control systems. Understanding of Bode plots enables designers to enhance system design for maximum performance and achieve desired specifications for implementations. Secondly, the Nyquist stability criterion is also important in testing systems' stability by their frequency response characteristics. According to the frequency response, if the Nyquist plot is constructed, engineers can comprehend whether the feedback system is stable or unstable without directly examining the systems' poles and zeros. This method is a synthesis of both magnitude and phase data, summarizing them as a single graphical representation of the behavior of the systems. Stability margins—phase margins and gain margins—can particularly be calculated from the Nyquist plot, providing key measures for predicting how changes in system parameters can affect stability. Lastly, frequency response analysis combined with Bode plots and Nyquist stability criterion is an exhaustive collection of techniques used for designing and analyzing stable control systems in various applications, including tokamak reactors and converter-based power systems (Sun, 2022; Tommasi, 2018).

### 4.1.2. Importance of Amplitude and Phase in Frequency Response

We need to understand phase and amplitude in frequency response to be able to analyze systems in the frequency domain correctly. Amplitude tells us how much the system is going to increase or decrease the signals at various frequencies, whereas phase tells us about the delay in the signals with respect to a reference. These two components together tell us about the dynamic behavior and stability of the system. Tools such as the Bode plot enhance this connection in that they display a visual plot of phase and amplitude across a frequency range. Using such tools to analyze allows engineers to predict how systems will behave in response to inputs, thereby making it easier to design and optimize control systems. Amplitude and phase are both essential to keeping systems online and performing well with dependability over a broad array of applications, including telecommunications, power systems, and beyond. Nyquist stability criterion is also focused on phase in the matter of control system stability. By looking at the phase and gain margin with Bode plots, engineers can determine how much phase shift or gain a system can tolerate without becoming unstable. Moreover, other research suggests that bad phase margins can lead to oscillations or instability and cause severe issues in power systems and control loops. For instance, it is noted by a study that the behavior of a centralized model predictive controller (MPC) is highly phase-margin sensitive because optimally tuned gains can provide improved transient responses compared to traditional controllers such as PI controllers. Therefore, an understanding of amplitude and phase interaction not only assists in designing a system but is also crucial for finding sound control measures in industry applications (Kumar et al., 2022). In real-world situations, for example, in the increasing application of renewable energy sources, amplitude and phase variations can have profound impacts on system performance. The unpredictable nature of wind and solar power generates fluctuations that need precise frequency response analysis to maintain grid stability. The synchronization of such variable resources with the grid is highly dependent on amplitude and phase shift prediction. Sophisticated tuning strategies, including the application of the pity beetle algorithm in PI controllers in synchronization systems, have been found to be effective in controlling amplitude and phase efficiently. Using numerical simulation and frequency response techniques like Bode plots, the methods under different grid disturbances have been found to be more efficient, stressing the importance of considering amplitude and phase for effective control and stability in modern power systems (Riyas et al., 2025) (Figure 4.3).

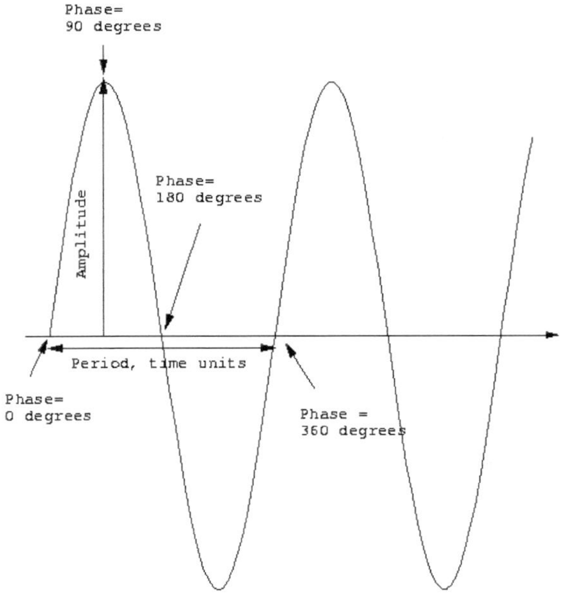


**Figure 4.3**. *Frequency response.*

*Source: https://www.homepages.ed.ac.uk/jwp/control06/controlcourse/restricted/course/advanced/figs7/sine.gif.*

### 4.1.3. Relationship Between Frequency Response and System Stability

The connection between frequency response and stability of a system is basic to control theory, especially in the study of linear time-invariant (LTI) systems. Frequency response simply outlines how a system reacts to sinusoidal inputs at different frequencies, and this tells us essential things about its response under dynamic states. Knowing this connection is important in building Bode plots, as they use graphical visualization to show system gain and phase over a spectrum of frequencies. These plots permit direct inference of the stability of an LTI system; gain and phase margins tell one how near a system is to becoming unstable. As the gain approaches the critical values, small changes can make a system oscillate, and therefore strict frequency domain analysis becomes important in forecasting stability. Hence, both frequency response information and graphical representations like Bode plots hold supreme significance in the design of stable systems and controllers. The application of the Nyquist stability criterion again emphasizes the connection between system stability and frequency response. It is the plotting of the open-loop transfer function on the complex plane and signifies the significant points that can determine the stability of a closed-loop system. In particular, the Nyquist plot illustrates how encirclements around the critical point ( –1, 0) are related to stability. If the encirclements agree with the number of poles in the right-half-plane of the open-loop transfer function, then the system is stable. Frequency response data collected by virtue of methods such as impedance measurement, as explored by recent research, corroborates this rule. For example, Mohammed et al. (2024) elucidates the importance of frequency-domain stability analysis of inverter-dominated power systems and how impedance-based methods yield vital stability information when conventional time-domain methods are not successful. In addition, the relationship between system stability and frequency response is demonstrated in renewable power systems, where there are varying grid conditions that need robust stability tests. The interaction between converters and power grids brings the requirement for impedance-based stability analysis to facilitate engineers in making suitable system performance predictions against fluctuating operating conditions. For instance, Arshad et al. (2024) emphasizes frequency scan-based analysis in stability determination for a large wind farm and asserts the applicability of frequency response analysis in complex situations. A firm understanding of gain and phase margins through frequency response techniques enables better controller design, such that systems are able to react to oscillations without degrading performance. The abundance of information yielded by frequency response analysis is thus essential to determining stability criteria for modern energy systems, where renewable resources are added to traditional grids efficiently.

## 4.2. BODE PLOTS: AN OVERVIEW

In signal processing and control systems, the Bode plots are a useful graphical tool for analyzing the frequency response of linear time-invariant (LTI) systems. In gain versus phase form plotted against angular frequency, the log-log output provides crucial information on the behavior of the systems across different frequencies. Specifically, Bode plots break down the relationship between system response and input signals into something easier to ascertain from stability margins and bandwidth. Decoupling the gain and phase information into two distinct plots makes it infinitely easier to analyze, and so Bode is immediately applicable to feedback systems in which stability and performance are priority number one. Apart from that, the two-graph technique allows engineers to visualize the impact of system parameter variations on system behavior overall, combining theoretical constructs and practical application rather effectively in frequency domain analysis. Construction of Bode plots is a step-by-step process, beginning from the transfer function of the system under consideration. The function should be written in a suitable form for analysis, ideally as the ratio of polynomials. Each zero and pole in the transfer function has a specific function in the frequency response of systems; therefore, their effects are important to plot appropriately.

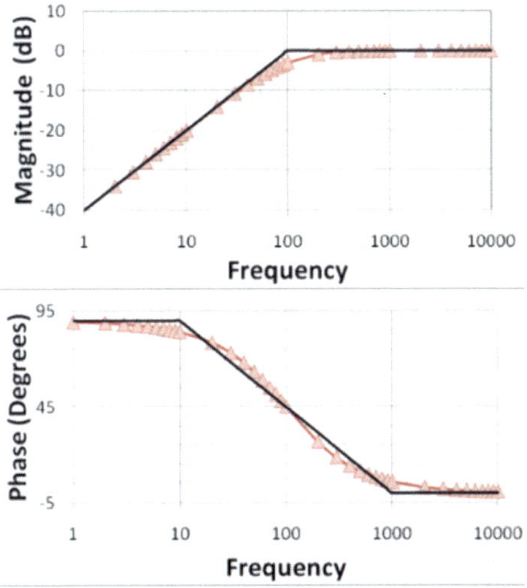


**Figure 4.4**. *Bode plot.*

*Source: https://upload.wikimedia.org/wikipedia/commons/b/be/High-Pass_filter_Bode_Magnitude_and_Phase_plots.png.*

Following this, phase and magnitude can then be independently calculated for a range of frequencies, plotted on logarithmic scales. Specifically, the gain plot indicates logarithm of absolute gain (in decibels) against the logarithm of frequency and the phase plot indicates the phase angle in degrees. This methodical technique not only simplifies the reduction of complex behaviors to manageable forms but also allows systematic comparison between theoretical predictions and experimental findings, thus facilitating engineers to design more stable and efficient systems. In real-world applications, Bode plots are generally applied to interpret stability criteria, especially the Nyquist stability criterion. The Bode stability criterion examines how gain and phase margins of a system influence its stability. For instance, an increase in phase margin often indicates greater stability, as discussed in the context of grid-connected inverters, where adaptive phase-locked loops significantly increase stability margins in dynamic grid environments (Meng, 2020). These observations point towards the importance of Bode plots in real-world applications, indicative of their place in not only predicting system performance but also enhancing control measures. Besides, the Bode plots gain and phase margins analysis permits engineers to make rational choices by adjusting the design to suppress undesirable oscillations and enhance system robustness in general (Bhattacharyya et al., 2009). This marriage of theory with practice renders the use of Bode plots importance in modern engineering disciplines irrevocable (Figure 4.4).

## 4.2.1. Definition and Purpose Of Bode plots

In frequency domain analysis tools, the Bode plots are useful graphical representations of a system's frequency response in magnitude and phase format. The plots consist of two separate graphs: one is utilized to represent the logarithmic magnitude of the transfer function against frequency, and the other to represent the phase shift for the same array of frequencies. The primary use of Bode plots is in providing a complete description of the way a system behaves against various frequencies. In viewing amplitude and phase data simultaneously, engineers are better positioned to design, analyze, and adjust systems for desired performance, particularly in control systems and signal processing. Such graphical displays facilitate improved system behavior understanding and simpler identification of potential stability issues. Bode plot design must be accomplished in an orderly fashion, typically beginning with the transfer function of the system, expressed as a ratio of polynomials. Polynomial terms project onto specific frequency response characteristics such as poles or zeros. By decomposing the transfer function into lower-order first-order and second-order parts, asymptotic approximations for phase and magnitude can be obtained. This enhances clarity and accuracy of the Bode plot and invites orderly comprehension of how the parts advance the overall system response. This type of treatment is necessary because it habituates engineers to simulate and predict the dynamics of systems under different states and, ultimately, lead to enhanced design choices and verification of stability requirements, such as those tested through Nyquist techniques solely based on frequency-response data (Sun, 2022). Bode plots are also appropriate for stability analysis, particularly control systems. The information documented in both phase and magnitude plots is critical for the calculation of gain and phase margins, and therefore determining a system's closeness to instability. A system may be deemed stable if these margins are within reasonable limits, which can readily be read graphically through the use of the Bode plot. This kind of analysis enables engineers to pre-tune system dynamics, which enables the robustness of a system across a range of operating conditions. In systems such as converter-based power systems, where frequency domain stability requirements are of utmost importance, the transition to frequency response allows for a broader understanding of system component interactions and thus the sustainable nature of Bode plots in modern-day engineering applications (Messner et al., 2020).

### 4.2.2. Components of Bode Plots: Magnitude and Phase Plots

In frequency response measurements, Bode plots are a fundamental tool to graphically represent the response of linear time-invariant (LTI) systems. The plots consist of two related but individual components: magnitude and phase plots. The magnitude plot gives a graphical representation of how the system gain varies with frequency, typically in decibels (dB). The logarithmic plot gives an intuitive sense of how the system responds to various input frequencies. The Bode plot is supplemented by the phase plot, which illustrates the phase shift produced by the system over the same frequency interval, in degrees. Together, these provide enough information for engineers to make system performance judgments, particularly in control and signal processing, so that important frequency intervals where the system may exhibit undesirable behavior can be determined. The relationship between phase and magnitude plots on Bode plots is critical for system performance as well as stability analysis. In accordance with the Bode stability criterion, stability is provided when the phase margin is sufficiently large so that the net phase shift never crosses $-180°$ at the frequency where the gain crosses unity. These plots are central to the definition of damping behavior and dynamic response of a system, indicating how changes in system parameters affect stability margins. As an example, in feedback systems, examination of the magnitude and phase plots can offer information on potential oscillation or instability as frequency increases. This connection is critical to the development of effective and stable systems, as it enables engineers to make logical modifications to maximize performance. Moreover, advanced uses of Bode plots extend them to adaptive systems and variable operating parameters. In particular, the use of adaptive phase-locked loops (PLLs) has been found to lead to dramatic improvement in stability margins, as observed in frequency domain analysis. For instance, there is evidence that adaptive PLLs can significantly enhance phase margins and, therefore, stabilize systems which otherwise would have oscillated in the event of varying load conditions (Meng, 2020). These developments reflect the importance of the magnitude and phase plots when designing sophisticated feedback circuits. By utilizing Bode plots, impacts of disturbances may be foretold and compensated in real-time and keep the systems in the top performance despite varying external conditions. This role is vital for equipment such as grid-connected inverters and other basic infrastructure, where stability is an essential ingredient in success of operation (Figure 4.5).

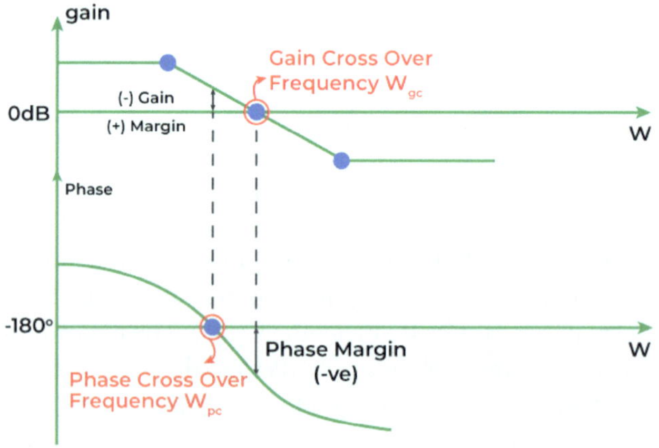


**Figure 4.5**. *Bode plots in control system.*

*Source: https://media.geeksforgeeks.org/wp-content/uploads/20230811100020/Bode-Plots-in-Control-System-1.png.*

## 4.2.3. Advantages of Using Bode Plots in System Analysis

The application of Bode plots in system analysis has some advantages when assessing system stability and performance. The graphical representations of magnitude and phase versus frequency assist engineers in viewing the way systems respond to inputs. One of the benefits is the clarity with which Bode plots indicate the gain and phase margins, crucial indicators of system stability. By analyzing these margins, the engineers can readily determine whether a system is likely to be oscillatory or unstable and require immediate measures to be taken to enhance performance. Further, the plots enable one to compare different control policies, with best system parameters readily optimized that are dependent on specific performance criteria so that robustness can be guaranteed in different operating conditions. Apart from stability insight, Bode plots facilitate the analysis of frequency response characteristics, which is essential in control system design. Their two-dimensional plot provides a beautiful image of a systems performance over a wide frequency range. This clarity is particularly helpful when working with systems that incorporate time delays, which can have a radical impact on performance. The closed-loop Bode plot, for instance, is an insightful tool to check delay margins, advising engineers on the compromise in delay compensation and tracking performance. As elucidated in recent studies, interpreting Bode plots in the context of stability analysis may give hints at the sophisticated dynamics between system performance and delay stability, which is an essential aspect in high-performance applications like big gas turbines (Su et al., 2025). Another convincing advantage of Bode plots is their easy compatibility with more complex analysis methods such as the Nyquist stability criterion and gain/phase margin analysis. By providing an intuitive visual guide, Bode plots are complemented by other analytic methods towards a fuller insight into system behavior. For instance, incorporating Bode plots within a multi-dimensional performance assessment tool allows for detailed investigation of performance characteristics, such as demonstrated by recent studies in the application of power electronic systems (Lin et al., 2024). Synergistic benefits of this method enhance the design process, in that engineers are able to measure stability margins and dynamic performance characteristics. Therefore, Bode plots do not merely exist as useful tools but are actually contributing to the broader context of frequency domain analysis and thereby contributing to the design and implementation of stable control systems.

## 4.3. CONSTRUCTION OF BODE PLOTS

Bode plots are made as a necessary tool in frequency domain analysis to graphically plot system response across a range of frequencies. A Bode plot is made of two separate graphs, one plotting the magnitude (in decibels) and the other the phase (in degrees) of the systems frequency response versus frequency on a log scale. This two-dimensional representation allows control system researchers and designers to analyze the performance and stability of control systems economically. These plots are typically designed starting from the transfer function of the system, where the frequency response can be found through substituting complex variables. Bode plots have uses extending beyond mere visualization; they are critical in control system design and tuning, as they allow for easy identification of key parameters such as gain and phase margins that decide the stability and responsiveness of a system (Messner et al., 2020). In practice, the construction of Bode plots involves a series of systematic procedures which consist of computation of the magnitude and phase contribution of individual discrete system poles and zeros. Each has an uncomplicated geometric effect on the global system response, and their individual Bode plots can be added directly. Asymptotic approximation also simplifies representation because linear segments would be able to be graphed in order to estimate system behavior in various ranges of frequency. Adjusting these plots involves learning how system parameters shift when component values change, which is typical when engineers tweak parameters to optimize them. Moreover, the ease with which Bode plots can be used to convey complex frequency response information makes them an important asset in learning environments as well as practical applications of engineering (Barcel, 2019). Ultimately, the knowledge gained through Bode plot generation not only aids in control system analysis and design but also opens it up to engineers of various disciplines. By observing the impact of frequency changes on system dynamics, Bode plots become a key tool to identify such problems as resonance or excessive phase lag that may jeopardize stability. They can also be used to aid in the choice of suitable compensators or controllers by observing the difference between desired and actual response. Through constructive iteration and creation of Bode plots, engineers gain an enhanced understanding of the behavior of systems, which significantly contributes to the efficiency of used control methods in different areas of engineering. Application of computer-aided software tools facilitates it further, allowing simulation and analysis of Bode plots in real time with a suitable correlation with contemporary requirements of engineering practices (Messner et al., 2020) (Figure 4.6).

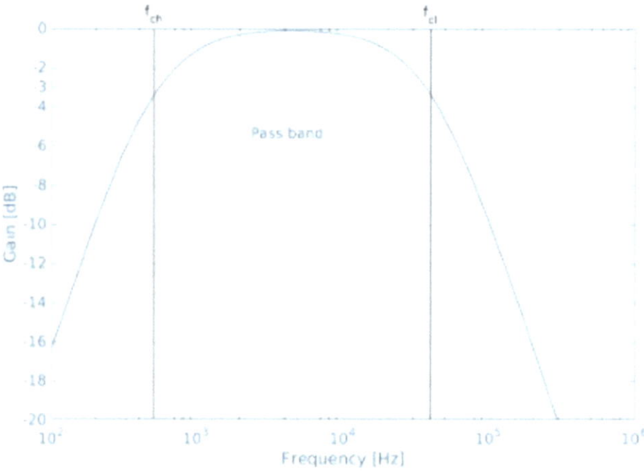


**Figure 4.6**. *Construction of bode plots.*

*Source: https://library.fiveable.me/_next/image?url=https%3A%2F%2Fstorage.googleapis.com%2Fstatic.prod.fiveable.me%2Fsearch-images%252F%2522Bode_plot_construction_magnitude_phase_plot_frequency_response_LTI_system_transfer_function_visualization%2522-figmatlab6-2.png&w=1080&q=75.*

### 4.3.1. Steps for Constructing Bode Magnitude Plots

Construction of the Bode magnitude plot begins with identifying the transfer function to be employed in building the plot. This is achieved by identifying the component elements of the system and their respective characteristics to form the overall transfer function H(s). The second step is to replace the complex variable s with jω where j is the imaginary unit and ω represents the frequency variable in radians per second.

This substitution allows one to compute the value of the systems response over a range of frequencies, both the transient response and the steady-state behavior of the system. By analyzing the principal factors of the transfer function such as poles, zeros, and gain constants, data gathered could be organized rationally into a comprehensible representation to enable better understanding of the system's behavior in the frequency domain. As soon as the transfer function is established and substituted for frequencies, the second most crucial step is to calculate the magnitude of H(jω). The magnitude can typically be computed in decibels (dB) using the formula 20 log10|H(jω)|. The individual components of the transfer function operate independently on the complete magnitude plot, so an additive procedure is feasible. Poles and zeros within the transfer function dictate the slope of the plot, typically a slope downwards for poles and an increasing slope for zeros.

There exists a requirement to break these contributions down over the logarithmic range of frequency, which does the job of keeping the large range of values manageable. Additionally, selecting significant points and their corresponding contributions does the job of making the plot accurately reflect the systems dynamic characteristics, which is particularly relevant when considering stability and response. The final step in constructing the Bode magnitude plot is to sum up all the individual contributions as a single continuous plot. The graphical plotting needs to be done with caution so that each contribution is plotted against its respective frequency ranges appropriately. The final Bode magnitude plot may include critical markers, that is, corner frequencies at which the slope of the plot changes, yielding important information about system behavior, particularly concerning stability margins. Checking the characteristics of the Bode plot can reveal essential information about the stability of the systems—especially in consideration of nonlinear effects in systems with dynamics based on converters, as shown in recent studies (Nduwamungu et al., 2024; Eckel et al., 2024). Through the understanding of how the magnitude plot displays the gain and phase relationships of a system, engineers can be more effectively informed in their decisions regarding controller design and optimization to meet desired performance demands (Figure 4.7).

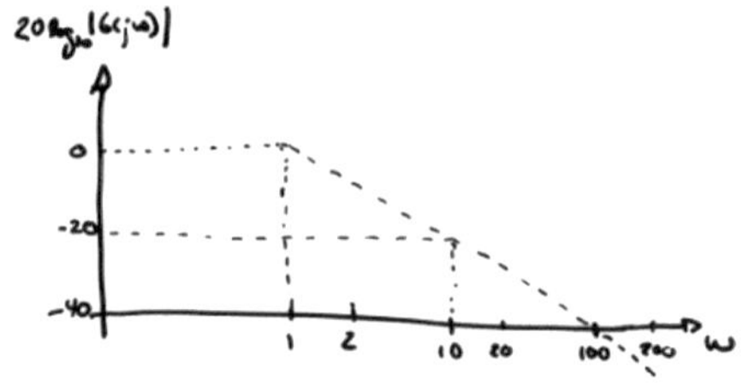


**Figure 4.7**. *Step-by-step bode plot.*

*Source: https://i0.wp.com/linearcontrol.info/fundamentals/wp-content/uploads/2019/05/bds1.png?resize=512%2C286&ssl=1.*

## 4.3.2. Steps for Constructing Bode Phase Plots

To build Bode phase plots, it is necessary to begin with the systems frequency response. The starting point for this is gaining the transfer function, which is mathematically a description of the response of a system output to inputs at different frequencies. This is normally obtained from the system's differential equations or determined from experimental data. Once the transfer function is ascertained, it is decomposed into its factors—commonly in the form of products of first-order and second-order factors. Decomposing into these factors is critical because it simplifies plotting the individual factors and then summing their phase contributions to obtain the overall Bode phase plot. The intricacies of such decompositions are key to how the system's dynamic behavior subtleties can be captured in the frequency domain. Having identified the transfer function, the next step towards constructing Bode phase plots is to calculate the phase contribution of each of these components across a range of frequencies. Every first-order or second-order factor contributes a particular phase shift, which may be calculated by conventional formulae: first-order factors contribute a phase shift of –90 degrees at high frequency, whereas second-order factors contribute a more complex phase behavior that must be investigated with utmost caution. The overall phase shift is determined by adding individual phase contributions for various frequencies, most often converted into logarithmic form. This convergence is particularly important for ensuring that the final plot entirely encircles the total phase response of the system to different frequency inputs, leading into further stability analysis. Lastly, Bode phase plots are a basic tool for system stability determination via calculations of phase margin. Once the cumulative phase response is plotted against the logarithmic frequency scale, the following crucial step is to have the phase margin, which informs us about the proximity of a system to instability.

Phase margin is the amount of phase lag at the gain crossover frequency—the frequency at which the system's gain is unity (0 dB). A good phase margin is equally critical to ensure during robust system design, as evidenced by research based on adaptive control models; such methodologies have taken significant steps toward enhancing stability margins and helping provide for higher quality response under changing demands (Meng, 2020). Additionally, approaches that utilize impedance analysis provide more clarity about why phase margins are sometimes at odds with system dynamic performance, as in more sophisticated stability evaluation frameworks (Lin et al., 2024). These techniques not only enhance the construction of Bode phase plots but also provide the foundation for effective control actions in engineering applications where stability plays a significant role.

### 4.3.3. Common Practices and Tips for Accurate Bode Plot Construction

Creating a good Bode plot involves a systematic process beginning with an accurate mathematical model of the system in question. The process begins with determining the transfer function, which is a characterization of the input-output behavior of a linear time-invariant system in frequency domain. It is required to take care that poles and zeros of the transfer function are accurately represented because they dominate the shape of the Bode plot. Extra care must be exercised in constructing the asymptotic approximations for each pole and zero because these will set the resultant frequency response behavior. By utilizing software tools, the process will be made easier through the provision of numerical values for various frequencies, and hence the ability to easily make magnitude and phase plots. Double checking all these components with analytical techniques or simulation ensures a solid foundation for further analysis, especially in cases involving complex systems (Goubej et al., 2021). In addition to ensuring mathematical accuracy, correct comprehension of the range of interest frequency is also essential in Bode plot construction. The range of frequency should span all the relevant dynamics of the system in respect to the bandwidth limitation by the application. This necessitates specification of low and high-frequency trends, which can be influenced by a number of factors such as system damping and resonances. With appropriate selection of the range of frequency, engineers will not draw wrong conclusions that would be due to lacking information. Logarithmic frequency scales enhance granularity of representation, particularly for systems with high dynamic ranges. Moreover, it should be ensured that there should be a sufficient number of points over the specified frequency range so that there is a smooth curve between the points on the Bode plot to ensure proper system stability and performance interpretations

(Arvid, 2017). Lastly, the graphical form of the Bode plot is extremely critical in communicating basic system behavior. Careful axis labeling, equal scales, and a distinction in color or line style for magnitude and phase can assist considerably. The inclusion of grid lines will serve to resolve more critical intersection points, which are crucial for stability analysis, such as gain and phase margin calculations. The inclusion of a legend to illustrate various system configurations or controller settings can be of significant use for the reader. After the construction of the plot, a check against earlier known benchmarks or simulation runs is a good idea in order to determine its accuracy against predicted results. All these collectively not only result in the construction of a good Bode plot but contribute to the overall interpretability and effectiveness of frequency domain analysis within engineering problems (Goubej et al., 2021) (Figure 4.8).

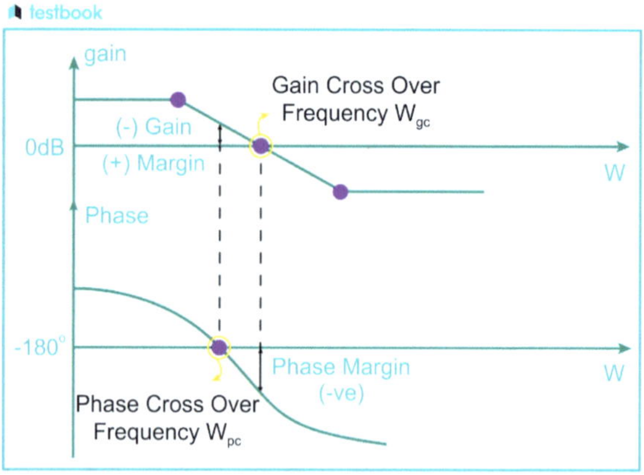


**Figure 4.8**. *Common practices and tips for accurate Bode plot construction.*

*Source: https://cdn.testbook.com/1718630926780-Bode%20Plot_1.png/1718630928.png.*

## 4.4. NYQUIST STABILITY CRITERION

The Nyquist Stability Criterion serves as a fundamental principle in the analysis of linear control systems, particularly within the framework of frequency domain techniques. This criterion facilitates the assessment of system stability by analyzing the frequency response of the open-loop transfer function. By plotting the Nyquist plot, which represents the locus of the complex-valued function as the frequency varies, engineers can determine whether the system will exhibit stable or unstable behavior. Specifically, the criterion asserts that the number of encirclements of the critical point ( –1, 0) on the complex plane indicates the stability of the closed-loop system relative to the number of right-half plane poles of the open-loop transfer function. Thus, the Nyquist Stability Criterion provides a powerful tool for designers in anticipating and mitigating potential stability issues before proceeding to implementation. In practical applications, the utility of the Nyquist Stability Criterion often extends into realms such as control system design and performance optimization. As systems develop increasingly complex dynamics—often arising from inherent time delays—validating stability in a systematic manner becomes essential. Notably, in systems with time delays, analyzing the Nyquist plot can reveal critical factors contributing to instability.

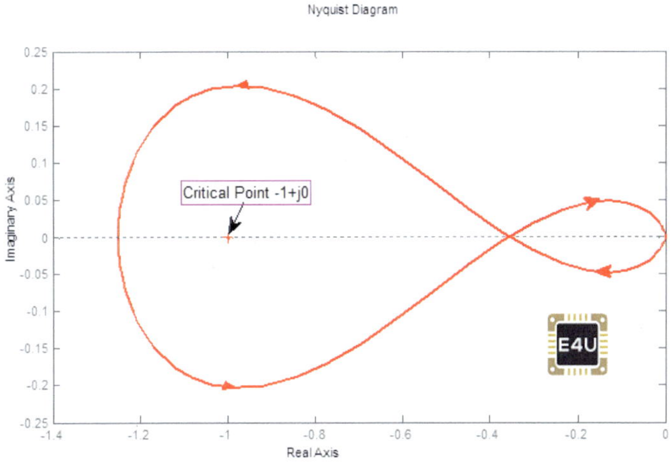


**Figure 4.9**. *Nyquist stability criterion.*

*Source: https://www.electrical4u.com/wp-content/uploads/What-is-Nyquist-Criteria.png.*

For instance, incorporating delay effects into the frequency response analysis allows for a more comprehensive understanding of the system's behavior. An empirical study on a heavy-duty gas turbine control system demonstrates how applying the small-gain theorem in conjunction with the Nyquist Criterion facilitates adequate delay margin assessments, ultimately ensuring stability in dynamic conditions (Su et al., 2025). This synergy highlights the role of the Nyquist Criterion in navigating complexities in real-world systems. Moreover, the integration of a multi-attribute evaluation approach extends the classical Nyquist Stability Criterion by quantitatively assessing stability margins in various control contexts. This method, known as the MAMD-CAS, evaluates not just the stability but also the dynamic responses across different timescales. The enhancement of the Nyquist Criterion through this multi-dimensional assessment allows for deeper insights into aspects such as phase margin and low-frequency oscillations (LFO). In this context, understanding the interactions between parameters becomes crucial for optimizing control systems. While achieving optimal dynamic performance, the MAMD-CAS framework aids in balancing trade-offs between stability margins and response times. This innovative approach not only reinforces the significance of the Nyquist Stability Criterion but also underscores its adaptability in addressing the increasingly complex dynamics of modern control systems (Lin et al., 2024) (Figure 4.9).

### 4.4.1. Definition and Significance of the Nyquist Criterion

The Nyquist criterion is a fundamental control theory and frequency domain analysis theorem, which enables us to determine the stability of a control system from its open-loop frequency response. More specifically, it states that stability of the closed-loop system can be determined from a consideration of poles of its transfer function using the context of a Nyquist plot of the system's complex frequency response. The elegance of the Nyquist criterion is that it relies on inherently different criterion for encirclements of the point $(-1, 0)$ of the critical point in the Nyquist plot, changes in system stability. Engineers can therefore predict how various parameters of a system influence its dynamic response and whether specific designs meet the desired stability margins for excellent control system performance. The significance of the Nyquist criterion extends from stability assessment to directly involve the concepts of gain and phase margins that are so important in robust control design. Measuring the degree of gain or phase margin the system possesses prior to destabilization, engineers can adjust system parameters to improve stability while permitting responsiveness. This focus on gain and phase margins acts to neutralize the adverse effects of system model uncertainties and real-world disturbances. Practically, achieving desirable gain and phase margins can contribute to increasing a systems robustness and flexibility to varying operating conditions, which aligns with industry demands for reliable control systems. The need is essential to processes in such complex engineering fields and therefore constitutes a standard part of system design and analysis. Along with this, the Nyquist stability criterion could be further narrowed down by using frequency domain techniques that enable stronger analysis of system responses under different circumstances. A case in point is the current research in the field, such as norm-optimal iterative learning control (NO-ILC) frameworks, and how it illustrates the application of frequency-domain analysis in measuring robustness against model uncertainty as well as handling system dynamics complexity. In this context, the Nyquist criterion is a vital tool that reconciles classical control principles with contemporary methods, compensating for deficiency in traditional approaches by considering variation in phase and model imperfections. The evolution of these techniques justifies the Nyquist criterions continued validity in today's engineering practice because it continues to be a critical influence in theoretical as well as applied settings (Edward et al., 2017; Ge et al., 2017).

### 4.4.2. Nyquist Plot and Its Interpretation

In control theory, the Nyquist plot is a significant method for testing the stability of systems by plotting the frequency response of a dynamic system on the complex plane. The real part of the systems response is measured along the main axis of the plot, and the imaginary part along the imaginary axis. By showing the frequency response of the open-loop transfer functions, the Nyquist plot enables engineers to examine the stability of a feedback control system through the encircling of critical points in the complex plane. Not only does it facilitate gain and phase margin analysis but also indicates potential stability peril resulting from system dynamics, especially under conditions with time delay. In practical applications, the use of Nyquist plots and Bode plots together can enhance analysis, providing a clearer understanding of system behavior in various ranges of frequency. Interpreting a Nyquist plot requires a clear grasp of its relation to stability criteria, particularly for intricate systems. One of the main principles here is the Nyquist stability criterion, where for a system to be stable the Nyquist plot must encircle the point of ( –1, 0) in the complex plane in a specified manner depending on the systems poles. The number of encirclements corresponds to the number of poles of the closed-loop system in the right half-plane. This is a critical relationship, as it offers explicit understanding of the manner in which parameter changes in systems could impact stability. Additionally, as indicated by researchers, the addition of Nyquist plots to Bode plot analysis, more so in systems prone to time delays, enhances stability analysis, allowing for complete evaluation of performance under varying operating conditions (Su et al., 2025). The application of Nyquist plots beyond mere theory is seen in the real world in the form of grid-connected complex systems. For instance, voltage-source converters (VSCs) connected to a weak grid can be analyzed for stability using Nyquist plots of open-loop transfer functions with precision. Upon such analysis, one can visualize how the interactions of feedback loops, such as phase-locked loop and outer control loop, can impact the overall system stability. Time-domain simulations have demonstrated a very high degree of correlation of Nyquist curve analysis results with real system responses, showing the applicability of the approach in engineering real-world problems (Wu et al., 2023). The above discussion concludes that the efficiency and flexibility of Nyquist plots make them precious tools in frequency domain analysis, which guides engineers towards designing and analyzing complex control systems (Figure 4.10).

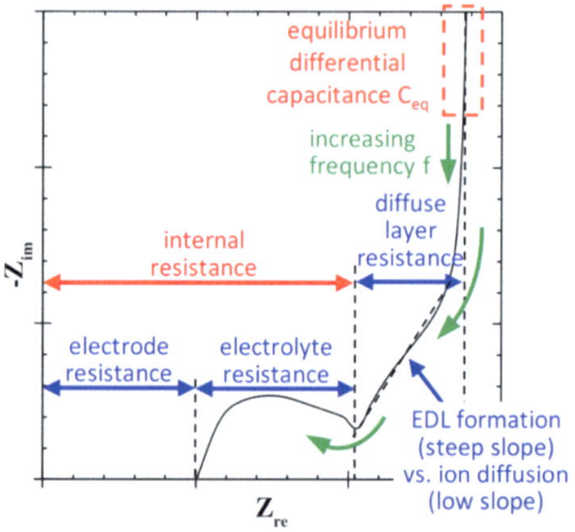


**Figure 4.10**. *Physical interpretations of Nyquist plots for EDLC electrodes and devices.*

*Source: https://www.seas.ucla.edu/~pilon/images/supercapacitors/EIS-Interpretation-TOC.png.*

## 4.4.3. Application of the Nyquist Criterion in Control System Stability

Application of the Nyquist criterion for stability of control systems remains a foundation for frequency domain analysis. With the application of Nyquist plots, engineers can graphically represent the relationship between the open-loop transfer function and control system stability. The criterion establishes a one-to-one correspondence between encirclements of the point ( –1, 0) in the complex plane and the number of right-half plane poles in the open-loop transfer function. This critical interaction alerts the designer of potential stability issues so that parameters of the system can be readjusted beforehand. The strength of this method comes to light when applied to systems that contain time delays because such delays have a tendency to introduce complexity that can damage stability. For instance, in heavy-duty gas turbine control systems, the Nyquist criterion plays a paramount role in delay performance and stabilizing efficient systems, stressing the importance of frequency domain analysis in actual applications (Su et al., 2025). Even more enhancing the Nyquist criterion's usefulness is its combination with novel analytical aids, such as integrated assessment systems for stability margins. The MAMD-CAS methodology, for example, quantifies numerically the stability of control systems using analysis of the phase margin and its relationship to dynamic performance criteria. This multi-dimensional analysis allows engineers to assess not just the qualitative character of stability as typically shown graphically within the Nyquist plots, but also to ensure that a preferred stability margin is achieved. It is understood that the relationship between stability margins and control parameters can yield critical information about system performance, as noted where there is a conflicting dynamic performance with stability. Hence, the application of the Nyquist criterion coupled with quantitative assessments helps facilitate easy understanding of stability in advanced control systems (Lin et al., 2024). The practical significance of the Nyquist criterion extends beyond theory to real-world situations, illustrating its importance in system design and observation. Through the correct application of Nyquist plots, engineers are able to identify possible instabilities before systems are sent to principal applications. For instance, in the assessment of a hardware-in-the-loop test for a gas turbine control system, the Nyquist criterion can help designers in the choice of appropriate controller parameters that involve delay margins without sacrificing desired tracking performance. This interaction between analytical analysis and experimental testing lays a foundation for robust control design methods under the guarantee that resulting systems will operate robustly under conditions of uncertainty. Such is actually the character of this technique, which not only justifies the theoretical postulates of the Nyquist criterion but also demonstrates its irreplaceable role in achieving stable and efficient control in more developed engineering systems (Su et al., 2025).

## 4.5. GAIN AND PHASE MARGINS

In frequency domain analysis, gain margin and phase margin are two key stability measures of control systems. Gain margin is a measure of how large the system gain can be increased before the system is on the verge of instability, and phase margin is a measure of additional amount of phase lag that can be tolerated before the system becomes unstable. They can be easily graphed in a graphical manner on Bode plots, and the gain and phase curves provide valuable information about system behavior. The stability of a system is directly related to these margins; poor gain or phase margins lead to undesirable oscillations and possibly fatal instability. Since systems generally are susceptible to demand changes or perturbation from the outside world, achieving adequate gain and phase margins is important for robust design and operation. Because Bode plots allow direct visualization of gain and phase margins, the engineer can easily pick out critical frequencies at which gain crosses 0 dB and phase crosses

–180° by plotting the logarithmic gain and the phase shift on a semi-log paper against frequency. These convergence points need to be established to find the gain and phase margins because deviations from these standards imply there is potential instability. Use of Bode plots in combination with the Nyquist stability criterion will also be able to provide insight into stability margins. Specifically, by understanding plots, it becomes easier to intuitively determine the effects of system parameter variations on stability and thus enable the engineers to make informed design choices that maximize performance as well as safety in control systems. Thus, Bode plots and corresponding gain and phase margins are key tools in the frequency domain analysis arsenal. Furthermore, gain and phase margin implications extend beyond the realm of theory; they are significant in applications, especially in time-delayed systems. Time delays occur in most control systems and will have substantial influences on both gain and phase margins to create unanticipated responses. Research has indicated that proper estimation of such delays using closed-loop Bode plots can yield valuable information regarding stability margins, as illustrated in Su et al. (2025). Trade-offs between good gain and phase margins and desired dynamic performance are equally crucial to successful system design. As referred to in Lin et al. (2024), some dynamic performance measures interdependencies and stability margins interdependencies can cause these design tasks to be difficult. For this purpose, exhaustive understanding of gain and phase margins under frequency domain analysis guidance enables engineers to masterfully manage these problems (Figure 4.11).

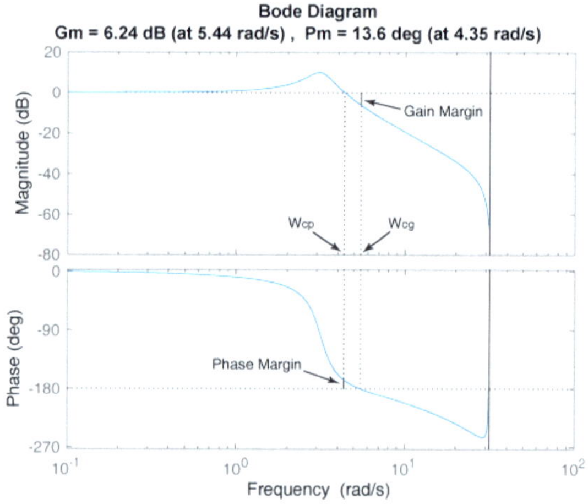


**Figure 4.11**. *Gain margin, phase margin, and crossover frequencies.*

*Source: https://www.mathworks.com/help/control/ref/margin1.png.*

### 4.5.1. Definition of Gain Margin and Phase Margin

Gain margin and phase margin definitions are vital to know in control system analysis, particularly in the frequency domain. Gain margin is referred to as the degree of increase in a system gain prior to losing stability. It is measured in decibels (dB) and is obtained from the systems open-loop frequency response. A positive gain margin indicates that the system is able to maintain an increase in gain without inducing instability, and a negative gain margin predicts impending instability. Gain margin thus becomes a valuable stability criterion in control system design, and system parameter changes and gain setting can be based upon it by engineers. Phase margin, however, quantifies the degree of additional phase lag at the gain crossover frequency at which a system is on the verge of instability. It is calculated based on the open-loop transfer function's phase angle at the frequency that has a gain of one (0 dB). A positive phase margin indicates that there is a margin against oscillations and instability; a negative phase margin indicates that the system will most likely be oscillatory or even diverge. Therefore, systems with larger numbers of phase margins are typically more robust and insensitive to variations in system parameters, and hence phase margin is naturally a consideration in control design for achieving commanded transient response and stability requirements. The combination of the gain and phase margins can provide more information regarding system stability and performance. High gain margin, for instance, may indicate a system's immunity to gain change but also for slow response times. Improvement in the phase margin at optimum may be at the cost of reduced stability if not regulated properly. Their connection to the Nyquist stability criterion makes them significant in analyzing control systems. Methods such as Bode plots enable such margins to be simply viewed and therefore enable engineers to examine trade-offs between gain, phase, and total system response. More contemporary developments, such as the sophisticated multifaceted analyses such as MAMD-CAS, enhance the capacity to measure and better delimit these margins to use in day-to-day applications, as seen in stability studies in complex systems (Su et al., 2025; Lin et al., 2024).

### 4.5.2. Importance of Gain and Phase Margins for System Stability

The gain and phase margin analysis is intrinsic in control system stability determination under the frequency domain analysis paradigm. The margins indicate the amount of gain or phase shift a system can endure before becoming unstable. Specifically, gain margin is the amount of gain increase that can be added before the system will be unstable, and phase margin is the additional phase lag that can be added before causing oscillatory behavior. These parameters cannot be overly emphasized because they give system designers valuable information about system robustness, so that instabilities can be forecasted and identified before they take place in real-time applications. Adequate examination of such margins in Bode plots and Nyquist diagrams allows engineers to predict system behavior in uncertain conditions, thereby laying a fundamental basis for the development of stable systems. In real-world situations, using gain and phase margin analysis can produce significant gains in system reliability and performance. For instance, for a heavy-duty gas turbine control system, like analyzed in one research, control loop stability is generally directly influenced by time delays and how these affect feedback interconnections.

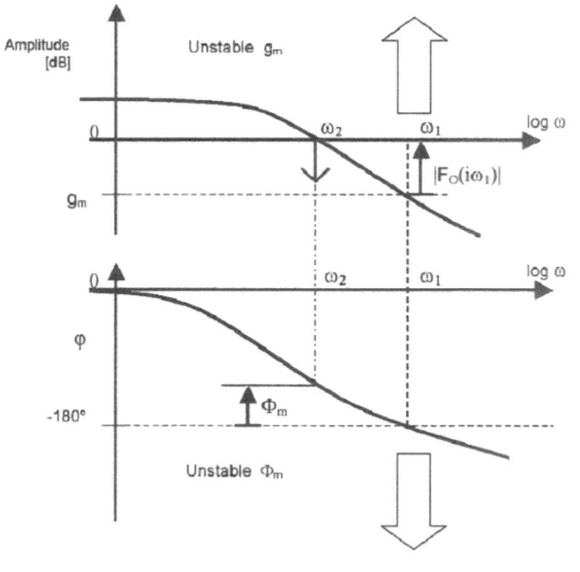


**Figure 4.12**. *Phase margin ( Φ m ) and gain margin (g m).*

*Source: https://www.researchgate.net/publication/26493103/figure/fig2/AS:310025265401868@1450927191975/Phase-margin-PH-m-and-gain-margin-g-m.png.*

The calculated delay margins, which are very critical in maintaining system stability, are shown to result in degrading performance and ultimately lead to instability in dynamic systems for low margins. By employing frequency domain analysis in the type of Bode plots, such margins can be graphically represented and associated with tracking performance by engineers, which is avoided through prevention of delays and system stability problems. This indicates the real-world application for knowledge of gain and phase margins in the avoidance of operation instability, and thus they are central aspects of control system design. Also, in the case of complex systems, such as a grid connected to a wind farm, the importance of gain and phase margins is even greater. For such cases, analyses in both the time and frequency domains are required in order to analyze overall system dynamics. While Nyquist plots are able to provide a clearer indication of stability with complex poles and zeros, gain and phase margins are still crucial in validating stability conclusions from frequency-domain methods. Stability being measurable in terms of these margins ensures the robustness of the system under varying operating conditions. As an example, as shown in grid stability analysis, the correct use of frequency response principles guarantees manufacturers to design for dynamic change without undue risk of instability. Gain and phase margins are thus still valuable tools in the assurance of control system performance across a wide range of applications, cementing their role in contemporary engineering solution (Figure 4.12).

### 4.5.3. Methods for Calculating Gain and Phase Margins

Gain and phase margin are both useful to determine in order to evaluate the stability of a system in the frequency domain. Gain margin, by its own name, is the extent to which system gain can be increased before the system becomes unstable, and can be derived from Bode plots. Precisely, it consists of determining the frequency at which the phase angle is –180° and then determining the gain at this frequency. Conversely, phase margin shows additional phase lag accepted before the system will collapse by finding the frequency where unity gain is reached, typically at 0 dB, and taking a measurement of that phase. Both margins are significant measures of system performance, enabling systems to cope with variations in component behavior or external disturbances without collapsing. The Nyquist stability criterion extends further to define improved procedures for gain and phase margin analysis, especially in systems complicated by multi-variable interactions. The criterion employs the Nyquist plot that displays a system's open-loop transfer function frequency response. From the plot, and more particularly from encirclements of the critical point in the complex plane, engineers are able to ascertain system stability. This analysis can be directly applied to gain and phase margin calculation, with specific regard to the marginal frequencies of Nyquist plots where marginal stability is found. Such structure conforms to the eigenvalue solution for small-signal stability in power-electronics-based systems, where the impedance matrices can represent the systems response against perturbations. For cases where the classical Bode or Nyquist methods become cumbersome, the calculation of gain and phase margins from such sophisticated methods brings clarity and ensures sound stability comprehension. To further make calculation and comprehension of gain and phase margins easier, recent developments in impedance modeling have emerged as a helpful approach. The impedance circuit model attempts to model power converters in terms of discrete circuit components so that control parameter implications on system dynamics can be readily visualized by designers. Through gain and phase margin tracing for various positions of impedances, this technique illustrates the interactions as well as the coupling effects among the control systems so that stability analysis is more intuitive. For instance, as outlined in more complex instances of grid-forming voltage source inverters, such a modeling approach not only facilitates ease of analysis of gain and phase margins but also leads design adjustments based on the observed interactions across several dynamic operating modes. By doing so, these innovations in modeling software can potentially enhance considerably the reliability and intelligibility quality of frequency domain analyses in control systems, especially when traditional methods are inadequate (Li et al., 2020; Wen et al., 2017).

## 4.6. PRACTICAL APPLICATIONS OF FREQUENCY DOMAIN ANALYSIS

Frequency domain analysis in laboratory work has important applications in many areas of engineering, particularly in control systems. By learning the response of systems to different frequencies, engineers develop a broad insight into system behavior under varying conditions of operation. This provides valuable knowledge when faced with issues like time delay effects, which have negative consequences on the performance of systems, causing instability. For instance, frequency domain tools can be utilized to model the load/speed control loop of a heavy-duty gas turbine such that its stability can be examined in detail. Together with employing the small-gain theorem, stability conditions can be derived, which illustrates how frequency response tools like Bode plots render it simple to calculate delay margins to make systems stable regardless of inherent delays (Su et al., 2025). This is evidence of how frequency domain analysis can enhance control system performance and reliability. Another point of note is the application of gain and phase margins to relative stability analysis of systems. Frequency domain analysis provides indications of how variations in system parameters affect these margins, which are such crucial indices of system robustness. Through a simple DC servo motor model, it is shown that although parameter changes cannot alter the gain and phase margins, these can effectively control the overall system dynamics. Such stability of margins underscores the importance of incorporating delay margins as an additional performance criterion. Through Kharitonov's theorem, engineers can effectively determine delay margins in interval systems and ensure thorough analysis of stability in practice application (Hote, 2021). The combination of these margins provides a more stable method of system analysis and system design for robustness against uncertainties in real-life applications. The use of frequency domain analysis in practice application extends beyond traditional control systems to include all areas from telecommunication to robotics. For example, in telecommunication, the frequency response analysis allows telecommunication engineers to optimize signal processing and communication systems for communication integrity over various frequencies. Similarly, in robotics, understanding the dynamic response of robotic systems by using Bode plots directs the development of control methods that are able to react towards changing environments. This cross-disciplinary application emphasizes the versatility of frequency domain analysis among various engineering fields and its pivotal role in technology advancement. Last but not least, the insight obtained from frequency response theory, Bode plots, and stability conditions is priceless for engineers seeking to innovate and enhance system performance in most fields (Figure 4.13).

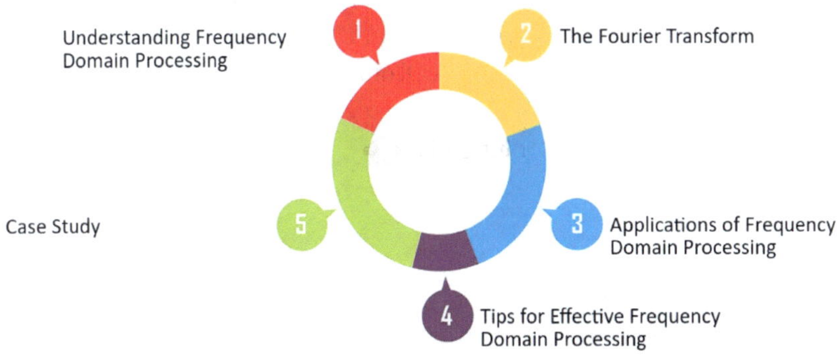


**Figure 4.13**. *The power of frequency domain analysis.*

*Source: https://fastercapital.co/i/Navigating-the-Frequency-Domain--FFCS-for-Advanced-Processing--Introduction-to-Frequency-Domain-Processing.webp.*

### 4.6.1. Use in Electronic Circuit Design

Frequency domain analysis is central to stability and performance in electronic circuit design. By analyzing the frequency response of a circuit, designers are best able to understand how the system reacts to different frequencies, more important in technologies such as DC microgrids. In DC microgrids, constant power loads might induce instability through negative incremental impedance behavior. Therefore, learning the principles of frequency response allows engineers to design more robust control methods that eliminate such instabilities. Bode plots, a graphical form of gain and phase margins, are found to be powerful tools in the exercise, allowing designers to perform qualitative stability analyses of circuits and also specify best parameters for desired system performance. Construction of the Bode plot is more than mere technical drill; it is a conceptual bridge between theoretical observations and practical applications to circuit design. Through systematic plotting of the gain and phase shift of a circuit versus logarithmic frequency, designers can pinpoint important transition points that herald approaching stability issues. These plots also highlight the gain and phase margins and give important insight into how close a system is to losing stability. In the application of DC microgrids, where reliability and efficiency are the requirements, the utilization of Bode plots can provide sound choices on component selection and control strategy. Thus, frequency domain analysis with the help of Bode plots guides designers through the complexities involved in modern electronic circuits such that systems are in an optimal state under varied conditions. The Nyquist stability criterion further complements the information provided by Bode plots such that more insight is gained regarding feedback systems in electronic circuit design.

**Figure 4.14**. *Use in electronic circuit design.*

*Source: https://www.autodesk.com/products/fusion-360/blog/wp-content/uploads/2021/12/shutterstock_2043923687.jpg.*

This criterion, through a contour integral method, allows system stability to be determined from open-loop frequency response. Applied to DC microgrids, it is of great importance as it aids in ascertaining the stability boundaries which arise while dealing with such recalcitrant loads as pulsed power loads. Because these problems emphasize the necessity of accurate impedance specifications, making the Nyquist criterion a part of design processes becomes indispensable.

A precise comprehension of the aforementioned principles provides engineers with the capability to create stabilization techniques that offset potential vulnerabilities, safeguarding the operation of electronic circuits. Thus, the combining of Bode plots and the Nyquist stability criterion provides a stable platform for analysis and designing stable electronic circuits in the realm of modern applications, including those new to DC microgrids (Nduwamungu et al., 2024; Tehrani, 2022) (Figure 4.14).

## 4.6.2. Application in Feedback Control Systems

The application of frequency domain analysis in the analysis of feedback control systems is crucial in ensuring performance and stability, especially when dealing with modern electrical systems. The primary tool employed in this type of analysis is the Bode plot, which enables engineers to plot the gain and phase characteristics of a system versus various frequencies. Engineers are able to obtain information regarding the response of a control system to different inputs from these plots. For instance, through the inspection of the Bode plots in systems like the Onshore Wind Power All-DC Generation System (OWDCG), stability parameters can be ascertained that are of utmost importance for small disturbance stability, an element emphasized in recent studies that study interplay among subsystems (Wang et al., 2024). This finer understanding aids in ascertaining control schemes to ensure maximum system reliability and responsiveness, particularly important for the integration of renewable energy. Furthermore, Nyquist stability criteria form the basis of feedback control systems in terms of a graphical approach towards checking stability in systems under the effects of feedback loops. The Nyquist criterion specifically analyzes how the frequency response of a system will have an interaction with the feedback path and how engineers can predict how variations in system parameters will create stability or instability. As has been demonstrated by the analysis in DC microgrids, in which constant power loads (CPLs) exhibit negative impedance characteristics that can make the system unstable, it is extremely critical to study gain and phase margins (Nduwamungu et al., 2024). These margins give us a quantitative approximation of how much gain or phase displacement a system can tolerate before instability is reached and thus allow pre-emptive adaptation in controller design that retains stability even as operating conditions vary. Ultimately, the marriage of frequency domain analysis with feed-back control systems reduces the design and optimization process to simplicity and facilitates handling the intricacies of electrical systems more easily. The interlink of terms such as Bode plots and Nyquist criteria guides solutions in responding to issues encountered in next-generation applications, including the recently emerging technologies in wind energy systems. Through the analysis of feedback mechanisms using the above-mentioned tools, engineers are able to counteract instability risks and enhance operational performance. As indicated in the OWDCG schemes and methods to DC/DC converters, a clear understanding of impedance behaviors and their effects on control variables is pivotal in offering insightful guidance for optimal design choices ensuring stable and efficient performance in diverse applications (Wang et al., 2024; Nduwamungu et al., 2024).

### 4.6.3. Role in Signal Processing and Communications

Signal processing and communications are fundamentally two interconnected disciplines that apply frequency domain analysis in order to maximize the performance of a broad spectrum of systems. Through the conversion of signals to the frequency domain, engineers are able to utilize tools like Bode plots to present a system's frequency response, providing useful information regarding how systems will perform under varying operating conditions. This graphical representation not only makes it easy to identify the resonant frequencies at which amplification occurs but also aids in the design of filters to selectively eliminate unwanted signals.

The application of Bode plots thus fills the gap between theoretical investigation and engineering implementation, making them a necessary tool for use in the design and optimization of communication systems. Besides Bode plots, the Nyquist stability criterion is also important in ensuring system stability within the frequency domain. The criterion provides a graphical method for assessing system stability by checking the open-loop and closed-loop gain and phase margins relationship. A system stability is a direct function of its ability to operate under varying conditions in a reliable manner, a primary requirement in communications networks where breakdowns cause significant data loss. By employing the use of the Nyquist criterion, the engineers can identify proper gain and phase margins necessary to guard against oscillations and thus ensure reliability in signal processing capability in communications. The addition of state-of-the-art tools such as the interactive programming developed for passive low pass filters also represents a critical step in frequency domain analysis.

The collaboration of tools such as Arduino and LabVIEW has simplified real-time visualization of circuit activity when one is changing parameters, allowing a hands-on experience in understanding system dynamics. This evolution not only facilitates pedagogic practice but also allows engineers to conduct iterative testing on the behavior of systems. As attested to by the development of an interactive front panel for analyzing RC circuits, these methods illustrate the union of theoretical concepts with practical application, adding to the significance of frequency domain analysis in signal processing and communication's dynamic aspect (Gonz et al., 2022; Sanjit, 2021).

# SUMMARY

Frequency domain analysis research has produced immense insight into system behavior complexities in control engineering. Through application of frequency response concepts such as the utilization of Bode plots and the Nyquist stability criterion, the engineer is well placed to analyze and ensure system stability in both theoretical and real situations. The capacity to plot Bode plots enables the direct interpretation of gain and phase margins, which are crucial in robust performance in various operating points. These analysis tools are crucial for detecting the potential instabilities and for optimizing control strategies, hence maximizing overall system performance and safety. Furthermore, incorporation of time delays in control loops is a major challenge that has the potential to destabilize the system. As demonstrated in Su et al. (2025), the assessment of delay margins using closed-loop Bode plots is highly valuable, particularly for complex systems such as gas turbine control systems. Understanding the compromises between an adequate margin of delay and desired tracking performance showcases the nuances involved in making decisions for engineers while designing effective control systems. The result of this analysis has practical use in a variety of domains, including aerospace and energy management systems, where optimum performance is paramount. The value of Nyquist plots, as stressed in Panah et al. (2025), is also illustrated through their potential to effectively investigate systems with more intricate dynamics, such as non-minimum phase behavior. Nyquist plots' ability to present a general overview of stability characteristics without any constraints like in the case of Bode plots further improves the process of stability analysis. In the connection between both frequency domain and time domain evaluations, engineers are capable of verifying transient responses and steady-state ones, eventually yielding decisions in terms of design. The controversy surrounding these tools and methods highlights their critical contribution to contemporary control systems engineering and demands them to be continuously integrated and utilized in designing adaptive and strong systems.

In the field of frequency domain analysis, the concept of frequency response is a foundation building block of control theory. This method allows researchers and engineers to discover how a system responds to varied input frequencies and thereby describe system behavior across a variety of conditions. Frequency response analysis is essentially all about discovering the outputs that come from a system when it is exposed to sinusoidal inputs. The most important tool in such analysis is the Bode plot, which is a graphical representation that captures both magnitude and phase information across a set of frequencies. The visualization facilitates the identification of critical dynamics like resonances and bandwidth, essential in the design of stable control systems that meet specified performance specifications. The construction of these plots requires close attention to detail so that they represent system properties correctly and serve the foundation for their subsequent application of control measures. The Nyquist stability criterion is a valued tool in the frequency domain analysis, particularly in the determination of the stability of feedback systems. This condition looks at the encirclement of critical points within the Nyquist diagram, thereby indicating poles in the right-hand side of the complex plane. Connecting the open-loop frequency response to closed-loop stability, engineers can check for stability margins, which indicate how much gain or phase difference will exist before the system is unstable. Notably, extensive knowledge about stability margins gives assurance of system robustness against disturbances. Recent advances in techniques, such as the MAMD-CAS, offer quantitative evaluations of stability and aid parameter design, ensuring an optimal dynamic response with stability margins as it resolves the dynamic performance-stability margins relationship (Lin et al., 2024). Such extensive knowledge is critically important for stabilizing complex systems, particularly in the case of power electronics applications. The gain and phase margins then contribute to the analysis of system stability by defining how much gain or phase can be varied before the system loses stability. Gain margin defines by how much gain needs to be increased before the system begins to oscillate, and the phase margin provides a measure of how much phase delay can be tolerated before instability sets in. Techniques for increasing such margins, such as application of adaptive phase-locked loops (PLLs), have been significant improvements in maintaining stability in systems subjected to varying grid conditions. For instance, an adaptive PLL module can boost the stability margin of inverters connected to a grid significantly, as demonstrated by research indicating an increase in phase margin by more than 78% in specific conditions (Meng, 2020). Not only do these innovations stabilize systems as well, but they also enhance overall performance indices, that is, reducing total harmonic distortion (THD), thereby highlighting the integration of theory and practical application in frequency domain analysis.

# REFERENCES


1. Arshad, M., Beik, O., Manzoor, M. O., Ateeq, W., & Jalal ud Din, M. (2024). Stability analyses of Bison Wind Farm using frequency scan. *IECON 2024–50th Annual Conference of the IEEE Industrial Electronics Society.* https://www.semanticscholar.org/paper/f90fa414edc3da2b6a0940c10edd9dc24fbc2fbc (accessed on 19th September 2025).
2. Bauer, W., & Baranowski, J. (2020). Fractional PIλD controller design for a magnetic levitation system. Electronics, 9, 2135–2135. https://doi.org/10.3390/electronics9122135 (accessed on 19th September 2025).
3. Beleca, V., Pleşa, C., Onet, R., & Neag, M. (2021). A novel approach to the stability analysis of conditionally stable circuits. 2021 International Semiconductor Conference (CAS). https://www.semanticscholar.org/paper/fc37ca4e441ddb7ec2c367c8ed22ef54b7c21c10 (accessed on 19th September 2025).
4. Bhattacharyya, S., Datta, A., & Keel, L. (2009). Linear Control Theory. https://www.semanticscholar.org/paper/942c5f2b44fa6712c2a1f119c932285e00de6e33 (accessed on 19th September 2025).
5. De Tommasi, G. (2018). Plasma magnetic control in tokamak devices. *Journal of Fusion Energy, 38*, 406–436. https://doi.org/10.1007/s10894-018-0162-5 (accessed on 19th September 2025).
6. Eckel, C., Babazadeh, D., & Becker, C. (2024). Classification of converter-driven stability and suitable modeling and analysis methods. *IEEE Access, 12*, 53056–53073. https://doi.org/10.1109/access.2024.3388098 (accessed on 19th September 2025).
7. Ge, X., Stein, J. L., & Ersal, T. (2017). Frequency-domain analysis of robust monotonic convergence of norm-optimal iterative learning control. IEEE Transactions on Control Systems Technology, 26, 637–651. https://doi.org/10.1109/tcst.2017.2692729 (accessed on 19th September 2025).
8. González-Galindo, E. A., Carrillo-Moreno, A. O., Domínguez-Romero, F. J., & Pérez-García, J. (2022). Block programming to develop interactive activities in the analysis of passive filters using an RC circuit for distance classes. *Journal of Computational Systems and ICTs*. https://doi.org/10.35429/jcsi.2022.21.8.19.29 (accessed on 19th September 2025).
9. Goubej, M., Königsmarková, J., Kampinga, R., Nieuwenkamp, J., & Paquay, S. (2021). Employing finite element analysis and robust control concepts in mechatronic system design—Flexible manipulator case study. *Applied Sciences, 11*, 3689–3689. https://doi.org/10.3390/app11083689 (accessed on 19th September 2025).

10. Hancock, E. J., Ang, J., Papachristodoulou, A., & Stan, G. B. (2017). The interplay between feedback and buffering in cellular homeostasis. *Cell Systems, 5*, 498–508.e23. https://doi.org/10.1016/j.cels.2017.09.013 (accessed on 19th September 2025).
11. Hote, Y. V. (2021). Stability analysis of time-delayed linear and interval systems: Application to DC servo system. 2021 IEEE International Conference on Automation/XXIV Congress of the Chilean Association of Automatic Control (ICA-ACCA). https://www.semanticscholar.org/paper/6f68cc2fc467bfc9be53ed49482ecee390355d32 (accessed on 19th September 2025).
12. Keemink, A. Q. L. (2017). *Haptic Physical Human Assistance*. https://doi.org/10.3990/1.9789036544085 (accessed on 19th September 2025).
13. Kumar, V., Sharma, V., Arya, Y., Naresh, R., & Singh, A. (2022). Stochastic wind energy integrated multi-source power system control via a novel model predictive controller based on Harris Hawks optimization. Energy Sources, Part A: Recovery, Utilization, and Environmental Effects, 44, 10694–10719. https://www.semanticscholar.org/paper/492224cb6ba50cc251552dab367f1f2bc0147080 (accessed on 19th September 2025).
14. Li, Y., Gu, Y., Zhu, Y., Junyent-Ferré, A., Xiang, X., & Green, T. C. (2020). Impedance circuit model of grid-forming inverter: Visualizing control algorithms as circuit elements. IEEE Transactions on Power Electronics, 36, 3377–3395. https://doi.org/10.1109/tpel.2020.3015158 (accessed on 19th September 2025).
15. Lin, G., Liu, J., Rehtanz, C., Li, Y., Wang, P., & Zuo, W. (2024). A comprehensive stability assessment system for EV DC charging station based on multitimescale impedance model. *IEEE Transactions on Transportation Electrification, 10*, 938–961. https://www.semanticscholar.org/paper/83ac8079815adf58aafe6720e95c622d5239e3bb (accessed on 19th September 2025).
16. Lyu, J., Cai, X., & Molinas, M. (2015). Frequency domain stability analysis of MMC-based HVDC for wind farm integration. *IEEE Journal of Emerging and Selected Topics in Power Electronics, 4*, 141–151. https://doi.org/10.1109/jestpe.2015.2498182 (accessed on 19th September 2025).
17. Mazaheri Tehrani, H. (2022). *Identification and Nonlinear Stability Analysis Using Black-Box Models in DC Smart Microgrids*. https://doi.org/10.20868/upm.thesis.70221 (accessed on 19th September 2025).
18. Meng, H. (2020). Stability analysis of grid-connected inverter with adaptive PLL. *2020 5th International Conference on Automation, Control and Robotics Engineering (CACRE)*. https://www.semanticscholar.org/pa

per/1b155d592941ee5ba362b0342a9f7db86d25151c (accessed on 19th September 2025).

19. Messner, W., & Tilbury, D. M. (2020). A case study in software instruction over the World Wide Web: The Michigan CMU control tutorials for MATLAB. https://doi.org/10.18260/1-2--6441 (accessed on 19th September 2025).
20. Mitra, S. K. (2021). Hendrik Wade Bode [Pioneer in CAS]. IEEE Circuits and Systems Magazine, 21, 90–92. https://doi.org/10.1109/mcas.2021.3092588 (accessed on 19th September 2025).
21. Mohammed, N., Zhou, W., Bahrani, B., Hill, D. J., & Blaabjerg, F. (2024). PRBS-based impedance measurement tool for stability analysis of black-box EMT models in PSCAD. *2024 4th International Conference on Emerging Smart Technologies and Applications (eSmarTA)*. https://www.semanticscholar.org/paper/51395744783b32c18554dc5f169f3d4f7311c86d (accessed on 19th September 2025).
22. Najdek, K., Nalepa, R., & Lis, R. (2021). Selection of output voltage compensators gains in two cascaded boost converters with input filters by means of the D{\mathfrak{D}}D-decomposition technique. *Energies*. https://www.semanticscholar.org/paper/0be13f503cae8491adeb66be9483173cb7c333d8 (accessed on 19th September 2025).
23. Nduwamungu, A., Lie, T. T., Lestas, I., Nair, N. K. C., & Gunawardane, K. (2024). Control strategies and stabilization techniques for DC/DC converters application in DC MGs: Challenges, opportunities, and prospects—A review. *Energies, 17*, 669–669. https://doi.org/10.3390/en17030669 (accessed on 19th September 2025).
24. Orellana Barceló, M. (2019). *Advanced Digital Resonant Control Techniques for Grid-Connected Voltage Source Converters*. https://doi.org/10.5821/dissertation-2117-128858 (accessed on 19th September 2025).
25. Riyas, P., & Lakshmanan, S. A. (2025). Optimal tuning of PI-based LF for three-phase SRF PLL synchronization system using pity beetle algorithm under grid abnormalities. *Scientific Reports, 15*. https://www.semanticscholar.org/paper/7a303c3cb84899e9d1d8763c7259493091f830b7 (accessed on 19th September 2025).
26. Su, Y., Chen, Y., Liu, Y., & Fang, F. (2025). Frequency domain delay performance analysis of heavy-duty gas turbine control system: Theory and experiment. IEEE Transactions on Industrial Electronics, 72, 2001–2011. https://www.semanticscholar.org/paper/7e43f2de8d8f72f4a6724af93b34b279a0a42ec1 (accessed on 19th September 2025).

27. Sun, J. (2022). Frequency-domain stability criteria for converter-based power systems. *IEEE Open Journal of Power Electronics, 3*, 222–254. https://doi.org/10.1109/ojpel.2022.3155568 (accessed on 19th September 2025).

28. Wang, J., Yan, J. D., Jiang, L., & Zou, J. (2015). Delay-dependent stability of single-loop controlled grid-connected inverters with *LCL* filters. *IEEE Transactions on Power Electronics, 31*, 743–757. https://doi.org/10.1109/tpel.2015.2401612 (accessed on 19th September 2025).

29. Wang, T., Li, F., Yin, C., & Jin, G. (2024). Small disturbance stability analysis of onshore wind power all-DC power generation system based on impedance method. Energies, 17, 1459–1459. https://doi.org/10.3390/en17061459 (accessed on 19th September 2025).

30. Wen, B., Burgos, R., Boroyevich, D., Mattavelli, P., & Shen, Z. (2017). AC stability analysis and *dq* frame impedance specifications in power-electronics-based distributed power systems. *IEEE Journal of Emerging and Selected Topics in Power Electronics, 5*, 1455–1465. https://doi.org/10.1109/jestpe.2017.2728640 (accessed on 19th September 2025).

31. Wu, G., Wang, S., Zhao, B., Wang, D., Yuan, C., Ma, Q., Zheng, S., et al. (2023). Research on the active power control stability of the renewable energy connected to weak grids. *2023 6th International Conference on Power and Energy Applications (ICPEA)*. https://www.semanticscholar.org/paper/23b15e5a686464a9f72ff14a3e0aa3250ddc22f4 (accessed on 19th September 2025).

CHAPTER 5

# STABILITY OF CONTROL SYSTEMS

## Contents



## INTRODUCTION

Control system stability is a fundamental concept in the engineering field because it has ramifications on system performance and reliability. Control systems manage the dynamics of many applications, ranging from industrial processes to robotic motions. In this case, understanding how to analyze and establish stability becomes crucial. Stability means that a system will return to a desired state after being subjected to disturbances, and without it, even systems of high complexity become unstable. Stability analysis encompasses various methodologies and frameworks for dealing with various kinds of systems and situations. As technology evolved and the complexity of control systems increased, more accurate strategies are required to analyze and provide stability in a broad variety of applications. Recent research, given prominence in interval type-2 (IT2) systems analysis, introduces new understanding to stability analysis of advanced control systems. Introducing polynomial fuzzy event-triggered controllers offers the potential for control system stability even when there are uncertainties like time delays and asynchronous premise variables. The new approach to employing IT2 membership functions provides a more accurate representation of parameter uncertainty and provides a robust method to stabilize system behavior. As energy efficiency and communication bandwidth become increasingly larger demands in modern networks, developing event-triggered control methods that ensure stability and maximize resources is of prime concern. Such a shift not only indicates the ongoing evolution of control theory but also shows its effectiveness in real-world scenarios where the optimization of performance with minimum cost is of utmost importance (Li et al., 2024).

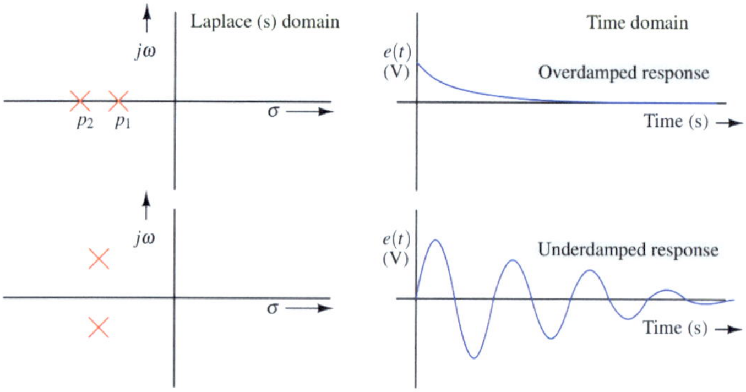


**Figure 5.1**. *Stability of control system.*

*Source: https://analogcircuitdesign.com/wp-content/uploads/stable_system-1-1-e1697558072417.png.*

Moreover, the integration of renewable energy resources, that is, wind power generations (WPGs), into traditional stability studies marks a shift in thinking in the analysis of control systems. Traditional analyses do not pay attention to the dynamics of WPG systems and therefore possess the risk of mismatches in frequency stability analysis. Engineers can do a significant better job in the stability prediction for large power networks using methods that recognize their specific responses. Probabilistic frequency stability analysis evolution that incorporates the dynamics of WPGs based on sophisticated methods such as multi-interval sensitivity and multi-element low-rank approximations underscores the importance of a new perspective in control system stability (Wang et al., 2024). Not only does the new understanding enhance the system's reliability, but it also facilitates the transition towards more sustainable sources of energy (Figure 5.1).

## *Importance of Stability in Control Systems*

Control system stability is necessary to ensure that a system performs its desired function effectively and reliably. A stable system can respond to disturbances or changes in operating conditions without resulting in large oscillations or divergent behavior. This is particularly critical in applications involving safety, including aerospace and automotive systems, where instability results in catastrophic failures. If the system is not properly stabilized, it can go into an unstable state, not only affecting its performance but also the safety of the users. Therefore, offering a solid framework for the analysis and preservation of stability is critical, as it promises system reliability and makes the user more confident in several applications, such as industrial automation and robotics. Stability analysis not only avoids undesirable system behavior but also assists engineers in designing uncertainty-resistant and environmental-disturbance-resistant control systems. Techniques such as the Lyapunov method provide a global examination of stability, providing information on the possible effect of perturbations on system behavior. In more complex systems, such as those occurring in nonlinear dynamic investigations, becoming more sophisticated. For instance, the behavior of coupled nonlinear systems, for instance double nonlinear damped spring pendulums, illustrates how small variations in parameters greatly influence changes in stability zones. This is important information in that it informs the design of the controllers that can effectively manage system performance under various operating conditions, hence optimal performance in all kinds of situations, as proposed in nonlinear stability analysis research (i.e., Amer et al., 2021). Beyond being technically important only, implications of stability extend far beyond technical interests alone; economic and operational implications are also entailed. In accurate control system industries, instability will lead to more downtime, expensive fix, and lost output, so there is a strong rationale for an active approach to stability management. Stability emphasis can also assist in the development of more advanced control methods that leverage technology, such as adaptive control or robust control, that will adjust dynamically with changing circumstances. Aside from that, stability analysis of systems like the newly developed dynamical system of three degrees of freedom under analysis through Lagranges equations offers knowledge on how different parameters affect performance in systems. The results indicate that not only does stability enhance lifespan in systems but also offers space for innovative and creative applications in a number of applications in the fields of engineering and robotics (Abdelhfeez et al., 2021).

## 5.1. MATHEMATICAL MODELS OF CONTROL SYSTEMS

The development of control systems mathematical models is significant in ascertaining the stability of control systems, particularly in vehicular suspension systems. A common procedure in this area is utilizing a mass-spring-damper system to model the dynamics of a sports car suspension. The model is capable of illustrating the interaction between mass, damping, and spring constants, and control techniques such as the Proportional-Integral-Derivative (PID) control can be applied. As highlighted in the research, the movement of the system to be underdamped and converge towards critical damping maximizes comfort on ride through the minimization of vibrations. By mathematically simulating such dynamics, engineers can calculate the impact of various parameters on the overall stability and performance of the suspension and ensure that ride quality is not compromised due to road-induced disturbances, which are effectively catered to by such intricate control systems (Haditiansyahuthor et al., 2025). In general control systems, evaluation of stability is particularly involved when nonlinear components are added. Mathematical modeling is of prime importance in identifying how such nonlinearities influence system dynamics and stability. Traditional linear control models are poor at characterizing the intricate behavior of nonlinear systems, and the result is that there are problems in predicting performance under variable conditions. Synthesis of accurate mathematical models not only is important for theoretical application but also for practical application, especially where full-scale experiments are out of the question owing to complexity. Specialists have proposed new methods that consider asymptotic and absolute stability, enhancing the stability of control systems governed by nonlinear elements. The approach can significantly impact the design and stability analysis of various control systems, ensuring operation in real-world applications (Yablochnikov et al., 2024). Furthermore, integration of mathematical models into control system design allows stability margins and failure points to be better understood. The models provide the foundation for simulation and analysis, and through their application, engineers can forecast the effects of changing parameters and environmental disturbances on system performance. By doing so, they can develop control strategies that enhance stability, such that it becomes sharper to give precedence to required performance levels regardless of variations in operating conditions. The modern research on stability analysis, as evident through the evolution of modeling techniques, is also pushing the boundaries of control system design even further. These all efforts will ultimately lead to stronger systems that are capable of operating in the complexities of real world settings and offer functionality along with safety in a variety of uses (Haditiansyahuthor et al., 2025; Yablochnikov et al., 2024) (Figure 5.2).

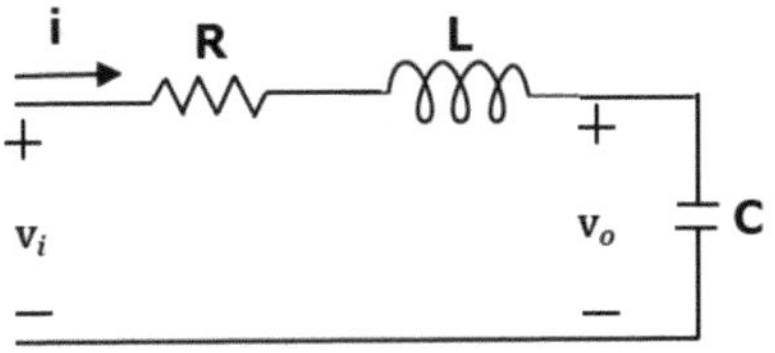


**Figure 5.2**. *Mathematical models of control systems.*

*Source: https://www.tutorialspoint.com/control_systems/images/series.jpg.*

## 5.1.1. Transfer Functions and Their Significance

The importance of transfer functions in control systems can scarcely be overstated, as they provide a mathematical model describing the system's input and output in the frequency domain. Through the transformation of dynamic equations into algebraic representation, transfer functions make it possible to analyze system behavior under various conditions easily. This adjustment is fundamental to calculating system stability since it allows engineers to calculate pole locations and frequency response characteristics. Pole locations—transfer function's denominator roots—are potentially indicative of whether a system is stable or not. Specifically, if the system has all poles in the left half-plane, then the system is stable. Hence, the understanding of transfer functions is crucial in the design of stable control systems that meet performance specifications such as rise time and overshoot. Transfer functions are also critical in the design and optimization of feedback control systems. They are used here to study how system dynamics and stability are influenced by different parameters. For instance, the feedback gain may be changed to modify the closed-loop transfer function in a way that system response properties can be improved through design. Study of open-loop and closed-loop system behavior using transfer functions can highlight model uncertainties and disturbances. Such an engineering application finds significance in engineering fields such as robotics, automotive systems, and aerospace, where stability and performance are critical. The analysis of control systems in terms of stability using techniques based on transfer function methods, such as Nyquist and Bode plots, enables engineers to design robust systems that can withstand uncertainties without sacrificing responsiveness or stability. The dynamic performance of control systems can also be characterized with transfer functions when they are subjected to time-varying inputs, further establishing their significance. Impulse and step response can be derived from the transfer function in time-domain analysis, which gives information concerning how a system reacts to sudden changes or constant input signals. This analysis of dynamic response is important in handling transient behavior and can be employed to guide control parameter tuning. In addition, transfer functions allow the design of advanced control methods such as PID control where their parameters are tuned in relation to the frequency response characteristics of the system. Lastly, the entire judgment provided by transfer functions is a component of the overall purpose of delivering stable and efficient control system design, thereby making their relevance particularly great in the case of stability in control systems (Krishna et al., 2019; Guillermo et al., 2018) (Figure 5.3).

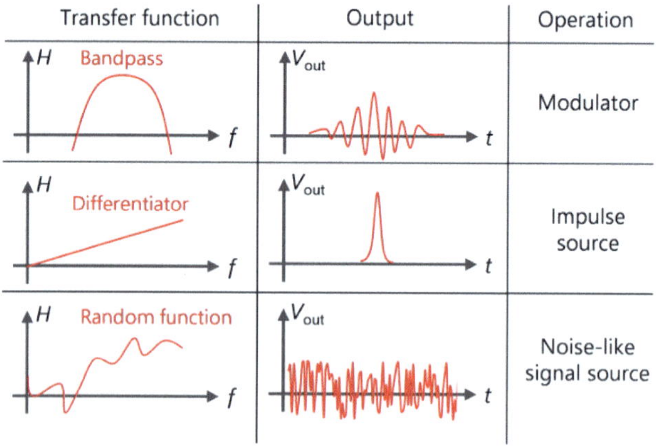


**Figure 5.3**. *Illustration of three examples of transfer functions.*

*Source: https://www.researchgate.net/profile/Mohammad-Samizadeh-Nikoo/publication/361654144/figure/fig2/AS:11737286668943337@1656850129307/Illustration-of-three-examples-of-transfer-functions-and-their-corresponding-output.ppm.*

## 5.1.2. State-Space Representation

State-space representation is a core framework employed in control system study that well represents dynamic systems behavior. This methodology is characterized by the formation of a system dynamics set of first-order differential equations by means of state variables. State variables encode all information regarding systems' behavior history such that future states may be determined. Utilization of state-space models allows engineers to capture the complex interactions among systems that are not necessarily revealed by reduced input-output representations. By focusing attention on state variables, designers can rigorously impose control actions that realize maximum system stability, thereby ensuring optimal performance under various operating conditions. State-space representation is not confined to linear systems, and non-linear dynamics can be treated with modifications and extensions. While LTI systems can be well-represented in such a framework, it becomes challenging when non-linearity is added, something that has driven substantial amounts of research. One new direction is the use of Markov parameters derived from the characteristic polynomial of a system, as seen in recent work. Using techniques such as the Leverrier-Takeno algorithm for state-space formulation, researchers can determine such parameters efficiently so that stability can be analyzed rigorously. This versatility is among the strengths of state-space models that can serve not only as analysis tools but also as control design tools in complex real-world scenarios, even guaranteeing system reliability and responsiveness. In addition, state-space techniques make it much easier to design and test stability criteria in control systems. The use of Lyapunov functions in conjunction with state variables makes it an efficient means to test for stability. Recent advances in the utilization of the coefficient vector as a natural state variable hold much potential for constructing stability analysis frameworks in the absence of known system dynamics or causal variable interactions. By bridging trajectory and state space, this approach simplifies complex analyses, ultimately leading to robust control strategies that can adjust to the systems needs dynamically. Such developments are a revolutionary shift in control system design to greater stability and efficiency as indicated by the results of the research studies examining the mathematical roots of state-space representation (Yan et al., 2025; Fatoorehchi et al., 2024).

### 5.1.3. Stability Criteria in Mathematical Models

Mathematical model stability provides a basis for control system analysis because it dictates the behavior of systems when under disturbances and equilibria. Sufficient requirements need to be provided in an attempt to assess the stability of dynamic systems, particularly those involving renewable energy. For instance, a 100% grid-integrated system of renewable electricity simulated with grid-following and grid-forming machines validates the need to use multiple stability metrics to assess the performance of such systems in varying conditions. This renders small-signal analysis vitally significant since it examines the qualitative behavior near the equilibrium points. By selecting appropriate control mechanisms such as droop control and virtual synchronous control, engineers are able to make significant alterations to the small-signal stability. Thus, the stability criteria enhances the design processes of modern power systems to facilitate reliability in the event of disturbances, as provided in Song et al. (2025). Apart from the utilization of external control techniques, the intrinsic nature of the mathematical models matters in system stability evaluation. The effects of model parameters on system dynamics must be studied with caution since these effects are at the core of predicting the systems behavior under various operating conditions. Sensitivity analysis becomes a crucial tool through which researchers are able to experiment with how minute changes in parameters affect the overall stability of the system. Methods employed to minimize sensitivity to the environment contribute to control systems resilience.

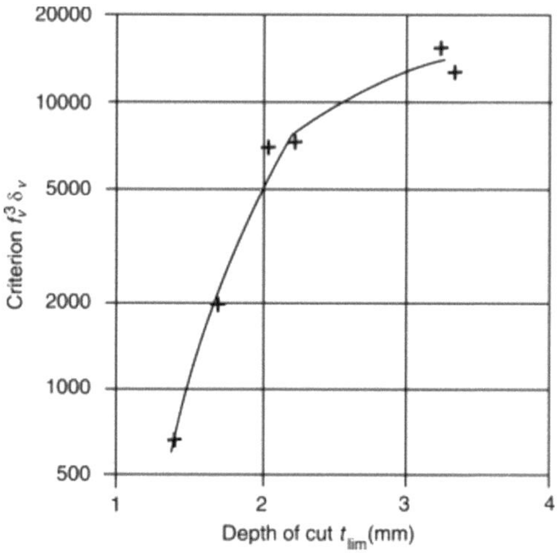


**Figure 5.4**. *Stability criteria.*

*Source: https://ars.els-cdn.com/content/image/3-s2.0-B0122270851001818-gr8.gif*

These methods favor stability through assurance of effective control actions regardless of variation of ambient conditions. Model convergence and actual implementation are critical, as seen in other engineering applications, like machining of rigid components, where stability criteria are used to optimize control structures, as noted by Sulyukova et al. (2025). Lastly, integral quadratic criteria development is a more advanced approach to the assessment of stability in control systems. It is a framework by which the performance and quality of control can be maximized so that engineers would be able to predict how modifications of control structures can lead to improved system responses. In technological fields involving applications like axisymmetric part machining, considerations regarding cutting force dynamics and precision in processing play central roles. Systems can achieve a higher level of precision by embedding positive feedback mechanisms to counter static errors. Additionally, the approaches discussed in this chapter enhance control systems' adaptability and therefore make them robust to various sets of application demands. Lastly, incorporating these stability conditions into mathematical models is paramount to the evolution of the control systems field since they form the basis for designing effective and stable technological solutions (Figure 5.4).

## 5.2. TYPES OF STABILITY

Understanding the various types of stability is relevant for the effective design and analysis of control systems. Stability in control systems is broadly categorized into three broad categories: absolute stability, relative stability, and asymptotic stability. Absolute stability refers to the capability of the system to return to equilibrium following a disturbance, whereas relative stability quantifies the rate at which the system stabilizes relative to a desired level of performance. Asymptotic stability captures not just the fact that the system returns to equilibrium, but that it does so within a reasonable time period, often to a unique equilibrium point to which the state of the systems converges in the long term. Identification of the appropriate type of stability is crucial to ensure that engineers can ensure sustained system operation in whatever condition and also design controllers that meet specific performance specifications. Another critical aspect of stability is the functioning of control action under diverse operating conditions. For instance, for applications involving vibration reduction, a quasi-zero stiffness vibration isolator (QZSVI) provides an excellent example of special stability characteristics being applied in complicated systems. These isolators are especially utilized in conditions where low-frequency vibration has the capability to disrupt sensitive instrumentation, for instance, aerospace or microelectronics. The low stiffness of QZSVI makes it possible for high vibration attenuation levels, showing effective integration of control philosophy and design. Through the use of negative feedback control techniques such as composite negative velocity control and cubic velocity control, the QZSVI can be optimized to provide a performance level that brings about a level of stability at around 94.8%. Through the use of simulations and analysis procedures, the study highlights the complex interaction between stability types and control methods, supporting effective system designs in real-world applications (Taher et al., 2025). Furthermore, hybrid systems including diverse converters present unique stability challenges, particularly in renewable energy-dominant power systems. The interaction between grid-following (GFL) and grid-forming (GFM) converters is crucial, depending on which stability analysis should be comprehensive. While traditional research is inclined to limit analysis of transient stability to each of the individual converter types, their interactions, particularly in hybrid systems, are more and more crucial to know. The control design developed herein to mitigate loss of synchronization (LOS) in such systems can further enhance stability to allow these converters to cooperate more effectively to support system resilience. Through the use of combined models containing both forms, this approach not only gives theoretical insight but also aligns with practice in contemporary power systems today. Thus, as the complexity of control systems increases so does the need to investigate and integrate various forms of stability (Xu et al., 2024) (Figure 5.5).

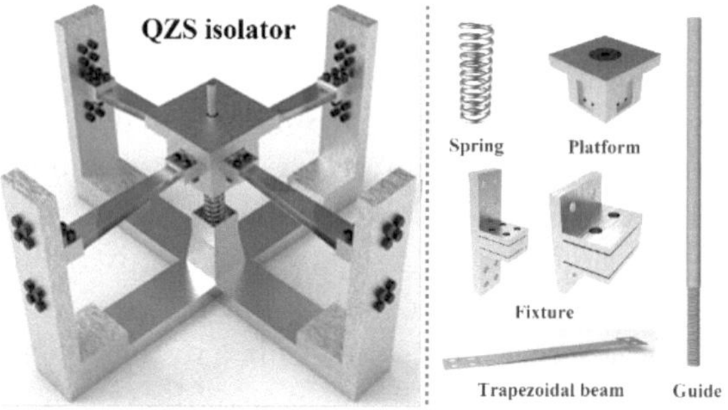


**Figure 5.5**. *Quasi-zero stiffness.*

*Source: https://www.mdpi.com/machines/machines-11-00512/article_deploy/html/images/machines-11-00512-g001.png.*

## 5.2.1. Absolute Stability

The absolute stability theory is a fundamental cornerstone for control system analysis and design, particularly nonlinear dynamics control systems. Absolute stability refers to the assurance that a control system will be stably operating independent of any external perturbations. This concept is even more pertinent when considering systems where there are added soft sensors and sophisticated algorithms such as gated recurrent neural networks. In recent work, it has been shown that traditional methods could inadvertently break underlying stability conditions, prompting the need for novel approach developments. For instance, a proposal to introduce the Lure-Postnikov gated recurrent neural network (LP-GRNN) into the Lure-Postnikov framework has been made to enhance stability at the expense of prediction accuracy on difficult benchmarks like NASA CMAPSS (Hilgert et al., 2025). These advancements underscore the ongoing advancement in control system design and the urgent need for guaranteed stability evaluation techniques. The need for absolute stability transcends its status as a theoretical idea; it has practical implications to control system design in a wide range of applications. Designers must evaluate the effect of nonlinear components in a system, as these components have the ability to significantly affect dynamic performance and behavior. As emphasized in current works, it is very challenging to quantify the stability of systems with nonlinearities internal to them due to complexity in modeling and experimental verification. Traditional mathematical modeling techniques have a tendency to fail in capturing the complexities of these nonlinearities, making the necessity to analyze other analytical methods.

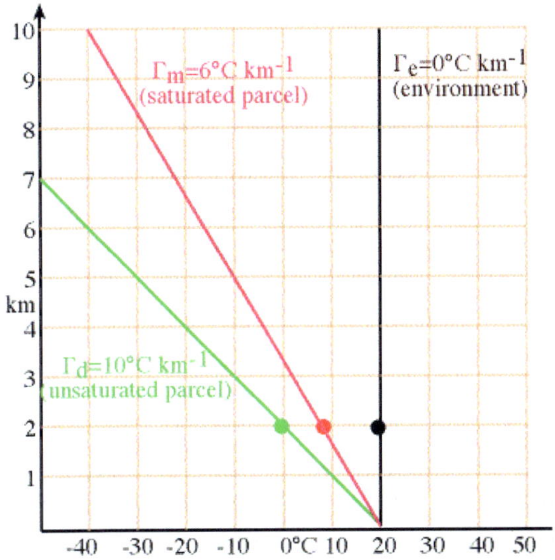


**Figure 5.6**. *Criteria for absolute stability.*

*Source: https://apollo.nvu.vsc.edu/classes/met130/notes/chapter6/graphics/absolut_stable.free.gif.*

In an effort to overcome this issue, new approaches have been proposed in recent years that enable engineers to identify both the asymptotic and absolute stability, providing us with valuable insights to create control systems that are effective in real-life applications (Yablochnikov et al., 2024). Under the absolute stability paradigm, the application of linear matrix inequalities (LMIs) as a practical tool for certifying global asymptotic stability in nonlinear control systems has been increasingly noticed. The integration of LMIs in the context of control design enables designers to articulate stability conditions in a mathematically tractable manner. By combining various components of a control system, that is, controllers and plant, into a unified standard nonlinear operator form (SNOF), one can translate difficult stability problems into a manageable set of LMIs. Not only does this facilitate the stability analysis, but it also enhances the confidence of designers in their designs. For instance, case studies in linearized systems have been able to validate this method in an effective manner, establishing its benefit in ensuring closed-loop performance and closing the gap between modern soft-sensor technology and strong stability assurances (Hilgert et al., 2025). Developing trustworthy procedures for assessing absolute stability is, thus, an important research domain in control systems (Figure 5.6).

### 5.2.2. Relative Stability

The concept of relative stability of control systems is crucial in determining the performance of a system in the face of perturbations and varying conditions. Stability measurement is a comparison of robustness between different configurations in relation to how they handle the varying environment. According to relative stability, scientists can determine the performance of a control system in varying conditions. For instance, the worth of assimilating satellite observational data to enhance atmospheric analysis proves relative stability. Dramatic enhancements to ozone concentration estimates were observed when satellite data were integrated into reanalysis models, and this reduced discrepancy between various systems. These findings identify the importance of data assimilation methods in achieving increased uniformity and reliability in control systems under various operating conditions, again demonstrating the emphasis laid on relative stability as a measure of overall system performance and responsiveness (Sekiya et al., 2025). Adaptive control methods also hold a key role in achieving maximum relative stability in specific applications, particularly industry. One example is the regulation of electric drives on hot rolling mills, where fluctuations in supply voltage can lead to dynamic deviations in drive behavior. By using an adaptive control system that adjusts magnetic flux in line with real-time conditions of voltage, engineers are able to absorb subsequent speed fluctuations and tension anomalies. Mathematical models of such systems transparently illustrate how precise tuning of control parameters significantly improves stability in operation. This versatility guarantees that irrespective of unstable grid conditions, the system remains stable and efficient, clearly showing that relative stability can be optimized by the integration of sophisticated strategies into control schemes (Druzhinin et al., 2025). In summary, the research of relative stability in control systems demonstrates the intricate relationship between system responsiveness and external variables. As realized in numerous applications, from atmospheric observation to industrial control, the measure undertaken to cope with uncertainty heavily impacts system performance. The integration of high-resolution data assimilation techniques and adaptive control methods accentuates the need for maintaining stability in control systems, particularly in dynamic environments. As the demand for sustainable and reliable systems grows, further investigation in relative stability will be instrumental in guiding the development of innovative solutions to address needs in the modern world across various fields (Sekiya et al., 2025; Druzhinin et al., 2025) (Figure 5.7).

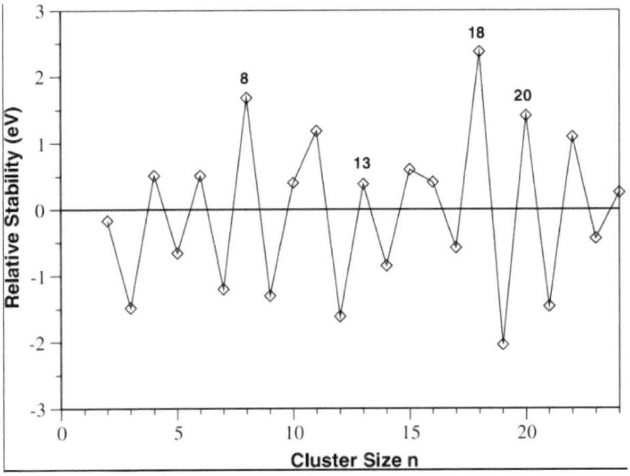


**Figure 5.7**. *Relative stability.*

*Source: https://www.researchgate.net/profile/Abhijit-Mookerjee-2/publication/227092944/figure/fig3/AS:668907497082886@1536491383542/Relative-stability-2En-En-1-En-1-2En-as-a-function-of-cluster-size.png.*

## 5.2.3. Asymptotic Stability

The concept of asymptotic stability is key to control systems, particularly in the design and analysis of feedback mechanisms. Asymptotic stability is that state of a system for which, when perturbed away from an equilibrium point, the systems paths will be drawn towards that equilibrium point as time approaches infinity. This feature is crucial for guaranteeing a control system can revert to a preferred state after it has been subjected to fluctuations or disturbances. Several methods of analysis, such as Lyapunov's direct method, are usually utilized to determine the asymptotic stability of nonlinear systems. Through constructing a proper Lyapunov function, engineers can achieve conditions under which the system is stable, ultimately demonstrating how slight deviations from equilibrium do not produce unlimited behavior, thus validating the validity of the control system for use in real-world. Asymptotic stability's applicability moves beyond theoretical models to real-world practice, where control systems typically encounter dynamic and uncertain environments. A straightforward example is found in designing robust controllers like the robust integral of the sign of the error (RISE) controller. Recent developments in RISE control techniques have aimed at achieving not only asymptotic stability but exponential convergence of the tracking error even in the existence of disturbances and actuator saturation. Patil et al. (2025) introduces new design methods that apply projection algorithms and auxiliary filters to enhance the stability of the RISE controller in saturation. This is a big step in control system design to obtain more robust controllers with closed-loop stability and effective management of the challenges brought by nonlinearities and external environment disturbances. Besides, asymptotic stability in networked control systems has defined the difficulties in multi-variable communication and control. In systems where variables are on differing time scales, stability under successful communication is a prevailing challenge. Research documented in Wang et al. (2025) suggests new methods such as dual clock mechanisms to remedy this issue with a view to obtaining semi-global practical asymptotic stability. This approach allows for efficient transmission scheduling, both high- and low-speed system dynamics with no loss of stability. The presentation of these advanced methods indicates the evolving context of control system design in which asymptotic stability not only ensures performance reliability but also allows adaptation of control policies to meet the demands of more advanced systems (Figure 5.8).

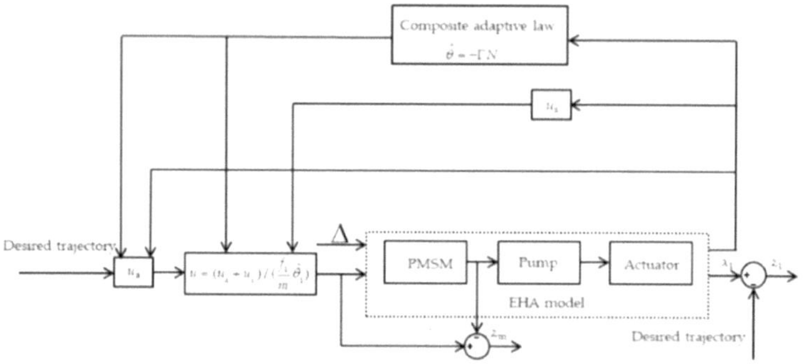


**Figure 5.8**. *The flow diagram of robust integral of the sign of error (RISE).*

*Source: https://www.researchgate.net/publication/367862855/figure/fig2/AS:11431281213142974@1702956485625/The-flow-diagram-of-robust-integral-of-the-sign-of-error-RISE-based-composite-adaptive.tif.*

## 5.3. STABILITY ANALYSIS TECHNIQUES

Analysis of stability of control systems is a fundamental necessity to make a system robust and reliable for various applications. Although there are numerous techniques available, Lyapunov's direct method is one of the most flexible and strong methods. The technique uses a Lyapunov function, a scalar function, to achieve a way to establish stability based on energy considerations. By showing that this function diminishes with time, it is possible to conclude the convergence of the system state towards an equilibrium point. These are especially useful in nonlinear systems where usual methods fail since they give a more generalized framework for stability analysis. Since stability ensures proper operation and reaction of the system in changing operating conditions, the choice of method of analysis directly affects design and optimization of control techniques of engineering systems. Modelling and control of DC microgrids have become issues of concern in recent studies, reflecting a move towards clean energy solutions. The necessity for effective utilization of renewable energy sources has driven the research toward advanced stability analysis techniques dedicated to such conditions. An understanding of control strategies is critical to efficiently stabilizing DC microgrids, which exhibit unique modes of operation and complexities. For instance, methods covering the hierarchical control structure of such microgrids—encompassing primary, secondary, and tertiary levels of control—are needed to improve performance. The emphasis is on ensuring grid stability and power quality alongside minimizing control system complexity. Lastly, the implementation of systematic stability analysis approaches is at the core of maximizing the efficiency and robustness of renewable energy systems to address dynamic energy demands (Bhunia et al., 2023). Piecewise learning architectures have also further enriched stability analysis approaches, particularly in stabilizing nonlinear systems with unknown dynamics. Traditional model-based reinforcement learning is generally plagued by limitations in learning complex behaviors across large domains. The addition of piecewise models that learn and control local segments of the state space, however, allows researchers to achieve robust stability guarantees even under uncertainty. Such a benefit is aided by rigorous approaches that find common piecewise Lyapunov functions, which are used as a stability certificate or as a potential counter-example at analysis time. These advancements, as seen in several case studies, indicate that this method yields a more conservative estimate of the region of attraction compared to conventional methods, with potential for use in a wide range of engineering systems (Farsi et al., 2022). In that respect, combining these new techniques not only deepens theoretical knowledge about stability but also constitutes a revolutionary departure in engineering practical control systems (Figure 5.9).

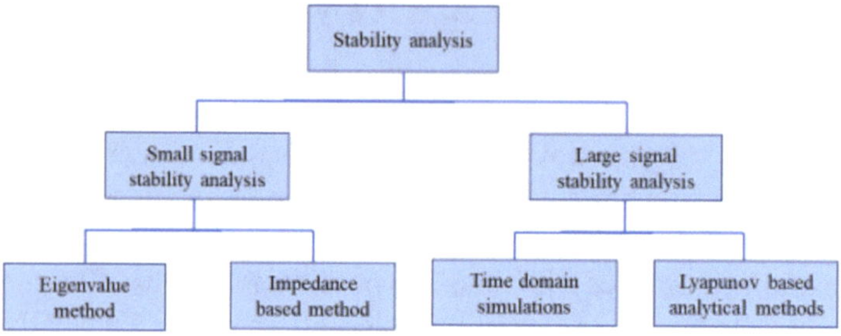


**Figure 5.9**. *Stability analysis methods.*

*Source: https://ars.els-cdn.com/content/image/3-s2.0-B9780128194324000184-f07-01-9780128194324.jpg.*

## 5.3.1. Routh-Hurwitz Criterion

The Routh-Hurwitz criterion is a leading cause for determining the stability of linear control systems by using an orderly method of calculating the location of the roots of characteristic polynomials. The criterion arose as a result of needing to examine the stability of a system without necessarily obtaining the roots, which may be complex or time-consuming. By constructing the Routh array from the characteristic polynomial coefficients, control engineers can calculate necessary and sufficient conditions for stability based on the sign of the first column of the derived array. If all the elements of the first column are positive, then the system is stable; therefore, the Routh-Hurwitz criterion is a powerful tool for stability analysis for most engineering applications. Application of the Routh-Hurwitz criterion not only serves the purpose of stability analysis but also allows us to understand complex systems under various perturbations. The criterion has now been applied recently in handling complex-coefficient systems, thereby extending its scope in real-world applications where systems are typically perturbed by non-trivial external forces. As such, asymmetric couplings in dynamic networks such as power systems and chemical reactors under these circumstances can lead to instability and necessitate stringent analysis via the Routh-Hurwitz test. Practitioners have indicated ways in which this criterion has been successfully utilized to evaluate stability conditions in such complex networks for engineers to be able to forecast operating parameters critical to maintaining system integrity (Muolo et al., 2024). Also, the versatility of the Routh-Hurwitz criterion is demonstrated through its integration into other stability techniques, such as time-delay systems. In vehicle suspension dynamics, the double time-delay feedback control method has been validated for stability through this criterion in conjunction with iterative numerical tools such as particle swarm optimization. Optimizing through iteration enables precise tuning of feedback parameters for robust performance against perturbation. The integration of the Routh-Hurwitz criterion with modern computational techniques and its application across all branches of engineering indicates the significance of ensuring the stability and reliability of control systems (Ma et al., 2024) (Figure 5.10).

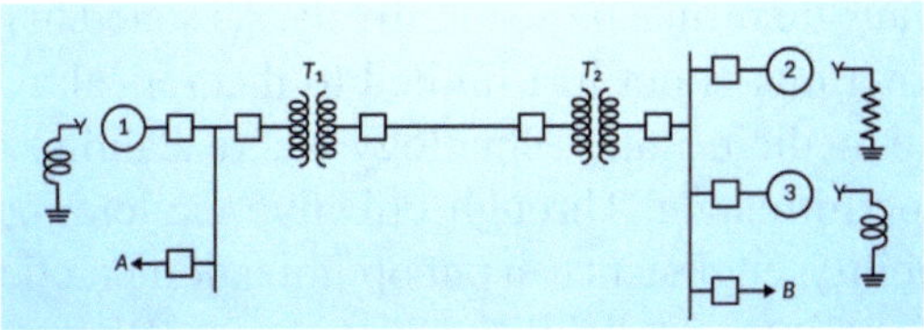


**Figure 5.10**. *Routh-Hurwitz criterion.*

*Source: https://www.jove.com/files/media/science-education/science-education-thumbs/16116.jpg.*

## 5.3.2. Nyquist Stability Criterion

Nyquist stability criterion is a core tool in the theory of control systems, and its purpose is to ascertain closed-loop system stability from the frequency response of their open-loop counterparts. The criterion utilizes the Nyquist diagram, which is a graphical plot of the complex frequency response of the open-loop transfer function. The study of the encirclement of critical points in the complex plane enables one to conclude the stability properties of the system. Specifically, the criterion is that the number of counterclockwise encirclements of the critical point ( −1, 0) equals the number of poles of the closed-loop transfer function in the right half-plane, and this asserts the system to be unstable. The Nyquist criterion thus not only provides information on stability, but also enables tuning of system parameters in order to achieve performance as desired in engineering applications. The Nyquist stability criterion is one of the best advantages of being in a position to analyze systems with time delays, which are present in most real-world applications. Its extension to include such delays has important consequences for control system design and analysis since it makes it possible to check the stability under more realistic operating conditions. As realized through the literature, stability conditions in positive delay time systems can be found that make it possible for engineers to perform efficient calculations of stable operating parameters. Practitioners can utilize such a scheme in order to provide means for control systems to be maintained stable under varying conditions, which is essential in numerous applications such as automation, robotics, and telecommunications. This adaptability shows the relevance of the Nyquist criterion in contemporary control theory and its application in complex systems with varying dynamics. Besides, the ease and openness of the graphoanalytic methods related to the Nyquist stability criterion enhance its applicability in practical cases. The easy accessibility of a Nyquist plot allows engineers to easily visualize the stability characteristics of a system and design and communicate results more intuitively. As stated in most research, the applicability of criterions is not just limited to theoretical analysis but can be traced to areas like medicine and technology where stability of a system plays a crucial role in performance. Through effective exploration and analysis of control system stability, engineers can put optimization practices into effect that increase the functionality and reliability of the system (Bieda, 2025; Umarova et al., 2024). Thus, achieving full control of the Nyquist stability criterion not only aids in increased understanding of control system behavior, but it also provides engineers with the tools needed to tackle real-world issues in any discipline.

### 5.3.3. Bode Plot Analysis

Bode plot analysis is a powerful tool in control system analysis that allows engineers to analyze the frequency response of systems effectively. Through the graphical display of the systems gain and phase shift against a frequency range, Bode plots allow the identification of important stability margins. These are the margins of relevance to the robustness of a control system across various operating points. Engineers can determine critical information by examining the phase crossover and the gain crossover frequencies, both of which, in turn, define the frequency range over which the system is stable. Not only does the graphical form facilitate system behavior analysis but also provides the basis for compensation design aimed at increasing stability and performance. Therefore, Bode plots are crucial in order to create good control schemes tailored to specific applications, ensure efficient system control even with the occurrence of disturbances. Stability analysis through Bode plots becomes increasingly critical when working with sophisticated systems, such as grid-forming inverters utilized in distributed power generation, as indicated in the stability analysis conducted in Narayansamy et al. (2025). These inverters operate in parallel as microgrids, and stability is pivotal to managing varying loads while ensuring efficient power delivery. The susceptibility of such systems, especially to asymmetrical operations or line inductance, suggests the need for profound analysis. Analyzing based on Bode plots enables engineers to explore the equivalent impedance and transfer functions of such systems to finally discover how well they can counteract the issue of instability. By employing damping networks that are specially designed based on passive techniques developed from Bode plot insights, control systems can be made more resilient and permit safe operation in different conditions. Besides its standard uses, Bode plot analysis is of utmost importance in novel technologies, for instance, solar tracker water pumping systems with respect to their dynamic performance and stability problems as explained in London (2025). The design and control of the systems are heavily based on accurate modeling of the transfer functions of individual components like stepper motors, PV cells, and pumps. Using Bode plots, the engineers are able to study the coupling between the solar tracking algorithm and the total system response, particularly for oscillating light intensity and water levels. This analysis is crucial in ascertaining if the pumps are operating optimally and also in ensuring the stability of the system when it is operating in dynamic conditions. In light of the growing importance of sustainable technologies, the role of Bode plot analysis in ascertaining the stability and viability of such emerging systems will continue to be pivotal in guiding future control systems engineering.

## 5.4. APPLICATIONS OF STABILITY IN CONTROL SYSTEMS

Applications of stability in control systems are extensive in ensuring the reliability and functionality of a vast variety of technological solutions. Stability of the system is the most critical area in engineering disciplines, including automotive engineering, where control systems control major operations such as braking and suspension. Development in magnetorheological fluids (MRF), as discussed in Fakir (2025), provides the opportunity to design smart dampers and actuators which could adapt dynamically their properties based on real-time feedback. This adaptability enhances riding comfort and driving stability with the focus on stability analysis in control system design. These innovations not only improve user experience but also enhance safety and reveal how an appreciation of stability dynamics directly influences applications in daily use within industries. In non-automobile applications, the ideas of stability are applied in robotics and aviation. Stability is crucial in robotics in ensuring the functionality of robotic arms and automation systems with resultful performance since efficient control translates to task effectiveness. Here, stability analysis minimizes oscillation or uncontrolled motion hazards, and overall dependability is enhanced. In aerospace too, stability plays a crucial role in flight control systems, where real-time control must be used to maintain altitude and path. The interrelationship between stability and controllability is more complex, particularly with external disturbances and system nonlinearities involved. An analysis of these interactions assists engineers in creating robust control systems that can manage uncertain operating environments to ensure mission accomplishment in terrestrial and extraterrestrial spaces. Also, the application of advanced mathematical theories, such as sub-Riemannian geometry, for stability analysis allows for new solutions to control systems with constraints. As explicated in Bala (2025), addressing singularities and understanding nonholonomic constraints provides a new method of achieving stability, particularly under circumstances where traditional solutions fail. A case in point is within constrained dynamical systems, where the existence and uniqueness of geodesics—a geometric model of stable trajectories—are of particular concern. Addressing these geometric properties informs the development of control tactics that are robust to disturbances and yet able to accommodate the complexity of operating environments. This cross-pollination of control and geometry not only enhances the scholarship but also provides insight into avenues for creating innovative applications where stability is critical, thereby guaranteeing that systems in diverse applications operate at their best under a multitude of conditions (Figure 5.11).

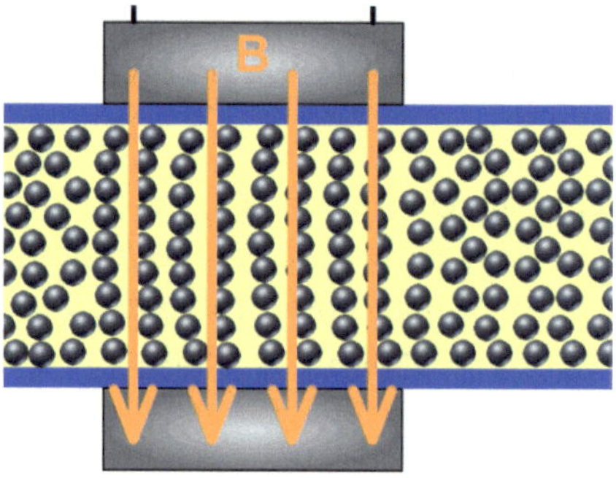


**Figure 5.11**. *Magnetorheological fluid.*

*Source: https://upload.wikimedia.org/wikipedia/commons/3/3e/MRF-Effekt-static-crop.png.*

## 5.4.1. Industrial Automation

Industrial automation plays a critical role in enhancing productivity and control system stability. Through integration of emerging technologies such as robotics and artificial intelligence, companies can automate processes and reduce errors. The complexity of modern manufacturing processes necessitates the use of reliable control systems with capabilities to adapt to varying conditions without compromising on performance stability. Automation provides real-time monitoring and constant adjustment of operating parameters for optimal use of machinery under different loads and conditions. Automation not only optimizes throughput but also maximizes the overall reliability of a system, which is crucial in applications like the automotive industry and pharmaceuticals where accuracy and consistency are vital to quality control. Wireless communication technologies transform industrial automation by addressing problems with wired systems. With processes needing real-time control applications, developing efficient wireless architectures is now crucial. Recent research focused on the optimization of wireless communication protocols, which reduced latency by up to 94.7% compared to conventional techniques. These innovations provide improved data transmission rates and response time of the system, which are critical to deliver stability in control systems. Also, the ability to achieve a transmission success rate of 99.3% with the presence of electromagnetic interference illustrates the ruggedness of wireless solutions as a viable option for complex manufacturing environments that are infested with conventional wiring constraints (Sahni et al., 2025). The spreading of the Industrial Internet of Things (IIoT) is evidence of industrial automation's way of evolving to provide solutions to the needs of the time, particularly connectivity and stability in control. Technologies like B5G and 6G provide unrivaled device density but are difficult to deploy, for instance, in the aspect of scarce spectrum resources. To ensure an effective communication in IIoT systems is a guarantee for working in a consistent fashion and minimizing latency, especially in real-time feedback applications. By adopting nonorthogonal multiple access (NOMA) techniques, industrial networks can enhance the dependability of the communication with low-latency demands. Research on throughput optimization with new frameworks reveals feasible solutions to power distribution and decoding faults, thus contributing to control system stability in industrial automation (Wang et al., 2025).

### 5.4.2. Aerospace Engineering

Aerospace engineering is practically related to the principles of control systems, mainly with the stability of various aircraft. As planes operate within intricate aerodynamics, stability becomes the prime factor for secure and reliable flight. Control system designs must take into account various factors, including the weight, shape, and velocity of the aircraft. Moreover, modern aerospace engineering uses advanced simulations to accurately model flight dynamics. The issues that disturbances such as turbulence and variations in payload cause introduce an additional level of sophistication to such systems. Engineers must then develop dynamic solutions that adjust based on real-time data, which renders stability not only a stationary need but one that is constantly monitored and managed. Aerospace engineers must therefore be well familiar with control theory to ensure the reliability and safety of their product within the aerospace sector are maximized. New control system design technologies offer apt illustrations of automation and machine learning's fundamental role in enhancing stability for aerospace vehicles. The advancement of new frameworks such as ControlAgent that couples large language models with control expertise facilitates an effective design procedure. This flow accelerates iterative refinement needed in control system design by prototyping and controller parameter tuning at high rates to meet harsh stability specifications. Simulating flight conditions and examining feedback, ControlAgent effectively mimics human approaches but with the advantage of not needing manual involvement. As a result, engineers can address increasingly complex design problems and leverage automated software to guarantee that stability and performance requirements are consistently met. This synergy of cutting-edge technology puts aerospace engineering at the forefront of developing sophisticated control systems for a variety of applications. Advanced aerospace vehicles require robust control systems with capabilities to maintain stability under varied operating conditions. A pertinent example is that of yaw damping control system employed in heavy commercial airplanes such as the Boeing 747. Because of the aircraft's weight and the forces it undergoes during flight, stability against yaw oscillation is extremely crucial from the point of view of safety as well as comfort. Application of an integrated approach that combines state-space modeling with closed-loop simulations has shown promising results towards greater yaw stability. Using stability derivatives, engineers can design controllers that dynamically adjust for changing flight conditions to ensure that the aircraft tends to return to equilibrium immediately whenever it is disturbed. The constant development of such robust systems underscores the importance of control theories in aerospace engineering, which allows engineers to offer more efficient and safer aircraft designs (José et al., 2024) (Figure 5.12).

**Figure 5.12**. *Aerospace engineering.*

*Source: https://happenings.lpu.in/wp-content/uploads/2019/03/aerospace.jpg.*

### 5.4.3. Robotics and Autonomous Systems

The integration of autonomous systems and robotics into control systems is key to the provision of perfect automobile performance. Stability is a fundamental attribute of control systems, which defines the reliability and safety of autonomous vehicles, especially as they traverse complex systems. Other control techniques have been used to enhance stability in robot cars, such as Proportional-Integral-Derivative (PID), Linear Quadratic Regulator (LQR), and Model Predictive Control (MPC), among the best. All have unique advantages in stabilizing vehicle dynamics under conditions, particularly when addressing critical parameters like yaw rate, slip angle, and steering angle. The comparative analysis of the approaches reveals that while MPC allows predictive optimization to lead to significantly enhanced performance, PID provides an easy and effective approach, due to which it is suitable for multiple purposes. Last but not least, the choice of control strategy may have a tremendous impact on the stability of autonomous systems, influencing both their responsiveness and safety while in operation (Yuvaraju, 2025). Multi-agent systems (MASs) illustrate the intricacy of stability guarantees in cooperative robots and autonomous systems. With increasing applications of these systems across spaces such as drone swarms and autonomous vehicles, the focus on stable motion among many agents is important. Utilization of the Lyapunov direct method for obtaining stability criteria makes it possible to derive more sophisticated control approaches, which permit effective cooperation among the agents. By converting continuous models to discrete systems, researchers are able to track formation shapes and enable real-time collision prevention. These innovations underpin the importance of cooperative behavior in MASs for achieving not only efficient task accomplishment but also stable operation under limited conditions. Simulation outcomes always validate the effectiveness of such control measures under actual conditions, emphasizing their capacity to revolutionize the field of autonomous systems by maximizing their cooperative potential (Qiang, 2025). The implications of stability in control systems extend beyond theoretical benefits; they are critical for real-world applications and safety standards in autonomous systems. As these technologies advance, so also do the challenges of providing stability in the midst of increasing complexity and uncertainty of operating environments. With the advent of artificial intelligence and machine learning, autonomous systems are now able to make decisions in real-time based on a multitude of variables. This dynamic character underscores the need for robust and responsive control methods that can react not just to sudden changes but also anticipate and predict anticipated instabilities in advance. Lastly, the ongoing research of control systems stability has the potential to make robotics and autonomous systems more efficient and adaptable, culminating in their eventual deployment in everyday life and prime applications in areas such as transport, medicine, and others, making them safer and more efficient (Yuvaraju, 2025; Qiang, 2025).

## 5.5. ISSUES OF ACHIEVING STABILITY

Control in systems is a problem that is inherently multifaceted in attaining stability in applications. Balancing system responsiveness against stability alone is one of the primary challenges, especially in those complex environments such as electrical power systems. One such case is integrating sophisticated control schemes such as Load Frequency Control (LFC), which requires fine-tuning for not only computational expense but also system dynamics. The use of techniques such as model order reduction techniques can also significantly enhance stability by minimizing computation requirements without compromising the quality of performance. A good case demonstrates that using a systematic procedure by model reduction in addition to tuned PID achieves enhanced performance, resolving the duality between complexity and solution engineering for actual control, remaining at the forefront of engineering studies in the area of control systems (Singh et al., 2025). The second major hurdle in achieving stability is addressing the uncertainties and variabilities that control systems inherently possess. Whether oscillating loads in electric power grids or unknown external environmental inputs to mechanical systems, such uncertainties can lead to instability if not effectively addressed. This necessitates resilient control approaches that are able to learn and adapt to changing conditions without discarding desired system performance. Research indicates that various control methodologies—ranging from traditional PID to high-performance optimization techniques—exhibit diverse advantages when tackling such uncertainties. Specifically, PSO-optimized PID controllers have been demonstrated to provide satisfactory performance in minimizing overshoot and facilitating rapid frequency recovery, which are indispensable for stabilizing interconnected power systems. These observations underscore the necessity of special approaches that can negate the adverse effects of uncertainties on system stability (Bano et al., 2024). Besides, the scalability issue in control systems is a very critical factor in seeking stability. As complexity in systems increases, interactions between more than one component bring unforeseen dynamics that make it difficult to evaluate stability. For instance, transitioning from single-area control systems to networked systems requires sophisticated methodologies capable of handling stability at the micro level but also maintaining consistency at the macro level. The interactions between various factors, for example, load fluctuations and network topologies, further complicate the difficulty of obtaining overall stability criteria. Recent research attests that multilevel control strategies are helpful in enhancing stability by addressing these complexities, thus making possible integration of heterogeneous applications in modern power systems. Lastly, the journey towards realizing stability in control systems demands ongoing research and adoption of new solutions tailored to emerging challenges (Singh et al., 2025; Bano et al., 2024) (Figure 5.13).

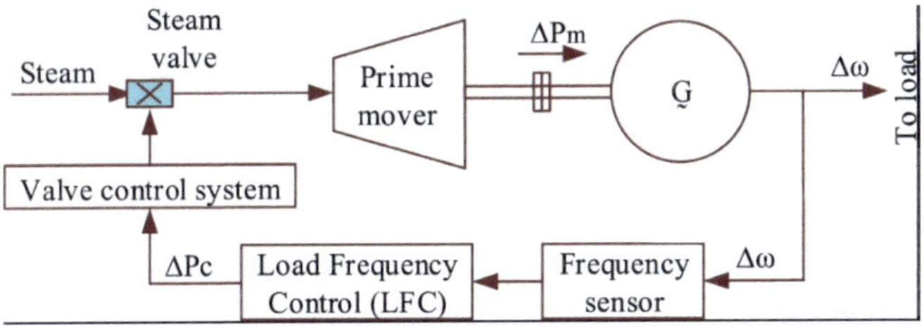


**Figure 5.13**. *Improving load frequency controller tuning with rat swarm optimization and proposing feature.*

*Source: https://media.springernature.com/full/springer-static/image/art%3A10.1038%2Fs41598-024-66007-y/MediaObjects/41598_2024_66007_Fig1_HTML.png.*

## 5.5.1. Nonlinearities in Control Systems

Control system nonlinearities are a primary impediment to researchers and engineers, primarily because standard linear control techniques tend to fall short in addressing complex real-world problems. Nonlinear systems can exhibit responses such as bifurcations, saturation, and hysteresis, which are inadequately characterized by linear models. Such phenomena are bound to produce system responses that are difficult to anticipate, so stability analysis becomes involved and demanding. Hence, understanding the role of such nonlinear characteristics in system performance is critical. For instance, systems with parameter uncertainty or bounded gain error pose possible stability challenges that require advanced methods, as seen from the generalized Cauchy–Schwarz inequality-based methods introduced in Ren et al. (2020). Traditional methods are generally insufficient, and innovative analytical models must be employed for efficient control and performance. One of the promising avenues towards addressing the challenges posed by nonlinearities is the use of fuzzy logic systems, particularly in network control environments. By using the Takagi-Sugeno fuzzy systems, the nonlinear dynamics can be represented more accurately and delays and uncertainties associated with the network communication can be included. By using a new Lyapunov-Krasovskii stability theorem, one can probe these nonlinear dynamics further, particularly when time-varying delays due to sensors and actuators are considered. In such scenarios, the traditional tools of stability analysis may yield too conservative estimates to avoid viable implementations. By employing multiple Lyapunov-Krasovskii functions, as discussed in Yoneyama (2017), the analysts can achieve less conservative conditions for a more stable and robust control system. Such design embraces the nonlinearities rather than rejecting them, bridging the gap between theory and practical implementability. Moreover, the integration of such advanced analytical tools in control system design reflects a trend toward a deeper insight into stability under nonlinearities. By addressing the challenges introduced by nonlinear qualities in a methodical way and honoring uncertainties, engineers can develop more resilient control techniques. The ability to quantify and control the influence of nonlinearities ultimately results in enhanced system performance, ensuring robust behavior over a variety of applications from aerospace to industrial automation. The advances in stability analysis result in a wealth of knowledge that informs future investigation and application. As these methods continue to evolve, they promise to hone our understanding of how best to control nonlinearities, inviting new architectures for control capable of shining in complex environments.

### 5.5.2. Time Delays and Their Effects

Time delays are a fundamental aspect that plays a major role in affecting the performance and stability of control systems. Time delays usually result from multiple sources, including sensor faults, actuator constraints, and communication delays. If a control system does not consider such delays, the resulting performance degradation may result in instability and inefficiencies. In actual applications, particularly those based on timely feedback of data for decision-making, disregarding the emergence of delays may cause destructive impacts. For instance, not responding effectively in a timely manner in response to system fluctuations or perturbations could manifest as oscillations or undesirable dynamical behaviors, eventually degrading the systems operating integrity. It is thus of utmost importance to comprehend the implications of time delays to make the design and implementation of robust control strategies that are stable and perform optimally. In addition, the integration of modern control methods is essential in fighting the problems introduced by time delays. Recent research identifies that intelligent control strategies, such as using neural network-based methods, can impart fault tolerance to delayed systems and systems with uncertainty. Such approaches enable more flexible and stable control actions, which are capable of compensating for disturbances typically associated with delays. For instance, the new neural network control approach presented in Wang et al. (2021) effectively solves the stability issue of fractional time-delayed systems by embedding fault tolerance. This evolution serves to show how novel control strategies can convert latent weaknesses into manageable problems, ultimately ensuring that control systems remain stable despite the occurrence of unforeseen delays or faults. Time delays in distributed control systems, particularly in microgrids, take on a new significance. As required by Lou et al. (2020), the interplay between delay-dependent stability and controller designs highlights the critical balance necessary for effective control of the grid. This study highlights that communication network delays, unless addressed, might strengthen operational risks and destabilize the system. Furthermore, the derivation of analytical equations that provide individual delay margins is of invaluable use in optimizing controller gains and pinning locations in such systems. With advancements in control systems, the identification of qualitative effects of delays enables engineers to design techniques that enhance stability but also maximize the overall system efficiency, leading to more efficient infrastructures that can support existing energy demands (Figure 5.14).

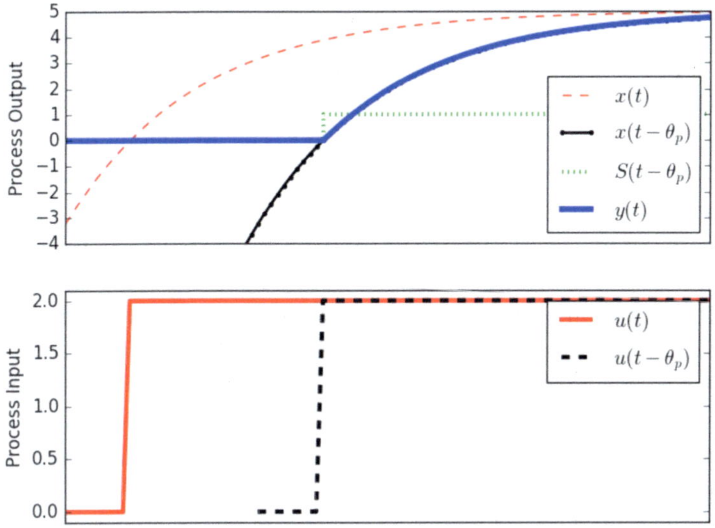


**Figure 5.14**. *Time delay in dynamic systems.*

*Source: https://apmonitor.com/pdc/uploads/Main/step_with_time_delay.png.*

## 5.5.3. External Disturbances and Uncertainties

Disturbances from the external world are major challenges in control systems that can destabilize system performance. They can arise from numerous uncertain sources, for instance, environmental conditions, operating conditions, and uncertainties internal to the system. For instance, a variation in wind speed or unexpected object in the path of quadrotor unmanned aerial vehicles would significantly impact navigational accuracy and stability. Overcoming these challenges requires advanced control solutions that not only boost robustness but can dynamically adjust to real-time variations as well. One such solution is the utilization of new control techniques such as adaptive neural network backstepping controllers that can dynamically counter disturbances while maintaining system integrity, as exemplified by system models that incorporate perceptive algorithms to counter unforeseen external interference (Li et al., 2023). Moreover, uncertainties in system dynamics render it difficult to make effective control strategies. The uncertainties may arise from changes in system parameters, sensor imprecision, or environmental impacts that are not possible to model accurately. In designing control systems, the designer needs to take account of such uncertainties at the design level to ensure robustness and stability under a range of operating conditions. By employing methods such as barrier Lyapunov functions and robust control mechanisms, constraints can be imposed such that it safeguards system operation while simultaneously allowing it to adapt to varying levels of input and output response uncertainty. Such methods facilitate stable control system performance even when varying external conditions exist, and this ensures attainment of control objectives under real-operating conditions (Panel I on Change C, 2023). The implications of exogenous perturbations and uncertainties extend beyond local control applications; they pervasively impact the broader context of systems engineering and management. The interaction between human activity and climate variability, in particular, underscores the need for systems that are not merely stable but also adaptive. Latest work by the Intergovernmental Panel on Climate Change highlights adaptation strategy incorporation as an imperative in addressing climate-related uncertainties whose cascading effects apply to ecological and socio-economic systems. As control systems are more coupled with more complex environments, such as those presented by climate change, stability will depend on robust frameworks that accommodate uncertainty as well as performance specifications. Hence, stabilizing control systems subject to external disturbances remains an important research and development area that is essential to mitigating threats and providing system robustness in a constantly changing international environment (Panel I on Change C, 2023).

## SUMMARY

Control system stability emerges as a critical factor in ensuring efficient and safe operations of various engineering systems. The application of complex control strategies, particularly in modern power systems, has highlighted the need for new solutions to stability issues caused by introducing new types of energy. One notable contribution in this direction is the formulation of a static voltage stability index, which may be employed as an index for assessment of new energy systems' static stability margin, thereby aiding in controlling voltage and frequency stability. Grid-forming control strategies are essential for improving the resilience of the system, ultimately working towards the support of pillars of stable control dynamics that power infrastructures for the modern era require, as discussed in the conclusions of Sun et al. (2024). Further development of control methodology, such as the event-triggered finite-time consensus control scheme, also contributes towards the topic of stability in control systems, particularly in multi-agent systems. This new design addresses the problems of unobservable velocities and external disturbances that have a tendency to make traditional control systems involved. With the use of a terminal sliding mode observer, the suggested approach guarantees that errors in velocities fall to zero in finite time, hence keeping control efforts responsive and effective. This decentralized control law allows for decentralized update in followers and hence enhances system robustness as well as the control of the update rate of control law. These are attributes testifying to the ongoing evolution of control methods and their contribution towards stabilizing complicated systems, as seen in Zhi et al. (2023). Lastly, the convergence of evidence from recent studies points to the significance of continued research in control system stability. Resolution of the newly emerging challenges for new sources of energy and adaptive control methodologies necessitates a commitment towards developing better ways of approach and exploring new solutions. With the changing landscape of the technological world, particularly in renewable energy systems, the influence of these studies extends well beyond the domain of theoretical discourse, with practical frameworks directly usable for real situations. Therefore, the future is in enabling interdisciplinary convergence and invention to ensure that the progress in control system stability stays in harmony with the requirements of the new energy economy. This path not only guarantees greater stability in control systems but also serves towards larger sustainability objectives critical to the next generation.

The stability of the control system is paramount to the reliability of operation of most engineering devices, particularly in cases where circumstances are dynamic and complicated. Central to this is the homeostatic balance, a principle which is omnipresent in as divergent as fields of physiology and biomedical engineering. The investigation of homeostatic balance not only promotes the understanding of the basic mechanisms shared by various physiological systems but also sets the grounds for designing artificial biological control systems. Adopting the closed-loop control paradigm involving negative feedback in its core, the proposed framework captures the underlying dynamics of multivariable physiological control systems. Such an approach acquires critical insight into the stability properties of systems to allow researchers to identify and investigate the intricate behaviors that define system performance and reliability (Procopio et al., 2025). Additionally, the study of stability is not only confined to biological systems but also technological applications, with particular emphasis on the employment of membrane systems in water treatment. Increased global demand for freshwater has made increased reliance on reverse-osmosis technology unavoidable, so robust control systems must be used to maintain fouling in check and boost the stability of operation. Using advanced diagnostic methods and adaptive control strategies, such as machine-learning algorithms, ensures enhanced monitoring and optimization of system parameters. Through addressing early foultant types and applying cascade prevention strategies, researchers have demonstrated outstanding enhancements in membrane systems stability. This not only increases periods between maintenance cycles but also decreases operation cost, pointing out the importance of stability in ensuring sustainable water treatment processes (Odnoralov, 2025). Fundamentally, the topic of control system stability points to increased understanding of both engineered and biological systems. Interconnectedness of homeostatic equilibria and technology is the demand for holistic models that will be able to react to different variables and perturbations. Examining dynamics of stability at the molecular and systemic levels allows researchers to benefit from understanding that frames theoretical and practical development. The continuous development of control systems employing continuous feedback processes will be at the core in problem-solving issues faced in diverse fields, guide designing and optimizing natural and man-made systems to improve performance and sustainability. The ongoing research in stability is therefore vital in leading effective solutions addressing urgent problems in a variety of fields, marking the significance of this area of study in the current research and application.

# REFERENCES


1. Abdelhfeez, S. A., Amer, T. S., Elbaz, R. F., & Bek, M. A. (2021). Studying the influence of external torques on the dynamical motion and the stability of a 3DOF dynamic system. Alexandria Engineering Journal, 61, 6695–6724. https://doi.org/10.1016/j.aej.2021.12.019 (accessed on 19th September 2025).
2. Amer, T. S., Bek, M. A., Hassan, S. S., & El-Bendary, S. I. (2021). The stability analysis for the motion of a nonlinear damped vibrating dynamical system with three-degrees-of-freedom. Results in Physics, 28, 104561–104561. https://doi.org/10.1016/j.rinp.2021.104561 (accessed on 19th September 2025).
3. Bahnasy, T. A., Amer, T., Almahalawy, A., Abohamer, M., Abosheiaha, H., & Elameer, A. S. (2025). A chaotic behavior and stability analysis on quasi-zero stiffness vibration isolators with multi-control methodologies. Journal of Low Frequency Noise, Vibration and Active Control, 44, 1094–1116. https://www.semanticscholar.org/paper/e05aa0c233c511befdf4176c20cacc0d26d716dc (accessed on 19th September 2025).
4. Bala, M. (2025). Singularities and metric structures in sub-Riemannian geometries with applications to control theory. *International Journal of Scientific Research in Science, Engineering and Technology*. https://www.semanticscholar.org/paper/b018cc80133297cfd63775eac407b79995285f53 (accessed on 19th September 2025).
5. Bano, F., Ayaz, M., Baig, D. Z., & Rizvi, S. M. H. (2024). Intelligent control algorithms for enhanced frequency stability in single and interconnected power systems. *Electronics*. https://www.semanticscholar.org/paper/344a2b6b2b762548abaf9f461f58e0de12cb9e7d (accessed on 19th September 2025).
6. Bhunia, A., Mathew, L., & Shimi, S. L. (2023). Various types of voltage stability control techniques of DC microgrid: A comparative analysis. *International Journal for Research in Applied Science and Engineering Technology*. https://www.semanticscholar.org/paper/35a5370325be43ebfa166d77dac58ab5443bc72e (accessed on 19th September 2025).
7. Bieda, R. (2025). On the stability of control systems with time delay. Pomiary Automatyka Robotyka. https://www.semanticscholar.org/paper/f87322f6312c0d8cd473ab77bf3ef686a728ac8e (accessed on 19th September 2025).

8. Druzhinin, V., Tytiuk, V., Kurliak, P., Chornyi, O., Sivyakova, G., Kalinin, A., & Busher, V. (2025). Electric drive excitation control for improved performance of hot rolling mill finishing groups. Energetika. https://www.semanticscholar.org/paper/7ba487695c3064df86bb8803f6e63e9040e7da66 (accessed on 19th September 2025).
9. Fakir, P. N. (2025). Review: Magnetorheological fluids and their automotive applications. International Journal of Scientific Research in Engineering and Management. https://www.semanticscholar.org/paper/df92eda61d0d6454e208a663eaac105a10380b29 (accessed on 19th September 2025).
10. Farsi, M., Li, Y., Yuan, Y., & Liu, J. (2022). A piecewise learning framework for control of unknown nonlinear systems with stability guarantees. https://www.semanticscholar.org/paper/9c99cc8bf08516844ea4703bc6b901fb046c9f7d (accessed on 19th September 2025).
11. Fatoorehchi, H., & Ehrhardt, M. (2024). A novel technique for stability analysis of linear time-invariant and nonlinear continuous-time control systems based on Markov parameters and Hankel matrices. *Transactions of the Institute of Measurement and Control, 46*, 1910–1923. https://www.semanticscholar.org/paper/e0633229e99bafc56f79a2b7082429b73cbd9834 (accessed on 19th September 2025).
12. Guo, X., Keivan, D., Syed, U., Qin, L., Zhang, H., Dullerud, G., Seiler, P. J., et al. (2024). ControlAgent: Automating control system design via novel integration of LLM agents and domain expertise. https://www.semanticscholar.org/paper/43bc723f6ec5e9dfabbd1a1ad8f9dc461935fc23 (accessed on 19th September 2025).
13. Haditiansyahuthor, A., Risdarmawan, F. A., Nabila, B., Wulandari, P., Samijayani, O., & Syahriar, A. (2025). Analysis of quarter-car suspension dynamics in sports cars with PID control. *Journal of Advanced Research in Applied Sciences and Engineering Technology*. https://www.semanticscholar.org/paper/69ba77b5e0ac65d6cfdd7643a8641f7d4125478d (accessed on 19th September 2025).
14. Hilgert, E., & Schwung, A. (2025). Stability analysis of soft sensor-based control systems—A Luré-Postnikov approach using a novel gated recurrent neural network. *ArXiv, abs/2505.11539*. https://www.semanticscholar.org/paper/ac7e79282bdcf87ab277a8eac507358931befb46 (accessed on 19th September 2025).
15. Intergovernmental Panel on Climate Change. (2023). *Climate change 2022 – Impacts, adaptation and vulnerability*. https://doi.org/10.1017/9781009325844 (accessed on 19th September 2025).

16. Krishna, S., Yim, D., Lakshmanan, V., Tirumalai, V., Koh, J. L. Y., Park, J. E., Cheong, J. K., et al. (2019). Dynamic expression of tRNA-derived small RNAs define cellular states. EMBO Reports, 20. https://doi.org/10.15252/embr.201947789 (accessed on 19th September 2025).

17. Li, J., Wan, L., Li, J., & Hou, K. (2023). Adaptive backstepping control of quadrotor UAVs with output constraints and input saturation. *Applied Sciences, 13*, 8710–8710. https://doi.org/10.3390/app13158710 (accessed on 19th September 2025).

18. Li, X., Wang, X., Liu, F., Lam, H., & Du, Y. (2024). Stability analysis and event-triggered control for IT2 discrete-time positive polynomial fuzzy networked control systems with time delay. IEEE Transactions on Fuzzy Systems, 32, 1346–1358. https://www.semanticscholar.org/paper/8a983bd80464386756c07f9a8b03181478581b58 (accessed on 19th September 2025).

19. London, J. (2025). *Control systems analysis of a 3-axis photovoltaic solar tracker for water pumping.* https://www.semanticscholar.org/paper/605a7efb48f2f65a2d5b8812e65ebb4ad8991a75 (accessed on 19th September 2025).

20. López-Doménech, G., Covill-Cooke, C., Ivankovic, D., Halff, E. F., Sheehan, D. F., Norkett, R., Birsa, N., et al. (2018). Miro proteins coordinate microtubule- and actin-dependent mitochondrial transport and distribution. *The EMBO Journal, 37*, 321–336. https://doi.org/10.15252/embj.201696380 (accessed on 19th September 2025).

21. Lou, G., Gu, W., Lu, X., Xu, Y., & Hong, H. (2020). Distributed secondary voltage control in islanded microgrids with consideration of communication network and time delays. *IEEE Transactions on Smart Grid, 11*, 3702–3715. https://doi.org/10.1109/tsg.2020.2979503 (accessed on 19th September 2025).

22. Ma, K., Ren, C., Zhang, T., Sun, B., Sun, Z., Li, L., & Zhang, Y. (2024). Dynamic response and stability analysis of vehicle systems with double time-delay feedback control. *International Journal of Control, 98*, 1085–1099. https://www.semanticscholar.org/paper/ee3e0153335f74e2d7bedf4a32f71fe2c9f8ed73 (accessed on 19th September 2025).

23. Muolo, R., Hastir, A., & Nakao, H. (2024). Stabilization of hyperbolic reaction-diffusion systems on directed networks through the generalized Routh-Hurwitz criterion for complex polynomials. 2024 SICE International Symposium on Control Systems (SICE ISCS). https://www.semanticscholar.org/paper/bc50b37c2bec1181643033eb5e02a7111428f6a4 (accessed on 19th September 2025).

24. Narayansamy, S., & Deepa, K. (2025). Stability analysis of grid forming inverters in master student control. 2025 7th International Conference on Intelligent Sustainable Systems (ICISS). https://www.semanticscholar.org/paper/7df8afd0988f4ca9fb4514a5ab43b54ca3d5d4af (accessed on 19th September 2025).
25. Odnoralov, A. (2025). Comprehensive analysis of physico-chemical and biological mechanisms of reverse osmosis membrane fouling with the development and optimization of preventive strategies to enhance the operational stability of membrane systems. *The American Journal of Engineering and Technology*. https://www.semanticscholar.org/paper/5585e8f8b33a8b5d908c6b7d1d3afc1586f90a9b (accessed on 19th September 2025).
26. Patil, O. S., Kamalapurkar, R., & Dixon, W. E. (2025). Saturated RISE controllers with exponential stability guarantees: A projected dynamical systems approach. IEEE Transactions on Automatic Control, 70, 4936–4942. https://www.semanticscholar.org/paper/a8e3b30738a5506d5b66e61d9e37e04063145597 (accessed on 19th September 2025).
27. Procopio, A., Tedesco, A., Lo Regio, F., Cesarelli, G., Donisi, L., Ricciardi, C., Merola, A., et al. (2025). A general framework for closed loop negative feedback multivariable physiological control systems: Existence, uniqueness, and stability of homeostatic equilibrium points. *IEEE Access, 13*, 17630–17651. https://www.semanticscholar.org/paper/df04e0ccfb901da1fdcb4660f3a726d546ded7fa (accessed on 19th September 2025).
28. Qiang, S. (2025). Differential dynamics modeling and simulation analysis of multi-agent cooperative motion. Journal of Advances in Engineering and Technology. https://www.semanticscholar.org/paper/73ed671bda78a8c66b4bab0f77141a0ede5eee5b (accessed on 19th September 2025).
29. Ren, Z., Wen, S., Li, Q., Feng, Y., & Tang, N. (2020). Stability analysis for nonlinear impulsive control system with uncertainty factors. Computational Intelligence and Neuroscience, 2020, 1–10. https://doi.org/10.1155/2020/8818794 (accessed on 19th September 2025).
30. Sahni, P., Prasad, S., Chandrappa, S., Pandalaneni, M., Thacker, C., & Vasmatkar, A. (2025). Performance optimization of wireless architectures for low latency communication in industrial automation. 2025 4th International Conference on Distributed Computing and Electrical Circuits and Electronics (ICDCECE). https://www.semanticscholar.org/paper/5e6b29667e650df2ce817ef9fe0884512395b48a (accessed on 19th September 2025).

31. Sekiya, T., Emili, E., Miyazaki, K., Inness, A., Qu, Z., Pierce, R. B., Jones, D. B. A., et al. (2025). Assessing the relative impacts of satellite ozone and its precursor observations to improve global tropospheric ozone analysis using multiple chemical reanalysis systems. Atmospheric Chemistry and Physics. https://www.semanticscholar.org/paper/2d6cc364f607a450cdbcfa822d028d1dd8c85251 (accessed on 19th September 2025).

32. Singh, A., Yadav, S., Tiwari, N., Nishad, D., & Khalid, S. (2025). Optimized PID controller and model order reduction of reheated turbine for load frequency control using teaching learning-based optimization. *Scientific Reports, 15.* https://www.semanticscholar.org/paper/e8dae474dfce7b83bebd39db386be04df7ecd6b8 (accessed on 19th September 2025).

33. Song, M., Ding, G., Ma, Y., & Ha, H. (2025). Analysis of the impact of control modes of grid-forming inverters in 100% renewable energy system on system's small-signal stability. 2025 8th International Conference on Energy, Electrical and Power Engineering (CEEPE). https://www.semanticscholar.org/paper/db888a5ff67134fc691a944db98fbf83903e9823 (accessed on 19th September 2025).

34. Sulyukova, L. F., Yakubjanova, D. K., & Djumayev, S. N. (2025). Mathematical modeling and control of non-rigid axisymmetric parts processing. International Journal of Innovative Science and Research Technology. https://www.semanticscholar.org/paper/48f10babf1f63eba557fb17a36053da9e00a7304 (accessed on 19th September 2025).

35. Sun, W., Liu, G., Yi, G., Wang, Q., Hui, C., Han, X., & Zhang, W. (2024). Research on static voltage stability enhancement for new energy station based on grid-forming control strategy. Frontiers in Energy Research. https://www.semanticscholar.org/paper/9439e80b43cdced667ebcea1690c5de3100b0988 (accessed on 19th September 2025).

36. Vieira, J. L., Dos Santos, J. A., Abur, K., de Souza, L. G. C., Azzolin, G., & Bellinaso, L. (2024). Development and analysis of a control system for yaw damper. *2024 16th Seminar on Power Electronics and Control (SEPOC).* https://www.semanticscholar.org/paper/36322f75f9972782e1f99703b62fdb8d698c268c (accessed on 19th September 2025).

37. Wang, B., Jahanshahi, H., Volos, C., Bekiros, S., Khan, M. A., Agarwal, P., & Aly, A. A. (2021). A new RBF neural network-based fault-tolerant active control for fractional time-delayed systems. *Electronics, 10,* 1501–1501. https://doi.org/10.3390/electronics10121501 (accessed on 19th September 2025).

38. Wang, W., Maass, A. I., Nevsi'c, D., Tan, Y., Postoyan, R., & Heemels, W. (2025). Stabilization of singularly perturbed networked control systems

over a single channel. https://www.semanticscholar.org/paper/4f3016ae31d59af5a5d1fda2e9c89509c3f17d06 (accessed on 19th September 2025).

39. Wang, X., Du, H., Feng, L., Zhou, F., & Li, W. (2025). Effective throughput maximization for NOMA-enabled URLLC transmission in industrial IoT systems: A generative AI-based approach. IEEE Internet of Things Journal, 12, 13327–13339. https://www.semanticscholar.org/paper/4b7bfd4fc615a88683ed4678fd3c20717145720a (accessed on 19th September 2025).
40. Wang, Z., & Bu, S. (2024). Probabilistic frequency stability analysis considering dynamics of wind power generation with different control strategies. IEEE Transactions on Power Systems, 39, 6412–6425. https://www.semanticscholar.org/paper/2249c7a0ab134df98d377ca828eabc4c318b40f9 (accessed on 19th September 2025).
41. Xu, C., Zou, Z., Yang, J., Wang, Z., Chen, W., & Buticchi, G. (2024). Transient stability analysis and enhancement of grid-forming and grid-following converters. *IEEE Journal of Emerging and Selected Topics in Industrial Electronics, 5*, 1396–1408. https://www.semanticscholar.org/paper/36a666331bd598b8410b1af43d2a373f0644dbcf (accessed on 19th September 2025).
42. Yablochnikov, S., Yablochnikova, I., Dzobelova, V. B., & Medvedeva, L. V. (2024). Aspects of assessing the absolute stability of control systems containing nonlinear components. 2024 Systems of Signal Synchronization, Generating and Processing in Telecommunications (SYNCHROINFO). https://www.semanticscholar.org/paper/119a2bce7cc67e5e25bc0dc5fed7a6fb536d48bb (accessed on 19th September 2025).
43. Yan, Y., Bao, J., & Huang, B. (2025). The intrinsic state variable in fundamental lemma and its use in stability design for data-based control. https://www.semanticscholar.org/paper/97c2f75b74e2d2cabd62f6f15d6f7a53a5073488 (accessed on 19th September 2025).
44. Yoneyama, J. (2017). Stability analysis of nonlinear networked control systems via Takagi-Sugeno fuzzy model. *IFAC-PapersOnLine, 50*, 2989–2994. https://doi.org/10.1016/j.ifacol.2017.08.665 (accessed on 19th September 2025).
45. Yuvaraju, M. (2025). Comparative vehicle stability analysis of indoor RC vehicle using control techniques. *International Journal of Scientific Research in Engineering and Management*. https://www.semanticscholar.org/paper/54b691d80db062be10bbb5a04065d9b547b84932 (accessed on 19th September 2025).

CHAPTER 6

# BLOCK DIAGRAMS AND SIGNAL FLOW GRAPHS

## Contents



## INTRODUCTION

The topic of systems engineering increasingly uses block diagrams and signal flow graphs in modeling complex systems and their interactions. The graphical tools are key methodologies that ease the analysis and design of dynamic systems in various applications such as telecommunications, automation, and control systems. Through decomposing a system into blocks, engineers can scrutinize the performance, stability, and behavior of the system under diverse conditions in a methodical method. This is particularly relevant in contemporary applications, such as the design of 5G networks, where the architecture relies so heavily on centralized cloud-computing paradigms-tuned setups, as cited in the increasing demand for signal processing systems to be flexible and optimized (Enrici et al., 2019). Awareness of the definitions and fundamental concepts on the topics of block diagrams and signal flow graphs is critical for practitioners and engineers. Block diagrams provide an abstract representation of a system in terms of blocks standing for the elements of a system and arrows standing for signal flow. Signal flow graphs provide a more complicated description of variable relationships but facilitate easier analysis on input-output relationships and system dynamics. For instance, the inherent attributes of discrete-time systems, as presented in classical texts, determine important parameters such as stability and causality, which link theoretical expressions with practical uses in system design (Jackson, 1985).

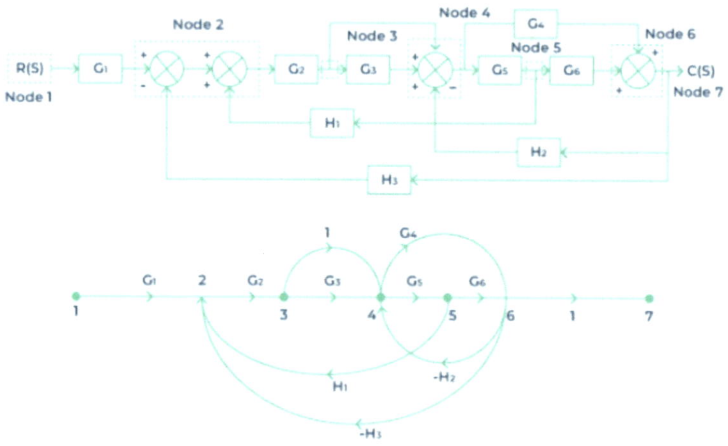


**Figure 6.1**. *Conversion of block diagrams into signal flow graphs.*

*Source: https://media.geeksforgeeks.org/wp-content/uploads/20240110152907/block-to-sfg.png.*

With improving technology, learning block diagrams and signal flow graphs is more than a classical interest. These graphs become particularly critical while optimizing feedback systems and enhancing data processing algorithms on various platforms. Moreover, as engineering problems continue to evolve, more effective analysis techniques are required. Techniques such as static data-flow analysis of SysML block diagrams and UML show how engineers can achieve significant reductions in unnecessary complexity while analyzing system performance graphs. So, as this chapter goes on to examine the intricacy of block diagrams and signal flow graphs, it tries to describe their pivotal position in contemporary engineering practice and their implications for emerging technologies (Enrici et al., 2019) (Figure 6.1).

## *Definition of Block Diagrams*

Block diagrams definition supports the system representation in engineering disciplines. A block diagram is simply a diagram that displays relations and interactions among a system by using blocks and arrows. Each block depicts a particular component or a function of a system, and the arrows depict the flow direction of signals or information among components. The ease of this description makes intricate systems easily comprehensible in terms of control or data flow without having to delve into mathematical equations or mechanism details. Block diagrams are also basic tools used by engineers and scientists when researching system behavior, supporting design, simulation, and optimization. Block diagrams are a vital tool in control systems and signal processing to provide insight into the transfer functions of each element. They can be used to visualize input/output relationships and assist in the determination of system performance features and system stability. Block diagrams can further be used in designing Signal Flow Graphs, even more specific diagrams with the interrelationship of signals and the routes involved. By converting the mathematical version of complex differential equations into a graphical form, block diagrams make it easy to simplify system analysis. The ability to decompose complex systems into understandable units simplifies the analysis load, as well as communication load, among designers and researchers dealing with system design and diagnostics. Additionally, enhancing the role played by higher-order interaction in complex systems has facilitated the use of block diagram representation. Researchers increasingly come to appreciate that traditional linear models may not be sufficient to accurately capture the dynamics of many applications, including traffic systems and ecosystems. Traffic flow forecasting models, for instance, have evolved to incorporate temporal dependencies as well as dynamic spatial relationships, typically shown in block diagram form for ease of understanding and accessibility. The application of such methods illustrates the resourcefulness of block diagrams for describing not only simple systems but also sophisticated systems that possess nonlinear characteristics and intricate interactions between various components (Battiston et al., 2020; Guo et al., 2019). Thus, block diagrams remain an essential framework for describing and interpreting the behavior of various systems in contemporary research and practice.

## *Definition of Signal Flow Graphs*

Signal flow graphs (SFGs) are a very useful representation in linear system analysis and design. Unlike block diagrams, which encase system functions in more pictorially unrelated form, signal flow graphs describe the signals going through nodes and branches such that a more integrated view of system dynamics is obtained. In SFGs, each node is a variable of interest, and an edge connecting nodes is the causal dependency between nodes, thus characterizing how output depends on various inputs. This formulation simplifies system behaviors to be more intuitively comprehensible, particularly when addressing systems' complex interdependencies. Therefore, signal flow graphs are particularly relevant in control system engineering, where component interaction must be understood to design and analyze stability appropriately, as also discussed in linear system interconnections (Enrici et al., 2019). Additionally, the mathematical modeling of signal flow graphs significantly enhances computational effectiveness in system analysis. By techniques such as Masons gain formula, engineers are able to directly get the transfer functions from the graph structure, making it possible to determine successful system behavior without the involved algebra typically required in more traditional techniques such as block diagram manipulations. Usage of SFGs is also specifically complementary with recent system design trends, especially in the context of greater use of cloud computing and distributed structures in signal processing, where digital signals travel over a range of hardware units. Such settings call for prompt and accurate analysis, and SFGs are particularly well-equipped for such purposes. It is stated that use of interconnected SFGs offers valuable information during static data flow analysis of complex systems, indicating their utility in modern applications (Enrici et al., 2019). Even though the advantages of signal flow graphs are of utmost importance, practitioners must also consider their limitations compared to block diagrams. For instance, signal flow graphs are especially well-suited for describing linear relations but become complex in the event of non-linear interactions or more than one feedback loop. Where such complexity is a factor, block diagrams may offer less complex representations, albeit at the cost of computational effectiveness. Yet, the use of SFGs or block diagrams is typically a function of the specific requirements of the system in question. Since engineering issues evolve—particularly in such fields as telecommunications, where 5G technologies and software-defined radios prevail—the ease of switching easily among these forms is required. Familiarity with them serves engineers well in selecting the best method to given problems within the universal paradigm of linear system analysis (Africa, 2019).

## *Importance of Visual Representation in Systems*

The function of graphical depiction within systems, particularly as block diagrams and signal flow graphs, cannot be exaggerated. These are potent abstractions that collapse intricate systems into manageable components. By means of the depiction of functionalities and interactions visually, such diagrams can aid in communication among stakeholders, whether engineers or non-engineering personnel. This is largely vital within interdisciplinary settings where an abbreviated, collective understanding of a systems architecture is critical. Effective visualizations facilitate quicker comprehension and simplicity of problem-solving because they clearly articulate the relationships and protocols involved. In system design and engineering, application of these representations is likely to reveal problems early during the development process and thus save time and funds. Visual representations are not only useful for communication purposes but also augment analytical abilities. They allow system behaviors to be seen through a structured organization that reveals interdependencies. For instance, block diagrams map out individual components and inputs and outputs, making it easier to see control loops and feedback mechanisms. These are needed in engineering disciplines where system dynamics are complex and potentially nonlinear. By plotting these relationships, engineers are able to employ techniques that ensure stability and optimal performance of the system. At this level, graphical representation is an immensely useful analysis tool, by which the optimization of many different situations or modification of the system design can be explored, thus resulting in improved outcomes. In addition, the creation of advanced computational tools has further elevated the role of graphical representation within systems. As scholars explore how machine learning and artificial intelligence are incorporated into system design, model interpretability becomes crucial. For instance, the rules of interpretable machine learning illustrated in the literature mirror the growing need for clear visualizations that explain what is happening precisely with intricate algorithms. Using techniques that operate on visual representations, described in Rudin et al. (2022), experts can perform high-stakes decisions according to a thorough image of the system being used. This is a trend towards more use of visual systems in order to fill the gap between advanced computational methods and actual usage, as a change that values transparency and knowability in system building.

# 6.1. HISTORICAL BACKGROUND

Historical evolution of block diagrams and signal flow graphs is intrinsically related to evolution in control theory and electrical engineering. Initially, there was a need for systematic description of dynamic systems with the industrial boom in the middle of the 20th century and its demand for effective system analysis and design. Block diagrams emerged as a way of simplifying complex systems into understandable visual representations that allowed engineers to express relationships and feedback loops clearly. Not only did this help make it simpler to understand, but it became the focal point of the design process, paving the way for the development of signal flow graphs, which give a better description of the dynamics of the system. Historical figures such as Norbert Wiener and Claude Shannon pioneered the early works that laid the foundation for the methods, enabling the paradigm shifts in the way engineers think and design control systems. The field continued to advance with the integration of mathematical concepts into the framework of block diagrams and signal flow graphs.

The stage was set when the engineers realized the potential of transfer functions and frequency response in system characterization. By the use of a consistent graphical approach, practitioners were able to effectively communicate and analyze system performance under varied conditions. The arrival of sophisticated computational software later accelerated this trend, as engineers have begun to tackle complex interactions once considered intractable. In addition, the arrival of learned machine learning could also be understood as an extension of this historical momentum, which involves the marriage of theoretical foundations and practical effort. As the record shows, this development is a sign of a gradual shift towards empirically founded methods to those based on sound mathematical foundations, thereby dramatically improving the field's analytical power (Laura et al., 2021). The late 20th and early 21st centuries have witnessed exponential leaps in terms of scope and complexity of control systems, which further called for the improvement of block diagrams and signal flow graphs.

The evolution of sophisticated computational techniques and data analysis methods has inspired fresh interpretations of these diagrams as dynamic tools to monitor and control systems in real time. The methodical approach of structuring information with block diagrams guarantees simplicity amidst complexity, while signal flow graphs heighten our senses regarding interdependencies within systems. Furthermore, the integration of biological information, exemplified through research in the field of neural circuitry, has further implications for such graphical models in relating them to biological processes with comparable feedback and transfer characteristics. It is this intersection of fields that emphasizes the flexibility of block diagrams and signal flow graphs as inherent tools across both engineering and biology, permitting continuing innovation and expansion (Louis et al., 2020).

## 6.1.1. Development of Block Diagrams

Block diagram evolution has had a significant influence on computer science, engineering, and systems analysis. Initially, they existed as tools to simplify complicated systems and facilitate processes to understand and explain. With the advent of digital technology, block diagrams expanded their application, including more complicated interactions and functions in systems. Their form allows the representation of inputs, outputs, and interactions among elements, which makes them indispensable in control system analysis and design. Additionally, development from traditional visual aids to digital platforms has enhanced their flexibility, with dynamic updates and changes that reflect shifting conditions in real-time, expanding their application beyond static representations to interactive simulation and performance assessment. Along with block diagram development, signal flow graph incorporation has provided a means to analyze system dynamics in more detail. Both tools are complementary system-description methodologies but have differing virtues. Block diagrams give a clean-cut presentation that focuses on signal flow and interconnection of various subsystems, while signal flow graphs portray feedback loops and effect of input changes on output more obviously. As signal processing and feedback control came into vogue, integration of the two paradigms enabled engineers and scientists to address problems more precisely and lucidly.

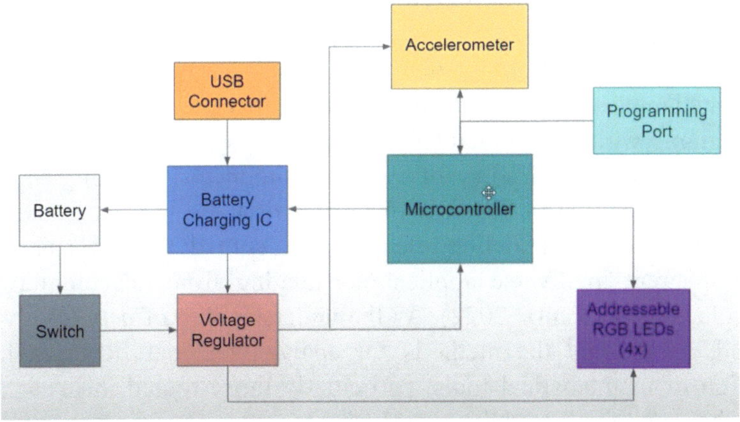


**Figure 6.2**. *System block diagram.*

*Source: https://predictabledesigns.com/wp-content/uploads/2021/09/SystemBlock-Diagram.jpg.*

This cross-use is an excellent instance of significant change in which graphical depiction and analytical accuracy blend, thereby enhancing the effectiveness of system depiction in theory and practice. Ongoing improvement in the design and utilization of block diagrams constitutes an indicator of larger patterns of technological enhancement and interdisciplinarity. With paradigms shifting toward more integration and complexity in systems in automation, practitioners have more leverage in the use of block diagrams' power to apply across areas such as artificial intelligence and robotics. Their development relies not only on traditional means but progresses through the use of a systematic method for adjusting to new techniques, as shown from the complementarity and contrast between graphical models and block diagrams. Subsequently, their role in education, particularly in the education of systems thinking and design engineering principles, continues to expand. Such an evolution not only highlights the relevance of visual aids in understanding complex systems but also the necessity for research on how to continually refine their design and functionality towards enhanced practical utilities in upcoming technologies (Figure 6.2).

### 6.1.2. Evolution of Signal Flow Graphs

The evolution of signal flow graphs (SFGs) represents a significant advancement in systems representation, particularly in engineering and control theory. Originally developed alongside block diagrams, SFGs serve as graphical tools for modeling complex interrelationships within systems by depicting signal pathways and interactions more dynamically. This approach emerged prominently in the mid-20$^{th}$ century, as engineers sought to develop more sophisticated methods to analyze and design feedback controls systems. Unlike block diagrams, which traditionally favor a linear presentation, signal flow graphs emphasize the interdependency of system components, allowing for a more intuitive understanding of the flow of information and energy. Consequently, as the need for precise modeling in engineering surged, the SFGs significance became apparent in diverse applications ranging from telecommunications to control systems (Kautz, 2022). As the understanding of network dynamics advanced, so too did the methods for analyzing signal flow graphs. The integration of mathematical tools, particularly those rooted in linear algebra and graph theory, facilitated the quantitative assessment of system stability and performance. Researchers began to explore the implications of feedback loops and cascading interactions, allowing for the examination of complex systems that could not be adequately captured by earlier methods. This analytical growth paralleled broader developments in related fields, where understanding higher-order interactions became key. The concept of higher-order networks began to gain traction, as interactions could not be merely represented through simple binary connections, prompting the need for more comprehensive frameworks

(Battiston et al., 2020). Such advancements underscored the necessity of evolving signal flow graphs to adequately encompass the intricacies of modern engineering challenges, positioning them as critical tools in both theoretical and practical contexts. The synergy between signal flow graphs and contemporary computational methodologies marked yet another evolution in their applicability. As the fields of artificial intelligence and machine learning burgeoned, the principles underpinning SFGs began to find relevance in increasingly complex systems. The transition to digital modeling enabled engineers to simulate and predict system behavior with unprecedented accuracy. Such technological advancements allowed for improved algorithmic approaches in optimization and control, with signal flow graphs acting as foundational elements in various software tools. This adaptation not only solidified the role of SFGs in traditional engineering disciplines but also broadened their scope in emerging fields. As interdisciplinary integration continues to advance, the evolution of signal flow graphs reflects an ongoing quest for enhanced understanding and control of complex systems, underscoring their enduring significance in both historical and contemporary contexts (Kautz, 2022) (Figure 6.3).

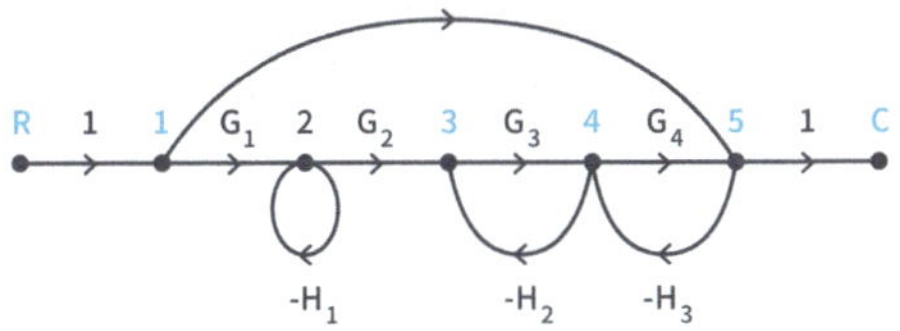


**Figure 6.3**. *Signal flow graphs.*

*Source: https://dwma4bz18k1bd.cloudfront.net/tutorials/Signal-Flow-Graphs-1.5-site.jpg.*

## 6.1.3. Key Contributors to the Field

The landscape of block diagrams and signal flow graphs has been shaped significantly by the contributions of many practitioners and researchers through the years. Notably, some of these early innovators like Norbert Wiener and Claude Shannon laid down the pillars of systems thinking and the theory of communication, on which the know-how of signal processing is based. Their innovative approaches provided the theoretical basis necessary for describing complex systems in terms of block diagrams. This pioneering work introduced fundamental principles such as feedback loops and transfer functions by which it became feasible to represent dynamic systems in a more palatable graphical representation. Therefore, these initiatives not only restructured the field of control engineering but further opened the doors for numerous applications in a range of disciplines from telecommunications to robotics, ultimately to an increasingly integrated perception of system behaviors. More recently, integration with sophisticated computational methods has propelled additional innovation in the design of block diagrams and signal flow graphs. Work on deep learning models has been particularly impactful, with the approaches here allowing new and hitherto unimaginable things to be done with the analysis of unstructured data. As the literature has found, the rapid application of these advancements within materials science is evidence of the potential of using sophisticated data-driven methods for system dynamics modeling and prediction. Additionally, the blending of these methods widens the scope of traditional block diagrams, enabling them to better represent intricate relations and interdependencies within systems. Researchers now examine these intersections increasingly, hypothesizing that the intermixture of conventional engineering techniques and modern computation techniques is critical in advancing the field. In addition, teaching methods towards greater student engagement and learning effectiveness have also found entry in the discourse regarding block diagrams and signal flow graphs. The inverted classroom approach has taken hold as a new pedagogy, supporting student-centered learning while maintaining rigorous academic content. With active learning through system analysis and design key concepts, students can engage with block diagrams in a setting that contextualizes application and problem-solving skills. Empirical evidence shows students in these settings develop a deeper sense of appreciation of the subject matter since they enjoy more control over their own learning (Mason et al., 2020). With the implementation of such pedagogical innovations, educators have enhanced the learning culture, ultimately enriching the control systems and signal processing fields through improved interaction and knowledge retention.

## 6.2. FUNDAMENTAL CONCEPTS

Basic ideas such as block diagrams and signal flow graphs are fundamental modeling and analyzing tools in system studies within fields such as control systems and signal processing. Block diagrams, being a visual representation of the elements of a system and their interconnection, facilitate the understanding of complex processes. A block may denote a specific function, and the arrows denote the flow of signals. This visualization enables simple analysis of dynamic systems by making it simple for engineers to visualize system dynamic and input-output relations. The explicit modeling of system functions not only enables stability analysis and frequency response but also enables the determination of controllers and feedback loops, which are instrumental in the optimization of systems. Using block diagrams, designers are able to systematically design and analyze various control systems. Signal flow graphs, however, present another graphic representation in terms of paths along which signals travel through a system. Signal flow graphs, unlike block diagrams, utilize nodes and directed branches to represent relationships among various signals more smoothly. Each node on the graph represents a signal, and the branches represent the mathematical relationship that governs their interaction. This method enhances the understanding of the system's dynamics since it allows the use of Masons gain formula, which plays a critical role in obtaining transfer functions. Signal flow graphs allow engineers to handle complex systems better, and hence, provide creative solutions in signal processing. Hence, signal flow graphs are essential in most areas of engineering, expanding the repertoire of analytical tools for the solution of such complicated signal-related issues. Knowledge about these fundamental concepts is crucial in making the most out of block diagrams as much as signal flow graphs in real-world applications. In telecommunication industries via automation, these graphical approaches enable the system that is at the core of society functions to be analyzed and designed. For instance, the intricate behavior of feedback control systems can be adequately captured using these representations and thus give rise to improved performance and reliability of the system. But the growing complexity of the systems introduces new challenges that require ongoing changes and adjustments of these fundamental concepts. As pointed out in recent research, predictive technology and machine learning development necessitates improved explanations of the underlying mechanisms, hence aligning with today's research on interpretability and system dynamics in engineering (Han et al., 2022; Tjoa et al., 2020). Through ongoing enhancement of these fundamental principles, engineers see block diagrams and signal flow graphs remain fundamental tools in a more advanced technology environment (Figure 6.4).

**Figure 6.4**. *Signal processing and machine learning.*

*Source: https://www.sintef.no/contentassets/151f67a6a457473eabb34c5de9e022ea/shutterstock_1236109669.jpg?width=890&height=500&quality=80&mode=crop&scale=both.*

### 6.2.1. Components of Block Diagrams

Block diagrams are a godsend for engineers and system analysis, employed to represent the important elements and relationships of a system graphically. The key elements of a block diagram are the blocks of system elements and arrows that symbolize the movement of signals or control between the system elements. Blocks are often marked so that the function that each signifies, for example, gain, filter, or summation point, can be recognized. Understanding these aspects facilitates clearer interpretation of the relationships between the system, and thus analysis, design, and communication for engineering teams is made possible. Representation in visual form makes these bring complex relationships into more manageable forms, making it invaluable while interpreting system performance or looking for changes that can enhance functionality. Another feature of block diagrams is how they represent feedback loops, which are crucial to identifying how systems react to changes in input over time. Feedback loops can be separated into positive and negative feedback, both of which influence system performance and stability in distinct manners. Positive feedback, for instance, may lead to exponential growth or unstable behavior, while negative feedback maintains system stability by opposing deviation from required levels of performance. These distinctions are key to control theory and graphically illustrated in block diagrams by structures connecting back from output to input. By analyzing these loops and their impact, engineers can predict system behavior under varying conditions and thus become part of achieving desired outcomes in design and operation. The addition of signal flow graphs to block diagrams provides an additional degree of complexity and richness to system analysis. Signal flow graphs offer another form of representation that emphasizes the directions in which the signals travel through a network of components and the interdependence of the variables. While block diagrams emphasize components and interactions among them, signal flow graphs emphasize the dynamic interdependencies and signal flow among them. The complementarity offers more effective system dynamics insight, especially in complex systems involving numerous simultaneous interactions. Both representations being analyzed give engineers valuable insights into system behavior and performance optimization so that structural as well as functional considerations are taken care of during the design process. This comprehensive critique is particularly relevant in areas that entail advanced modeling and simulation techniques, bridging the lights from insight to application across a broad array of disciplines, like control systems and signal processing, as a case in deep learning in image segmentation research where sophisticated models are analyzed in terms of how they work for detecting subtle details (Han et al., 2022; Minaee et al., 2021) (Figure 6.5).

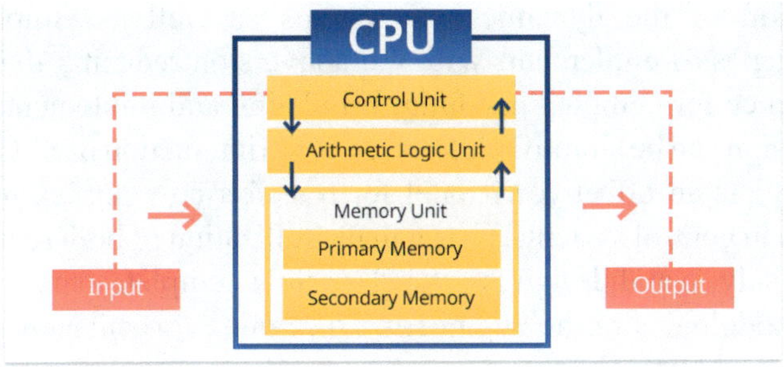


**Figure 6.5**. *Primary elements of a block diagram.*

*Source: https://www.21kschool.com/ge/wp-content/uploads/sites/9/2023/11/Block-Diagram-of-Computer-Featured-Image.png.*

## 6.2.2. Components of Signal Flow Graphs

Signal flow graph structure is a key component in control system analysis and design where the signals passing through various system components are illustrated graphically. SFG consists of directed edges and nodes in which nodes represent the system variables or signals, while edges illustrate the relationship between variables. Most vital in SFGs is the ability to show dynamic interactions, and hence an appreciation of how different components impact each other. The graphical representation not only allows for the simplicity in analyzing complex systems but also aids in transferring functions and system behavior with respect to stability and control. By describing mathematical relationships in a visual representation, SFGs can potentially uncover connections hidden in typical algebraic expressions. The second very crucial component of signal flow graphs is the novelty of introducing the idea of gains associated with each edge. These gains define how input signals are processed as they move through the network, effectively strengthening or weakening the signal at each step. This gain structure allows one to analyze the impact of various changes to the system, such as feedback loops or additional signal paths. As an example in the real world, one can examine the design plan of a three-stage amplifier through its SFG, where varied routes in the graph express varied levels of amplification and stability characteristics. The interactive nature of these gains emphasizes the importance of feedforward and feedback actions in developing wanted performance criteria, for example, bandwidth and phase margin. Besides, block diagram representation of SFGs is a fundamental operation that preserves the original system's properties but allows engineers to take advantage of the simplicity of block diagrams. This correspondence facilitates more intuitive comprehension of the dynamics of systems as well as simplification of control strategies to implement. With the conversion reducing the signal flow interdependence for complex handling, simulation and implementation phases of system design can be simplified. Therefore, the interaction of SFGs with block diagrams depicts an effective channel for transferring complex mathematical relationships in control systems. Through the utilization of both representations, the system analysis and design is carried out in a complete sense that supports diverse methodologies in the engineering discipline for enhancing clarity and efficiency in problem-solving.

### 6.2.3. Differences Between Block Diagrams and Signal Flow Graphs

Block diagrams are an elementary device within systems engineering and control theory for representing the functional relationships of a system. Block diagrams consist of blocks joined by arrows denoting the direction of signal transmission or control. Each block stands for some process or action and is a basic and intuitive way to describe system dynamics. Block diagrams are a top-down description to allow engineers to see the overall structure and interdependencies without being swamped with precise mathematical equations. Hierarchical organization ensures it is easy to communicate between members, supporting collaborative design and analysis processes. Nevertheless, their simplicity falls short in capturing complex interactions that are critical in complex systems and so there is a need for more sophisticated representations like signal flow graphs that give a clearer picture of the inner mechanisms and feedback loops of the system. Signal flow graphs, however, depict a more detailed picture of signal relationships in a dynamic system. In comparison with block diagrams, where the focus is on system elements and how they function, signal flow graphs depict signal delivery from nodes and edges, showing how disturbances propagate through the system. Each node is a variable, and edges are transfer functions between variables, allowing more precise study of system behavior. This graphical approach is most helpful in control theory since it makes it possible to apply techniques such as Masons Gain Formula in determining system stability and feedback loops. Therefore, signal flow graphs enable engineers to evaluate potential issues in system performance that otherwise would be hidden in a block diagram. Thus, while both diagrams are useful, they serve distinct purposes and allow for different levels of system analysis and design complexity. The most basic differences between block diagrams and signal flow graphs can be articulated in terms of their focus and application. Block diagrams are primarily utilized for the statement of general functional features of a system and are thus better suited for initial system design stages and intracommunicating among stakeholders. Conversely, signal flow graphs play an important role in in-depth analysis and optimization processes, particularly in feedback control systems where signal interaction needs to be fully understood. Furthermore, the mathematical definition used in signal flow graphs provides a platform for quantitatively extracting the behavior of a system, making simulations and controller design simpler. While block diagrams underlie the structure of learning systems, signal flow graphs extend the understanding of dynamic behavior, and evidence is given that a choice between the two has to be made with context and specific application of the engineering analysis in question.

## 6.3. APPLICATIONS IN ENGINEERING

In the realm of engineering, the utilization of block diagrams and signal flow graphs serves as a pivotal method for modeling complex systems. These graphical representations allow engineers to visually dissect and analyze dynamic systems, facilitating a clearer understanding of the interdependencies among different components. For instance, the newly developed web-based software UYAMAK exemplifies the integration of block diagrams in mathematical modeling, providing users the ability to design, simulate, and analyze systems with ease. This application offers a user-friendly interface that supports various mathematical operations and allows for the simulation of both discrete and continuous systems. Such capabilities are especially beneficial in education, where students can visualize complex dynamical systems and verify mathematical concepts, ultimately enhancing the learning experience through practical engagement with theoretical knowledge (Concha et al., 2025). Moreover, advancements in simulation software have further crystallized the essential role of block diagrams in engineering disciplines. The emergence of tools like DiaBloS allows for graphical modeling in Python, demonstrating significant progress in educational software that caters to dynamic system simulations. The algorithmic innovations in these tools prioritize the accuracy and efficiency of simulating system behavior, thereby affording students and professionals the opportunity to work with reliable representations of physical systems. By identifying execution orders of blocks within diagrams, such tools facilitate the transformation of abstract concepts into tangible analysis, crucial for engineers dealing with intricate variables across disciplines such as robotics, control systems, and network design (Torres-Torriti et al., 2024). Incorporating block diagrams and signal flow graphs not only aids in educational contexts but also enhances practical engineering applications. The versatility of these modeling tools finds usage in asynchronous computing and artificial intelligence, where understanding and manipulating multiple input-output relationships is vital. Systems that rely on the integration of countless subsystems can be better managed through the structured representation afforded by these diagrams. As the field of engineering confronts increasingly complex challenges, the demand for robust modeling techniques that incorporate block diagrams and signal flow graphs will only continue to grow. Such methodologies allow engineers to innovate and design systems that can adapt to the rapid technological advancements characterizing contemporary engineering landscapes (Concha et al., 2025; torres-Torriti et al., 2024) (Figure 6.6).

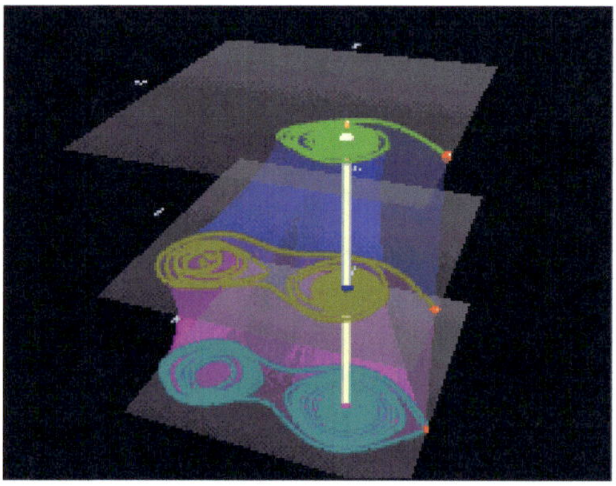

**Figure 6.6**. *Visualization of complex dynamical systems.*

*Source: https://www.cg.tuwien.ac.at/research/vis/dynsys/NDimViewer.gif.*

## 6.3.1. Use in Control Systems

Block diagrams and signal flow graphs find their prime importance in control systems in the visualization of system dynamics and feedback mechanism analysis. The methods enable visualization of complex control processes via simple models that determine the relation between system components. Each box in the figure is a dynamic system component, and arrows show signal flow and interaction and give a clear picture of how inputs are mapped to outputs. The graphical representation not only helps to describe the systems behavior but also enables the crucial step of system analysis, whereby engineers are able to ascertain stability and performance characteristics. As the complexity of control systems increases, such diagrammatic aids become indispensable to design efficiency and productivity, and are the core of modern control engineering techniques. Block diagrams and signal flow graphs have applications beyond representation; they form the core of control system analysis and design. For instance, in automated systems, transfer functions are obtained from the diagrams to quantify the input-to-output response of signals. These quantifications are required in order to implement many design approaches, including PID control and state-space form. Further, technological advancements of late demand an iterative design of systems, whereby digital signal processing is significantly reliant on algorithms minimizing signal transmission delay and latency (Enrici et al., 2019). As a result, engineers are better positioned to design fault-tolerant control systems that are not only operational but also adaptive to changing operating conditions and thus enhance system resilience. With the boom in digital technology and the sweeping transition to cloud-based systems, block diagrams and signal flow graphs have become all the more important. They are the very tools used for building software-based signal processing solutions in control systems and provide the means to efficiently analyze data flow and model complex behavior in systems such as Software Defined Radios. By applying sophisticated algorithms—such as the blocking combination of iterative and worklist approaches—analysts can maximize their analysis, expunging much of the extraneous computational overhead with enhanced performance (Enrici et al., 2019). As control systems evolve further, with increasingly greater demands for flexibility and throughput, disciplined use of these graphical representations will remain at the center of the advancements necessary to introduce increasingly more efficient and responsive systems into operation (Figure 6.7).

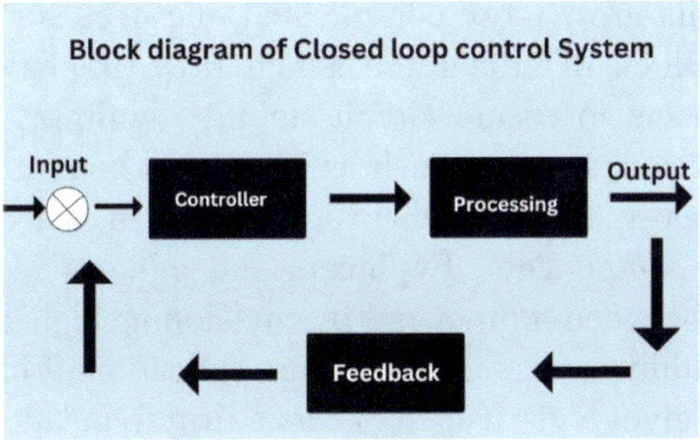


**Figure 6.7**. *Use in control systems.*

*Source: https://files.codingninjas.in/article_images/complete-introduction-to-control-system-2-1679994554.webp.*

## 6.3.2. Application in Electrical Engineering

Block diagrams and signal flow graphs are useful electrical engineering tools employed to concisely depict complex systems and interoperations. Block diagrams provide a general overview of a system, with major elements and functions drawn in the form of interconnected blocks. Block diagrams are simple to use in depicting system structure, with the use of amplifiers, filters, and controllers being depicted. By dissecting a complex circuit into easily understandable components, engineers can successfully analyze and debug systems. This is especially crucial in feedback control systems, where block diagrams allow the visualization of input-output relations as well as the foresight of potential stability issues or response time. Thus, the use of block diagrams promotes enhanced communication among engineers as well as learning scenarios where ideas must be made intelligible for pupils. Signal flow graphs complement block diagrams by adding signal processing's dynamic content. Signal flow graphs illustrate the movement of signals through a system and allow one to consider how a change in one area of a system influences others. Using nodes to symbolize variables and directed arcs to represent relationships, signal flow graphs allow engineers to use mathematical methods to infer system behavior. This method is beneficial in the study of differential equation-controlled systems, where signal flow graphs can provide structured handling of variables and interdependences. Such frameworks accommodate broad analysis of high-frequency behaviors, circuit stability, and performance indices in various applications, including telecommunications, signal processing, and control systems. The use of these visual representation tools into practical applications in electrical engineering gives the discipline a greater effect. For instance, as electronic circuits grow more complicated, the need for effective analysis becomes higher. Block diagrams and signal flow graphs can be employed by electrical engineers to ensure circuit stability, optimize performance, and improve system design. In areas such as high-level communication networks and renewable energy systems, the tools assist in managing interactions among numerous components. Engineers can address issues arising from interdependencies between components by envisioning higher-order complexity and interactions within a system. Lastly, the application of block diagrams and signal flow graphs gives electrical engineers a firm foundation for dealing with the intricate issues in today's technology, significantly enhancing their ability to solve problems (Louis et al., 2020; Battiston et al., 2020).

### 6.3.3. Role within Communication Systems

In the study of communication systems, block diagrams and signal flow graphs play a key role in the representation and analysis of complex interactions in the systems. Block diagrams assist in the simplified graphical description of signal processing systems by dividing them into building blocks that are representative of various functional components. Such functional components are capable of representing various operations such as signal amplification through to filtering. This visual approach allows engineers to visualize signal flows and build a sense of system behavior, creating better designs and better troubleshooting.

Moreover, this model of representation becomes increasingly valuable as one models more extensive systems, where system performance and stability can be significantly impacted by the interaction of numerous blocks. Signal flow graphs complement block diagrams with a different perspective on the same processes, emphasizing the signal flow in a network of nodes and associated paths. The method not only facilitates signal propagation visualization but also assists in the mathematical formulation of intricate systems. By illustrating signals as arrows and system variables as nodes, signal flow graphs highlight feedback loops and interactions that are not necessarily apparent through block diagrams.

**Figure 6.8**. *Role of communication systems.*

*Source: https://www.freyrsolutions.com/sites/default/files/blog-thumbnails/What-is-the-role-of-communication-systems-in-healthcare.jpg.*

This ability in graphing intricate connections is especially important in communication systems where dynamics can become nonlinear and multi-dimensional. Investigation of such higher-order interactions enhances the potential for maximizing communication processes and alleviates problems arising from the interaction among multiple signal pathways and elements, as suggested by the importance of network interaction levels in Battiston et al. (2020). In addition, the integration of signal flow graphs and block diagrams is a fundamental element in the latest advances in communication systems, including the application of technologies like transformers to analyze and process signals. The evolutionary capacities of the models, especially their fitness for burdensome work, mark the shift towards increasingly sophisticated approaches in communication theory. Signal transformation over multiple steps, made possible by the block diagram illustrations, follows the upgraded understanding of higher-order interactions provided by multiple nodes in a network. With researchers exploring these holistic methods, as emerges through the observations of Han et al. (2022), the potential for new communication system techniques goes higher, and the door opens for developments that will be able to optimize performance and the overall system efficiency (Figure 6.8).

## 6.4. ADVANTAGES OF BLOCK DIAGRAMS

Block diagrams are a system analysis and engineering tool of great utility with the ability to create simplified representations of too complicated systems otherwise. One of the most important advantages of block diagrams is their ability to reduce intricate processes to easily understood components. This abstraction enables engineers and analysts to conceptualize system behavior without being burdened with excessive detail. For instance, in fields such as control systems and signal processing, these diagrams assist in ascertaining system dependencies and interconnections, allowing for easier analysis and optimization of performance. By utilizing block diagrams, practitioners are better able to make more effective problem-solving and decision-making, which assists in guaranteeing clearer system dynamics understanding and how changes could be implemented to create improvement or innovation. Another significant advantage of block diagrams is that they are useful for communication among team members and stakeholders. Their graphical depiction eliminates the language barrier and technical jargons, thus making it easier to comprehend primary concepts and system structures by individuals with different backgrounds. In team settings, block diagrams provide a means of communication and feedback, thus supporting teamwork. In addition, this clarity ensures learning experience among new entrants within a domain or project because they can intuitively follow the direction of the system and comprehend the interdependencies at play. This open-minded and collaborative approach is particularly relevant to multidisciplinary projects where members must communicate efficaciously in an effort to bridge knowledge gaps and expertise. Moreover, block diagrams can best introduce even more sophisticated analytical methods, such as signal flow graphs or transfer function analysis. The simple visual framework provided by block diagrams provides a foundation upon which further investigations could be built, for example, simulation of nonlinearities and time behavior as presented in the literature for traffic flow forecasting (Guo et al., 2019). Through this formal process, complexity may be added in incremental layers with organizational consistency intact, making it easier to make clearer and more focused analysis. This layered approach not only saves time but also enhances the overall form of system design and assessment, as important relationships and interdependencies are established at the beginning stages. In this way, block diagrams become not only diagrammatic tools but also a strategic advantage in addressing full and productive systems analysis.

### 6.4.1. Simplification of Complex Systems

Streamlining complex systems is a key operation to effectively analyze and understand their behavior. For systems theory and engineering, block diagrams and signal flow graphs are the primitive instruments for such an operation. Through graphically representing complicated relations and interactions, they simplify such details into understandable elements such that engineers and analysts can visualize system dynamics without being overwhelmed by complexity. By employing such modeling techniques, large-scale processes and interactions can be identified, thus facilitating easier communication and deeper insight. This simplification not just facilitates the design of efficient control systems but also the identification of defects in operating systems. Lastly, by employing a simplified approach using neat diagrams, learning and application are facilitated, allowing different players, from new to old engineers, to utilize the systems. Further research into simplification effects unveils its vital role with respect to system performance and system efficiency. In the course of designing, designers utilize block diagrams to represent input-output relationships, with signal flow graphs rendering a graphical representation of signal flow through systems. Such graphic transparency not only creates a better understanding of system interdependencies but also makes debugging a simpler task in case of fault occurrences. For instance, when studying feedback loops and system response, these diagrams point out significant paths that can be studied further or revised. In this way, using such illustrations facilitates systematic analytical thinking, allowing experts to enhance complex systems on an ongoing basis. Secondly, as computing methods continue to improve, the simplification cycle easily integrates with modern simulation tools, thereby increasing accuracy with reduced computing effort. Besides, the function of simplification is not restricted to engineering but is present in interdisciplinary areas where complex systems are even more prevalent. For instance, the emergence of powerful data structures in natural language processing like transformers verifies the necessity for simpler models that can handle huge volumes of data successfully, as suggested in a recent publication (Han et al., 2022). Similarly, just as block diagrams allow engineers to conceptualize intricate mechanical relationships, computer scientists use similar paradigms to simplify and study linguistic structures. This cross-disciplinary application illustrates how the principles of simplification can operate across disciplines, more effectively preparing us for encountering and negotiating complexity. As scholars continue to push frontiers in understanding phenomena, reliance on simplified frameworks remains paramount to fostering innovation and forward thinking theoretical refinement across disciplines (Figure 6.9).

**Figure 6.9**. *Simplifying complex.*

*Source: https://fastercapital.co/i/Benefits-of-multimedia-in-blog-adding-visual-interest-and-boosting-engagement--Simplifying-Complex-Information.webp.*

## 6.4.2. Enhanced Communication Among Engineers

The importance of increased communication between engineers cannot be emphasized sufficiently, particularly for complex tasks involving designing and implementing block diagrams and signal flow graphs. With engineers combining to design complex systems, sound communication is the backbone in achieving project success. This is particularly important in multidisciplinary projects where various engineers bring specialized knowledge. Effective communication not only enables sharing of ideas but also enables all employees to be on the same page when it comes to project requirements and objectives. Therefore, a reduction in misunderstandings can mean fewer errors and greater productivity, in itself illustrating the core importance of communication in engineering endeavors. Enhanced communication ability can also leverage modern technologies facilitating real-time data and project update sharing, promoting team work, and greater project performance. The last area, digital tools, and processes have transformed the field of engineering communication. Digital twins are one such facility allowing engineers to create virtual representations of physical systems, which enables more interactive interaction in the design process. This digital twinning idea encourages improved communication via a platform in which engineers are able to virtually visualize and manipulate intricate systems without the potential breakdown in communications that could result from conventional approaches. As noted, the multimodal system's function of monitoring and regulating complex interactions has the potential to significantly facilitate coordination among engineers, especially when engineers utilize block diagrams and signal flow graphs during system development. This cooperation between technology and communication not only resolves problems but also makes a process of efficient decision-making, hence making an integrated and harmonious engineering environment that reacts better to issues. Improved communication modalities are secondly responsible for effective management of projects via engineering disciplines. By combining innovative communication paradigms, one is able to have better feedback mechanisms through which engineers can assess the effectiveness of a project on a cyclical basis. By acknowledging challenges in real time, teams are able to adjust their strategies and take full advantage of analysis tools. For instance, application of signal flow graphs is able to demonstrate that communication involves various inputs and outputs within a system. The graphical depiction is able to assist engineers in explaining ideas vividly and correctly and minimize miscommunication. As highlighted through current studies, understanding the complicated character of how systems communicate both internally and externally is vital for successful projects and ties back to the requirement for co-engineering interactions and the probable positive impact of better communication on upper-level project objectives (Christian et al., 2020; Rasheed et al., 2020).

### 6.4.3. Facilitation of System Analysis

Effective system analysis is crucial in the design and optimization of sophisticated technical systems, particularly when they involve efficiency and accuracy. Block diagrams and signal flow graphs are fundamental tools applied in such a scenario and provide visual methods of describing systems in a way that can be easily analyzed and designed. Through breaking down systems into their constituent parts, these diagrams enable one to visualize interactions and control behavior. Especially in the case of modern architectures, for example, cloud-computing paradigm-based centralized designs, there is an increased need for crisp visualization. As systems get complicated, having the ability to investigate data streams and execution dependencies is highly necessary in order to gain optimal performance and reduce the likelihood of error (Enrici et al., 2019). In exploring the application of block diagrams and signal flow graphs to system analysis, one has to examine their application across various technologies. For example, iterative and worklist algorithms explained in recent studies have been successful in performing static data-flow analysis, particularly for complex applications like Software Defined Radio systems and 5G channel decoders. These advances call for the utilization of tools that can efficiently produce software tailored for the particular needs of contemporary hardware. By applying visual modeling techniques, engineers can better analyze system performance and refine processes, which ultimately leads to huge reductions on unnecessary visits to control-flow graphs (Enrici et al., 2019). The information gained from such analyses not only enhances understanding but also stimulates system architecture and function evolution. Furthermore, the role of computational techniques in system analysis cannot be overstated. As mentioned in the literature, recent versions of texts such as Automatic Control Systems contain modern applications of control theory in addition to novel pedagogical methods. These books emphasize the application of block diagrams and signal flow graphs in control system analysis and design in general. Through analytical methods applied to practical problems, students and professionals can access the entire potential of these diagrams in solving actual problems. Stability analysis and frequency-domain analysis are now often treated together so that students can develop a clearer picture of control systems analysis. Through the synching of theory and practice, these learning methods develop a generation of engineers poised to handle rich system dynamics in an increasingly networked technological world (Figure 6.10).

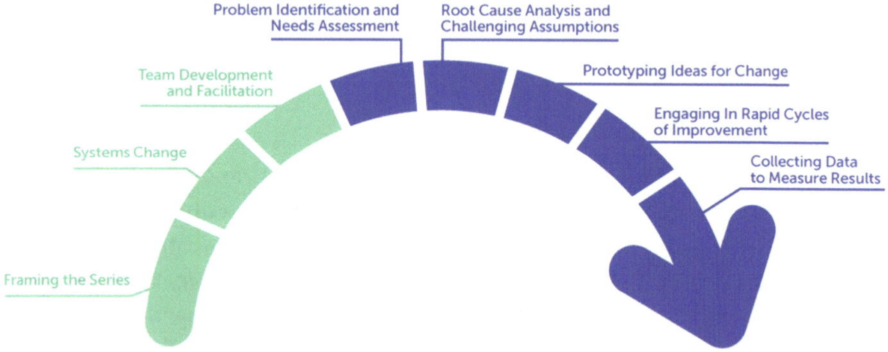


**Figure 6.10**. *Facilitation of system analysis.*

*Source: https://wested2024.s3.us-west-1.amazonaws.com/wp-content/uploads/2024/10/08131809/RelWest-Inquiry-Cycle-v2-2048x818.webp.*

## 6.5. ADVANTAGES OF SIGNAL FLOW GRAPHS

Signal flow graphs (SFGs) possess a number of advantages over traditional block diagrams, notably in terms of clarity and visualization of system behavior. Among the most compelling arguments for SFGs is the ability to convey complex interdependences among system variables in a straightforward manner. This visualization also makes it easier to inspect feedback loops and interactions and see how changes in one part of a system can affect others. By facilitating more intuitive understanding of dynamic systems, SFGs enable communication between team members to be more productive and presentations of system behavior easier to explain. The diminished representation level has a tendency to lead to quicker identification and rectification of issues, hence the acceleration of overall system analysis efficiency. Additionally, signal flow graphs promote systematicity to system modeling, particularly in linear time-invariant systems. In contrast to block diagrams that become haphazard and complex very quickly, SFGs are simpler to keep ordered by virtue of their new method of presenting nodes and connections. This property is very valuable while coding control systems or mimicking dynamic behavior because it minimizes the chance of missing something because of intricate visual representations. SFGs also facilitate the use of mathematical tools that simplify analysis, such as Masons Gain Formula. This equation, applied to computation of transfer functions from the graph, is especially beneficial for engineers who are eager to calculate required system parameters in an effective way, thereby once again projecting the role of SFGs in dynamic system analysis. In addition, utilization of signal flow graphs may lead to dramatic improvements in computational models and simulation techniques. With increasing reliance on automation in engineering use, SFGs can include algorithms that can specify complex systems with high accuracy and reliability. This is particularly crucial in fields where accuracy is critical in making decisions, such as traffic flow regulation and ecological observation. For example, in investigating dynamic traffic correlations in traffic data, techniques such as those outlined in more recent studies illustrate how attention mechanisms can be incorporated into graph-based models to gain insights into traffic patterns (Guo et al., 2019). In a similar vein, SFGs can be formulated for new paradigms in computation, making them more applicable in an ever-changing technological environment. Hence, the strengths of signal flow graphs exceed mere description and have impact on numerous applications in engineering practice.

### 6.5.1. Representation of Dynamic Systems

Dynamic system modeling lies at the basis of control theory and engineering to enable practitioners to model, analyze, and design complex systems economically. Along with other methods, block diagrams and signal flow graphs are excellent visual aids that simplify the dynamics of a system. System operations are encapsulated in block diagrams using linked blocks that symbolize various operations, and arrows show the direction of the movement of signals or information among these operations.

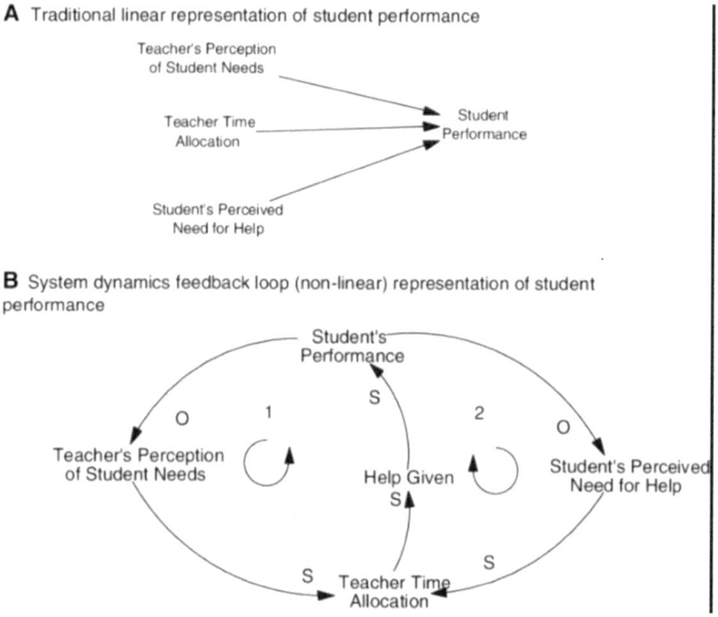


**Figure 6.11**. *Comparison of traditional and system dynamics representations.*

*Source: https://www.researchgate.net/profile/Gary-Hirsch/publication/6414923/figure/fig1/AS:601667925970979@1520460220694/Comparison-of-traditional-and-system-dynamics-representations-of-student-performance-A.png.*

This graphical representation simplifies the observation of system interdependencies and aid in troubleshooting and optimization work. As dynamic systems are represented, engineers can simplify complicated behavior into chunks that are easier to manage, thus making it more effective to come up with resilient solutions for real-world problems. Signal flow graphs constitute another form of representation that works well in tandem with linear algebra, emphasizing relationships and feedback paths in dynamic systems. Each node within a signal flow graph is a variable or signal, and directed edges are used to denote interactions and influences among variables. The use of such graphs is particularly useful when dealing with systems that involve both feedback and feedforward paths, allowing more complicated signal manipulation and combination to be achieved. The mathematical basis of signal flow graphs also facilitates the application of sophisticated computation techniques, such as Mason's Gain Formula, that can offer reliable information related to system behavior in various situations. Thus, signal flow graphs form a very versatile framework for comprehensive analysis of dynamic systems in academic and professional environments. Furthermore, the pairing of block diagrams and signal flow graphs can yield additional system dynamics insights through complementary strengths. While block diagrams can provide an intuitive sense of system structure, signal flow graphs can facilitate a deeper analysis of interactions and feedback loops. The combining of ideas formulated by both techniques provides a holistic view of system performance and allows for the application of efficient control methods. Such an integrative approach appeals to recent trends in engineering disciplines that employ computational tools to model and predict the behavior of dynamic systems in a broad array of applications—automations to aerospace engineering. The deeper researchers penetrate into the potential of such graphical representations, the more the paradigms for the design and comprehension of intricate systems become actualized, opening up doors to innovative solutions to existing challenges in science and technology (Figure 6.11).

### 6.5.2. Ease of Mathematical Manipulation

Mathematical manipulation within block diagrams and signal flow graphs plays an imperative role in reducing complex systems to simpler ones, enabling engineers to analyze and design control systems. The structured representation of signals and how they interact within the diagrams enables algebraic means to be employed. This role plays an imperative role when dealing with numerous components that interact with each other as it can transform complex equations into simple forms systematically. Through diagrammatic breaking down of system component interactions, engineers can readily visualize important parameters and interdependencies, which significantly contributes to the derivation of transfer functions and stability criteria interpretation. Moreover, the transparency provided through these graphical tools not only aids in better interpretation but also ensures precision in calculations so that errors that may originate from manual derivations are minimized to a mere minimum. Further, the ease of mathematical manipulation offered by block diagrams and signal flow graphs is enhanced by conventions used on them. Such conventions, for example, adding signal flow directions and block portrayal, provide an implied uniformity that facilitates the analysis process. Standard methods, like Masons Gain Formula, can be employed by engineers to derive important system behavior information without embarking on excessively intricate arithmetic.

The self-descriptive nature of the diagrams facilitates easy identification of signal routes and feedback loops, hence making the entire process more streamlined. This formal approach to visualization not only promotes efficient problem-solving but also facilitates teamwork since teams can easily communicate ideas and interpretations from a shared perception of the graphic representations. Apart from facilitating manipulation, block diagrams and signal flow graphs are also crucial for taking forward theoretical understanding of system dynamics. Their modularity itself allows one to study various configurations and modes of operation in a system through a simple change in the configuration of loops and blocks. Such sensitivity is particularly worth it's while in research and development environments, where rapid prototyping and simulation are priorities. Whenever engineers revise design, the diagrams serve as documentation and analysis aids that enable repeated learning and system improvement. As the computational equipment that employs these graphical methods becomes widespread, mathematical moves can be made with increasing efficiency and precision, advancing the boundaries of innovation in fields such as control theory and automation, as highlighted by some recent breakthrough in deep learning technologies (Choudhary et al., 2022) and the challenges posed by interpretability in complex systems (Tjoa et al., 2020) (Figure 6.12).

**Figure 6.12**. *Ease of mathematical manipulation.*

*Source: https://images.prismic.io/everway/Z7dplZ7c43Q3gB8K_equals-digital-solution.webp?auto=format,compress.*

### 6.5.3. Visualization of Feedback Loops

Visual representation of feedback loops is central to the dynamic behavior of systems represented by block diagrams and signal flow graphs. Feedback loops are circuits through which outputs get fed back to the system as inputs, ultimately influencing future outputs. Through this cyclic process, stabilizing and destabilizing effects can be introduced depending on the type of feedback.

Through the application of block diagrams, engineers and analysts can depict these relationships in a graphical form, thus easily identify points of inefficiency or instability in the system. By visualization, the complexity of feedback mechanisms is exposed, leading to insights that make possible the optimization of systems, troubleshooting, and designing improvements. Application of advanced techniques such as artificial intelligence and machine learning provides an added layer of sophistication when analyzing feedback loops. These technologies are complemented by their application to more sophisticated visualization methods that are able to simulate complex variables and predict system response under various conditions. For instance, recent advances in deep learning can be applied to analyze big data to detect subtle feedback patterns not easily identified with common analysis methods. Since the ability of deep learnings to process unstructured data was highlighted as a priority in current studies, its application for optimizing the presentation of feedback in block diagrams and signal flow graphs, particularly in such areas of complexity as safety and construction management, where feedback loops play a life-or-death role in project success (Baduge et al., 2022). In addition, with the introduction of big materials databases and high-technology visualization tools, development has been accelerated in understanding and controlling feedback loops in engineered systems. The databases allow for the exploration of atomistic predictions and other applications that are facilitated by detailed feedback analysis. The novel use of deep learning techniques to such materials data allows automation of feature extraction and pattern recognition, thus enhancing the quality of visual portrayal (Choudhary et al., 2022). As these new techniques become combined with conventional techniques in block diagrams and signal flow graphs, the possibility of system optimization through enhanced feedback visualization is not only highly promising but also imperative to satisfy current and future needs in all fields of engineering and the sciences.

## 6.6. LIMITATIONS AND CHALLENGES

The application of block diagrams and signal flow graphs brings with it a multitude of challenges and limitations to efficiently be applied in the analysis of complicated systems. One of the primary issues is that such diagrams cannot easily represent complicated systems, particularly those with many interconnected components. As the size of the diagrams increases, they tend to be very simplistic and will sacrifice valuable information required in proper analysis. As such, system designers and engineers typically struggle to achieve a balanced point between detail and clarity since each modification to a block diagram or signal flow graph will have a dramatic impact on what is known about system behavior. This challenge is amplified in todays' applications demanding complex modeling, where a simple visual representation fails to capture temporal dynamics or state transitions of the utmost importance, resulting in an inferior design process and decision-making. In addition to this, the application of software packages to aid generation and analysis of block diagrams and signal flow graphs has inherent challenges. The majority of available software packages are discovered to possess weaknesses, including exorbitant charges for licensing and a requirement for considerable staff training, serving as a barrier to widespread use in both education and professional setups. Moreover, the incompatibility issue can be triggered because expert software is often incompatible with existing reliability analysis software, hence forming knowledge silos and data silos. The barriers can have an effective workflow, where the users often switch between different platforms to assess system performance and reliability. Resolving such issues is crucial since effective software solutions are required to facilitate correct model creation and enhanced coordination within system engineering. Also, the dependability of such complex systems modelled by such diagrams is largely in doubt. Because the models have thousands of elements—each having multiple states of operation—there is intricacy in getting things right while defining reliability indicators. And those are matters pertaining to today's software systems, reliability block diagrams can depict huge areas of states, all for different operational scenarios and failures. The computers developed to calculate these quantities often lack usability and computation capabilities, as most legacy tools tend to present old-fashioned interfaces and narrow functionalities (Seniv et al., 2024). The creation of better systems that utilize sophisticated algorithms can counter such issues, resulting in higher speed and accuracy levels in reliability estimates. Lastly, overcoming these challenges and weaknesses is necessary for the effective use of block diagrams and signal flow graphs in engineering work today.

## 6.6.1. Complexity in Large Systems

Understanding the intricacies of complex systems needs a robust methodology for dealing with complexity, particularly when employing block diagrams and signal flow graphs. They are powerful graphical aids which enable engineers and system designers to dissect intractable interactions into manageable chunks. As systems become larger, interpiece dependencies can expand exponentially, leading to design and analysis issues. Block diagrams simplify the task by abstracting the functions and interrelations between system functions, while signal flow graphs provide an insight into data and signal processes. Such structuring is invaluable in the prevention of risks associated with high complexities, particularly in fields where performance and safety are of the utmost concern. By the advanced case modeling, techniques that integrate tools like Information Flow Hierarchical Dynamic (IFHD) and Inhibitor arcs Hierarchical Colored Petri Nets (IHCPN) illustrate how complexity can be dealt with systematically. The techniques not only ease the modeling of complex systems but also enhance validity and safety during design. For instance, with a Unified Modeling Language (UML) model, designers can derive a functional model of the system and subsequently simplify it using transformation rules that minimize model complexities. The efficacy of this strategy has been proved notable; simulations indicate that such models fulfill important performance demands and conserve development time and resources, thereby enabling engineers to focus on higher-order system design rather than being overwhelmed by minute details (Zhao et al., 2020). The past of system architecture, particularly in fields like telecommunications, places emphasis on managing complexity effectively. In consideration of increasing requirements for increased performance attributes—such as minimal latency and maximum throughput—designers are required to move towards centralized designs based on cloud-computing models. These developments downgrade the role of digital signal processing within various units of hardware, making optimized model compilers necessary in analyzing signals efficiently. Improved interactions among system components are facilitated through static data-flow analysis techniques, especially in UML Activity and SysML Block diagrams. The application of blocking algorithms has demonstrated stunning reductions in inconsequential visits to models, which ultimately led to a better method of handling the intricacy of big systems (Enrici et al., 2019). With the implementation of such methodologies, engineers can enhance both the reliability and the operating efficiency of complex systems (Figure 6.13).

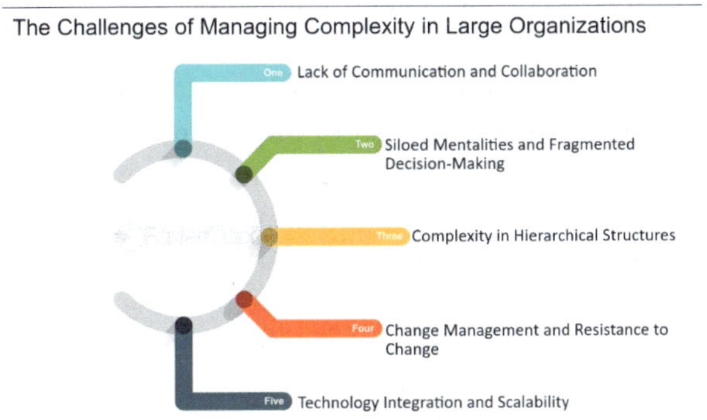


**Figure 6.13**. *Managing complexity: Balancing asset size and organizational efficiency.*

*Source: https://fastercapital.co/i/Managing-Complexity--Balancing-Asset-Size-and-Organizational-Efficiency--The-Challenges-of-Managing-Complexity-in-Large-Organizations.webp.*

### 6.6.2. Potential for Misinterpretation

The overlap between block diagrams and signal flow graphs in engineering is a set of possibilities for analysis and design; however, this detail is also a significant source of potential misinterpretation. These graphical tools are invaluable to the process of systems visualization, particularly in complex fields such as signal processing and control theory. When the stakeholders, including the non-technical stakeholders and engineers, look at these diagrams, varying levels of comprehension might lead to varying interpretations. An elementary block diagram, for instance, might hide required dynamics, leading one faction to rule out required feedback loops or extend system behavior assumptions. It is therefore very important to have a shared understanding, particularly in cross-disciplinary environments where miscommunication could threaten the integrity and results of projects.

Furthermore, the inherent abstraction in block diagrams and signal flow graphs can reinforce assumptions that might not necessarily be true to actual conditions. Although these technologies are designed to simplify complex systems into forms that are more manageable, their representational characteristics can occasionally provide an illusory sense of simplicity. This can make it more difficult in areas that demand strict interpretation, such as automated systems and medical technology. As depicted by current advances in artificial intelligence, the need for transparency and accountability has become increasingly critical; misinterpretation not only hinders technical accuracy but can also have significant ramifications in real applications (Tjoa et al., 2020). Where good decision-making hinges on accurate understanding, misinterpretation can have significant reach, so plain communication and sound education systems are all the more essential. Lastly, the danger of misinterpretation is also heightened by the limitations inherent in the visual presentations themselves. Block diagrams and signal flow graphs, while helpful at portraying relationships, are susceptible to a paradigm that conceals underlying complexity. For example, when designing solid-state UV emitters, the complexity of their structure and working problems cannot always be transferred to the graphical models used in their simulation (Amano et al., 2020). Since these diagrams are being referred to by inexperienced and experienced practitioners both, the risk of forgetting such crucial aspects, such as thermal stability or material limitation, is significantly raised. Thus, it is required that visualizations be complemented with rigorous contextual analysis so that the users are not lost in the critical perspective of information displayed. This itself not only avoids misinterpretation but also provides a more complete image of the involved systems, thereby enhancing the educational value of block diagrams and signal flow graphs (Figure 6.14).

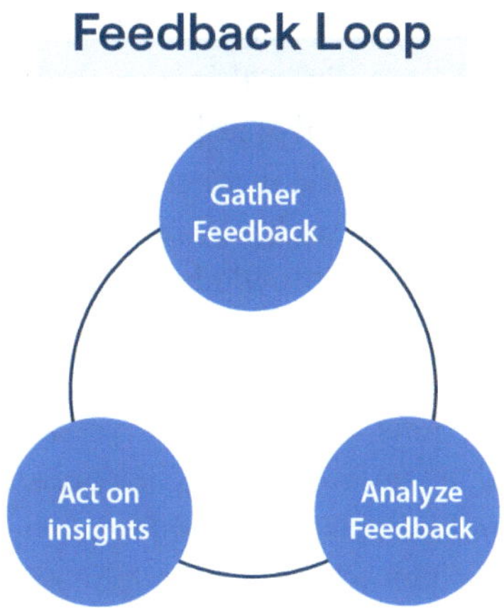


**Figure 6.14**. *Feedback loop.*

*Source: https://cdn.botpenguin.com/assets/website/Feedback_Loop_60dcf71d12.png.*

## 6.6.3. Limitations in Nonlinear Systems

Nonlinear systems present great challenges in control theory primarily due to their complex dynamics and behavior. Unlike linear systems, which have expected patterns and can be studied using established techniques such as block diagrams and signal flow graphs, nonlinear systems often show phenomena such as multiple equilibria, limit cycles, and chaotic behavior. These complexities rule out the application of traditional linear approaches, resulting in difficulty in stability analysis and system response prediction. Furthermore, complexities that accompany nonlinear relationships imply that solutions to problems are not necessarily feasible in closed form, and numerical methods which may not only be computationally intensive but also sensitive to initial conditions, thus providing uncertainty in system outputs. Also, constraints of nonlinear systems are compounded by issues which these create in practical applications, where disturbances and external inputs could lead to unpredictable outcomes. A very severe point is that these systems are difficult to model accurately, since even minor inaccuracies in depiction of nonlinearity can have a profound impact on performance metrics. For instance, in engineering control systems, the presence of nonlinearities such as friction or saturation would lead to system failure or degradation if not taken into account. This becomes a significant consideration in fields such as robotics and aeronautics where performance is coupled with safety. Therefore, designers resort to simplifying linear approximations, thus making concessions on system representation for the sake of ease of implementation and design. Apart from this, the analytic methods applied in order to study nonlinear systems are also at least intuitive calling for advanced understanding of advanced mathematical concepts, and hence beyond the reach of many practitioners. Techniques such as Lyapunov methods or bifurcation analysis are those that require deep knowledge and consciousness of complex mathematical frameworks that are definitely not within easy reach. With the improvement in nonlinear dynamics research, researchers are increasingly probing higher-order interactions within these systems, possibly providing a way of understanding their behavior. In fact, it was stated that it is possible to enhance predictability and increase the comprehension of complex systems beyond the simplest pairwise interactions by modeling higher-order interactions (Battiston et al., 2020). These advances can eventually lead to the building of robust mechanisms to account for nonlinear systems' inherent limitations, thus enabling more effective applications across many fields.

# SUMMARY

In summarizing the extensive discussion of block diagrams and signal flow graphs, it is clear that these tools are not merely structural diagrams but working tools in modern signal processing. Their ability to divide complex systems into bite-sized visual forms enables engineers to study and design complex systems better in various applications. The use of block diagrams, particularly when augmented with advanced techniques such as hierarchical modeling, underscores the part that structure plays towards achieving operating effectiveness. For instance, by applying modeling techniques such as the combination of Information Flow Hierarchical Dynamic with Inhibitor arcs Hierarchical Colored Petri Nets, complexity is reduced significantly without compromising system performance and reliability, as cited in Zhao et al. (2020). In addition, the new trend of technologies and increasing demands for efficiency make the importance of adjusting block diagrams and signal flow graphs according to the recent breakthroughs more obvious.

The new trend of further system centralization and increased use of cloud computing infrastructures result in more powerful modeling needs to achieve optimal performance. The trend towards digital signal processing calls for not only software adjustment but also reengineering traditional modeling methods. Progress in computation techniques allows for more precise and agile analysis, allowing systems to address throughput and latency requirements. The modifications, as outlined in Enrici et al. (2019), allow for a good integration between data flow and engineers, making possible a better design and implementation of the system. Finally, results from the study of block diagrams and signal flow graphs confirm their pivotal position in mapping the path of advanced engineering systems. The discipline's ongoing evolution, embracing revolutionary modeling paradigms, can generate material performance improvements, reducing development time and cost and enhancing system performance. Ongoing research should continue to explore such methods with the aim of integrating emerging technologies and practices. So, block diagrams and signal flow graphs remain firmly relevant, secured firmly on solid foundations, sustaining essential engineering progress that relies on precision and adaptability in a fast-changing world.

The employment of block diagrams and signal flow graphs is essential in control systems engineering as a compulsory visual tool in understanding complex system interactions. Block diagrams facilitate the description of dynamic systems through the illustration of input-output signal relationship, system components, and feedback loops. In this graphical representation, it becomes easy to visualize information flow and what operations are executed by individual subsystems. Each block represents a specific function or operation, and there is the use of connections to indicate signal processing order. The abstraction is critical, particularly for engineers and researchers designing and analyzing control systems, as it encapsulates complex mathematical models in an easily comprehensible manner, making it easier to understand system behavior and interdependencies, which is especially critical in applications ranging from robotics to industrial automation (Syed, 2021). As opposed to block diagrams, signal flow graphs are more informative in the depiction of system dynamics with emphasis on direction of signal flow from one node to another as well as from one branch to another.

This facilitates precise portrayal of the interactions among system variables with utility in higher-end analysis such as application of Masons Gain Formula. Here, signal flow graphs provide a more precise analysis of system behavior by offering a chance to study the impact of each variable on the system response. The ability to represent recursive relationships and feedback loops within a signal flow graph provides better insight into system interactions. Along with it, the practice is complemented by mathematical modeling techniques that are a cornerstone of control system design, and this enhances the ability to predict the system's behavior under varied inputs and constraints (Syed, 2021).

Control systems engineering utilization of block diagrams and signal flow graphs is not merely beneficial; it constitutes a fundamental practice that shapes engineers' mode of thinking and problem-solving. By applying such tools, the analyzer and designer can enhance the design process and analysis with precision in modeling the system. The evolution of control system methodologies is also a testimony to their significance since they react to emerging technologies such as intelligent and automated systems. The close understanding that is achievable by using these graphical models combined with robust mathematical constructs makes it easier to design more sophisticated control strategies. Collectively, signal flow graphs and block diagrams not only allow for theoretical understanding but also provide engineers with the skills needed for application to actual control system problems, making them precious in a professional as well as an educational environment (Kuo, 1962).

## REFERENCES


1. Africa, A. (2019). Linear system interconnections, steady-state analysis and stability theory. *International Journal of Advanced Trends in Computer Science and Engineering*. https://www.semanticscholar.org/paper/b9941f9eedf6fb578cc661e74f9995052bf0b162 (accessed on 19th September 2025).
2. Amano, H., Collazo, R., De Santi, C., Einfeldt, S., Funato, M., Glaab, J., Hagedorn, S., et al. (2020). The 2020 UV emitter roadmap. *Journal of Physics D: Applied Physics, 53*, 503001–503001. https://doi.org/10.1088/1361-6463/aba64c (accessed on 19th September 2025).
3. Baduge, S. K., Thilakarathna, P. S. M., Perera, J. S., Arashpour, M., Sharafi, P., Teodosio, B., Shringi, A., et al. (2022). Artificial intelligence and smart vision for building and construction 4.0: Machine and deep learning methods and applications. Automation in Construction, 141, 104440–104440. https://doi.org/10.1016/j.autcon.2022.104440 (accessed on 19th September 2025).
4. Battiston, F., Cencetti, G., Iacopini, I., Latora, V., Lucas, M., Patania, A., Young, J. G., et al. (2020). Networks beyond pairwise interactions: Structure and dynamics. *Physics Reports, 874*, 1–92. https://doi.org/10.1016/j.physrep.2020.05.004 (accessed on 19th September 2025).
5. Choudhary, K., DeCost, B., Chen, C., Jain, A., Tavazza, F., Cohn, R., Park, C. W., et al. (2022). Recent advances and applications of deep learning methods in materials science. *npj Computational Materials, 8*. https://doi.org/10.1038/s41524-022-00734-6 (accessed on 19th September 2025).
6. Concha Sánchez, A., & Gadi, S. K. (2025). UYAMAK: Empowering model-based design and simulation through an open-source web platform. *Computer Applications in Engineering Education, 33*. https://www.semanticscholar.org/paper/3561ae41af9cf28054279d44fc9c80f2901bc578 (accessed on 19th September 2025).
7. Enrici, A., Apvrille, L., & Pacalet, R. (2019). Efficient data-flow analysis of UML/SysML diagrams for optimized model compilation of hardware-software systems. https://www.semanticscholar.org/paper/daf52da48138f44f7dd794c6870adf0e29344be1 (accessed on 19th September 2025).
8. Ghashghai, M., & Ghaznavi-Ghoushchi, M. (2024). Design and analysis of a new three-stage feedback amplifier utilizing signal flow graph domain inspection approach. *IEEE Transactions on Very Large-Scale Integration (VLSI) Systems, 32*, 1792–1800. https://www.semanticscholar.org/pa

per/0d80cb74bdb2e9f96feb693bfcf7e928e120ce77 (accessed on 19th September 2025).

9. Guo, S., Lin, Y., Feng, N., Song, C., & Wan, H. (2019). Attention based spatial-temporal graph convolutional networks for traffic flow forecasting. Proceedings of the AAAI Conference on Artificial Intelligence, 33, 922–929. https://doi.org/10.1609/aaai.v33i01.3301922 (accessed on 19th September 2025).

10. Han, K., Wang, Y., Chen, H., Chen, X., Guo, J., Liu, Z., Tang, Y., et al. (2022). A survey on vision transformer. *IEEE Transactions on Pattern Analysis and Machine Intelligence, 45*, 87–110. https://doi.org/10.1109/tpami.2022.3152247 (accessed on 19th September 2025).

11. Jackson, L. (1985). *Digital Filters and Signal Processing*. https://www.semanticscholar.org/paper/a0694336f38f5ea52510fee70668773314d7f468 (accessed on 19th September 2025).

12. Kautz, H. (2022). The third AI summer: AAAI Robert S. Engel more Memorial Lecture. *AI Magazine, 43*, 105–125. https://doi.org/10.1002/aaai.12036 (accessed on 19th September 2025).

13. Khan, S., Naseer, M., Hayat, M., Zamir, S. W., Khan, F. S., & Shah, M. (2022). Transformers in vision: A survey. ACM Computing Surveys, 54, 1–41. https://doi.org/10.1145/3505244 (accessed on 19th September 2025).

14. Kuo, B. (1962). Automatic control systems. *Universe: Technical Sciences*. https://www.semanticscholar.org/paper/5da18bdfdf68386c583b759dbcb2479ee6d06306 (accessed on 19th September 2025).

15. Leslie, H. A., van Velzen, M. J. M., Brandsma, S. H., Vethaak, D., García-Vallejo, J. J., & Lamoree, M. H. (2022). Discovery and quantification of plastic particle pollution in human blood. *Environment International, 163*, 107199–107199. https://doi.org/10.1016/j.envint.2022.107199 (accessed on 19th September 2025).

16. Mason, G., Shuman, T., & Cook, K. (2020). Inverting (flipping) classrooms–Advantages and challenges. https://doi.org/10.18260/1-2--19842 (accessed on 19th September 2025).

17. Minaee, S., Boykov, Y., Porikli, F., Plaza, A., Kehtarnavaz, N., & Terzopoulos, D. (2021). Image segmentation using deep learning: A survey. IEEE Transactions on Pattern Analysis and Machine Intelligence. https://doi.org/10.1109/tpami.2021.3059968 (accessed on 19th September 2025).

18. Mircea-George, P., & Pungila, C. P. (2024). Control flow graphs against malware: Methods of analysis and detection. 2024 26th International Symposium on Symbolic and Numeric Algorithms for Scientific Computing (SYNASC). https://www.semanticscholar.org/paper/a3485540e43b88710ff159620b986b8e52a8fdb0 (accessed on 19th September 2025).

19. Rasheed, A., San, O., & Kvamsdal, T. (2020). Digital twin: Values, challenges and enablers from a modeling perspective. *IEEE Access, 8*, 21980–22012. https://doi.org/10.1109/access.2020.2970143 (accessed on 19th September 2025).

20. Rudin, C., Chen, C., Chen, Z., Huang, H., Semenova, L., & Zhong, C. (2022). Interpretable machine learning: Fundamental principles and 10 grand challenges. *Statistics Surveys, 16*. https://doi.org/10.1214/21-ss133 (accessed on 19th September 2025).

21. Scheffer, L. K., Xu, C. S., Januszewski, M., Lu, Z., Takemura, S., Hayworth, K. J., Huang, G. B., et al. (2020). A connectome and analysis of the adult *Drosophila* central brain. *eLife, 9*. https://doi.org/10.7554/elife.57443 (accessed on 19th September 2025).

22. Schürch, C. M., Bhate, S. S., Barlow, G. L., Phillips, D. J., Noti, L., Zlobec, I., Chu, P., et al. (2020). Coordinated cellular neighborhoods orchestrate antitumoral immunity at the colorectal cancer invasive front. *Cell, 182*, 1341–1359.e19. https://doi.org/10.1016/j.cell.2020.07.005 (accessed on 19th September 2025).

23. Syed, S. (2021). Comparative analysis of CNN & RNN for voice pathology detection (Preprint). https://www.semanticscholar.org/paper/7af371b6637518491d3a2af6475e3880bcb569a2 (accessed on 19th September 2025).

24. Tjoa, E., & Guan, C. (2020). A survey on explainable artificial intelligence (XAI): Toward medical XAI. *IEEE Transactions on Neural Networks and Learning Systems, 32*, 4793–4813. https://doi.org/10.1109/tnnls.2020.3027314 (accessed on 19th September 2025).

25. Torres-Torriti, M., & Rojas-Sepúlveda, M. (2024). DiaBloS: Diagrams of blocks for systems modeling and simulation in Python. 2024 13th International Workshop on Robot Motion and Control (RoMoCo). https://www.semanticscholar.org/paper/7c35efee214133beefaae113cc8266e1f658419c (accessed on 19th September 2025).

26. von Rueden, L., Mayer, S., Beckh, K., Georgiev, B., Giesselbach, S., Heese, R., Kirsch, B., et al. (2021). Informed machine learning–A taxonomy and survey of integrating prior knowledge into learning

systems. *IEEE Transactions on Knowledge and Data Engineering*. https://doi.org/10.1109/tkde.2021.3079836 (accessed on 19th September 2025).

27. Zhang, Z., Yang, H., Li, J., Chong, S., Eshraghian, J. K., Yong, K. T., Vigolo, D., et al. (2024). Neuromorphic imaging cytometry on human blood cells. Neuromorphic Computing and Engineering, 5. https://www.semanticscholar.org/paper/0e066961 9a4ed1f756e64c19a5601807cdb5c185 (accessed on 19th September 2025).

28. Zhao, S., Li, J., Tan, M., Xu, C., & Cheng, X. (2020). Modeling and simulation of signal acquisition system based on inhibitor arcs hierarchical colored Petri nets: Taking dust signal acquisition system as an example. 2020 IEEE 9th Data Driven Control and Learning Systems Conference (DDCLS). https://www.semanticscholar.org/paper/7c964b4ef8934d0f3824376305eee6eae9032d34 (accessed on 19th September 2025).

# CHAPTER 7

# FEEDBACK ANALYSIS

## Contents



## INTRODUCTION

In control systems, the nuances of feedback mechanisms have a determinant impact on system performance and stability. Placement of feedback analysis within linear control frameworks provides a foundation for understanding system reaction to variations in inputs and outputs. Conventional control processes have a tendency to rely on pre-programmed algorithms, but feedback allows systems to dynamically adapt based on real-time knowledge, encouraging sensitivity and accuracy. This skill is particularly important in modern uses of engineering, where precision and reliability are extremely vital. The use of feedback principles encourages a deeper comprehension of system performance, enabling engineers to devise improved control measures that not only deal with current operating conditions but also anticipate future issues. Researching the principles of feedback in linear control systems involves analyzing various methods of simulation. One of the primary methods is inverse simulation that tries to retrieve input variables that may be used to achieve desired outputs within a given model. Classical inverse simulation approaches have long been criticized on grounds of computational expense, thus posing problems in high-speed applications. As noted by recent studies, other methods using feedback concepts provide a more efficient framework for nonlinear and multi-input multi-output models. \

These approaches, which are based on a feedback loop, generate systems that can invert indefinitely, allowing for higher robustness to uncertainties and disturbances. The emphasis on feedback, therefore, not only makes it easier to simulate the process but also the reliability of control systems when they encounter real-world applications (Anonymous et al., 2011). Further, the meshing of feedback analysis with advanced control techniques holds out great research and development avenues. The feedback principles are not only to be extended to classical linear models but also to encompass new emerging fields, such as quantum control. With the investigation of well-studied theories such as closed-loop learning control and feedback stabilization, researchers are able to discover new methods specifically tailored for complex dynamical systems. This study highlights the demand for robust methodologies that can cope with system uncertainties while meeting performance specifications. Through the use of the innovation in feedback methodologies, researchers can take advantage of the theory foundations of control systems to cater to contemporary issues and, in the process, advance the development of more efficient and resilient linear control systems (Belavkin et al., 2010) (Figure 7.1).

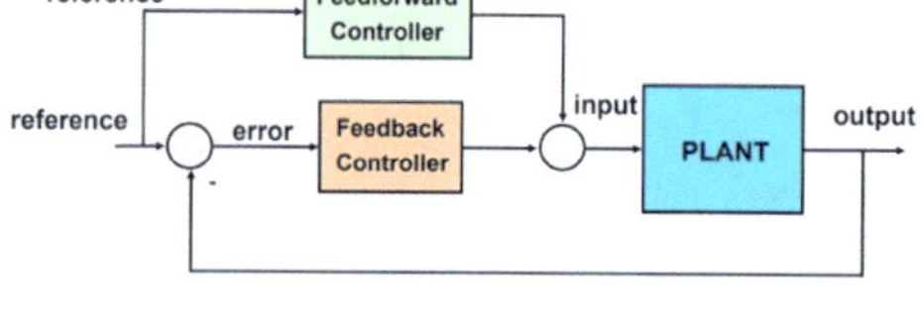


**Figure 7.1**. *Feedback control system.*

*Source: https://cdn.automationforum.co/uploads/2021/05/Untitled-26.jpg.*

## *Definition of Linear Control Systems*

A linear control system in the field of control theory is characterized by its adherence to principles of proportionality and superposition. Linear control systems obey linear differential equations where the output is a linear combination of input. The primary virtue of linear control systems is that they are stable and predictable and therefore can be easily designed by engineers based on an efficacious understanding of system behavior. The input-output relationship can often be characterized by transfer functions that provide a compact description of dynamic behaviors. The linearity inherent in the models enables classical control techniques to be employed since these types of systems can be thoroughly analyzed using techniques such as root locus and frequency response analysis, paving the way for more complex feedback mechanisms, which are pivotal in sophisticated control applications. Furthermore, the effectiveness of linear control systems also applies in their utilization in multi-input multi-output (MIMO) systems where feedback becomes paramount in enhancing performance and stability.

Recent studies have brought forth innovative definitions of phase response in MIMO linear time-invariant systems and significantly developed the theory of system stability under feedback. These developments are crucial as they allow for the development of new theorems, for example, the small phase theorem for feedback stability, that closely resemble established stability principles like the small gain theorem. These developments in theory are crucial for engineers to be capable of designing methodologies that not only predict the behavior of a system accurately but also allow fine tuning of the loops of feedback in systems with complication, thus optimizing overall performance and disturbance resistance (Chen et al., 2019).

Input-to-state stability is another crucial concept which arises in the investigation of linear control systems, particularly when considering feedback dynamics. This principle is worth its weight in gold when it comes to maintaining the desired performance level of the system in spite of disturbances. With the design of controllers such as linear DC motors by projective linear quadratic servo feedback, interaction between the input states and the controller assumes prominence. An understanding of what disturbance torques do to system stability requires a robust analytical framework that synthesizes both theory and practice implications. Through the use of input-to-state stability theory, engineers are able to properly describe the way disturbances affect system performance and guarantee that linear control systems operate accurately across different conditions. They are further confirmed by simulations, and hence the added practical element that links theoretical analysis with practical application (Doruk et al., 2016).

## Importance of Feedback in Control Systems

The value of feedback in control systems basically arises from the ability of feedback to enhance the performance and stability of a system. Feedback systems allow a system to self-correct and adjust its output to balance desired and actual performance disparities. It is imperative to maintain equilibrium in most engineering and technological processes, including robotics and automation manufacturing. Since the system responds to disturbance and changes, the application of feedback loops can dramatically enhance response time and precision to ensure the system operates optimally under varying circumstances. Especially in situations where uncertainty or environmental disturbances prevail, feedback is used as a stabilizing influence in a way that the system adjusts to accommodate its defined target states, which in turn ensures improved operation reliability. Feedback, in linear control systems, is crucial in implementing optimal control by regulating system dynamics. A linear model without feedback, for example, may be at steady state but will do so in a manner that it is not well suited for quick responses to disturbances or change in the load. Having mechanisms of negative feedback, such as in some kinetic models, enables the system to be more responsive. The mechanism by which an element, such as a protein, inhibits its own production is a fitting analogy. Through direct interaction with its product, the feedback mechanism facilitates a metabolic efficiency impossible without this regulation. The application of feedback, thus, brings an order of control at a high level that enhances their resilience and overall response, something required to enhance contemporary engineering practice (Kellman et al., 2018). Not just essential for stability purposes, feedback analysis also enables sophisticated simulations and modeling in dynamic settings. The concept of inverse simulation is one such best example of such importance as it utilizes feedback principles in deriving actual input histories from a known set of output responses. The methodology based on feedback makes simulation a simpler task while ensuring computational needs are eased and precision is maximized. As systems grow more complex, particularly in non-linear and multiple-input multiple-output systems, the feedback loop has a foundational contribution in attaining efficient control structures. The ability to iteratively adjust via feedback enables successful navigation within the complexities within dynamic systems. Hence, identifying and utilizing feedback principles is crucial for researchers and engineers towards achieving innovation and efficiency in complex control systems (Anonymous et al., 2011).

## Overview of Feedback Analysis

Feedback analysis in linear control systems is a fundamental component utilized to enable system performance and stability. Under mechanisms of feedback, engineers can observe the output of a system and adjust inputs based on the output in an attempt to reach required performance parameters. Feedback is a vital component in the control of dynamic systems whereby there are numerous variables to influence outcomes. Feedback analysis allows real-time correction of disturbances in a system, enabling the system to adapt to unforeseen changes. A testament to its importance is the development of such devices as the Disturbance Observer that proves the effectiveness of feedback in adjusting the effect of disturbances on a system's performance. Current attempts at making feedback analysis commonplace bear testament to its crucial application in engineering disciplines. Institutions like the California Institute of Technology have recognized the necessity of a two-track curriculum for control theory, dealing with conceptual understanding and analytical sophistication. When using a conceptual track, students from various science backgrounds are able to absorb the fundamental principles of feedback analysis without being weighed down by mathematical complexity. This approach not only enhances the applicability of control theory but also encourages interdisciplinary communication, with students having the option to learn from physical and biological sciences examples to make feedback principles more understandable. Better education of the framework for feedback analysis prepares future engineers for the tools necessary to make systems effectively designed and evaluated. With good control methods coupled with feedback analysis, linear control techniques are developed. Books such as the comprehensive review of Disturbance Observer-based robust control exhibit the progress in the tools in the last decades. With the use of unifying frameworks for analysis and synthesis, engineers can precisely apply the concepts to linear and nonlinear systems. The progress has serious implications for applications in real life, allowing engineers to stabilize the system and minimize disturbances in real-time applications. Additionally, such common knowledge begotten by these advances promotes a spirit of cooperation among the fraternity of engineers, one that inspires innovations that transcend traditional boundaries and broadens the reliability and performance of control schemes. On this basis, an introduction to feedback analysis not only becomes crucial to highlight its theoretical significance but also its key role in practical engineering solutions.

## 7.1. FUNDAMENTALS OF LINEAR CONTROL THEORY

Understandings of linear control theory are a prerequisite to the design and analysis of control systems, particularly in applications where performance and stability must be maintained. This theoretical point of view relies on the application of linear differential equations to model dynamic systems so that engineers can predict system response to various input conditions. The fundamental concepts are system representation both in state-space and frequency-domain domains, which plays a role in system dynamics and stability analysis. With the use of principles such as superposition and linearity, analysts can derive dominant performance parameters, which form the crux of effective feedback analysis. Such a basic understanding is most critical, as it serves to be the groundwork for more sophisticated control techniques capable of handling real-life problems associated with engineering applications. The core of linear control theory is the design and implementation of feedback mechanisms that directly influence the system response. Feedback is a corrective input that compensates for system outputs in the form of a difference between desired and actual performance.

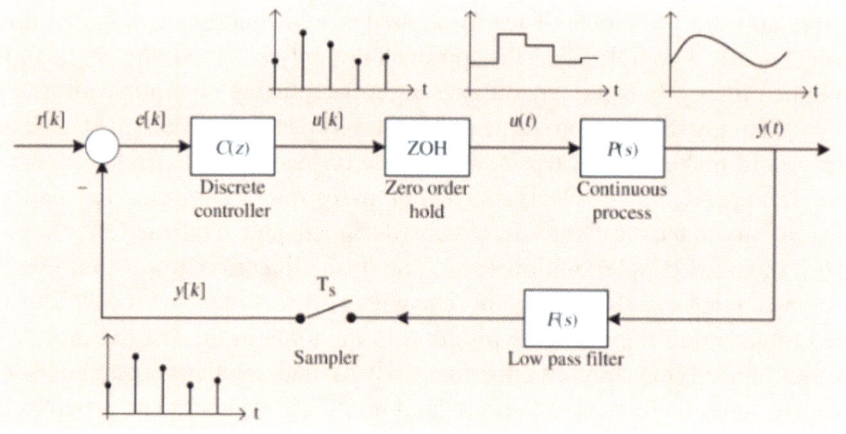


**Figure 7.2**. *Linear control theory.*

*Source: https://media.springernature.com/lw685/springer-static/image/chp%3A10.1007%2F978-3-030-96729-1_6/MediaObjects/128577_2_En_6_Fig6_HTML.png.*

This is best described with the assistance of Proportional-Integral-Derivative (PID) controllers that illustrate the manner in which feedback loops can mold system behavior. Specifically, while classical PID controllers are effective, they can be beset by overshoot and stability in controlling sudden changes in input. Literature indicates that the use of hybrid control methods such as the PI + CI controller can significantly enhance performance through resistance to overshoot and rapid stabilization of the system. As shown in recent research, the approach has been promising when applied in power converters, highlighting the significance of feedback analysis in control system optimization (Baños et al., 2019). Furthermore, practical applications of linear control theory cross into other engineering disciplines, particularly automation and robotics. Appropriate feedback analysis allows for improved command and control functionality, which is required for the operation of contemporary automated systems. For instance, in robotics, actuators must be tightly controlled to get desired motions, where a strong feedback mechanism is needed. Linear control theory-based methods can be modified to control trajectories and rectify errors in real-time applications. Research and development in this field continuously stress the inclusion of advanced control techniques in sync with evolving technology. Curiously, the faster evolution of computational hardware enhances simulation and modeling of complicated dynamics such that more sophisticated feedback strategies become possible in practical applications. Such advances harmonize with the ongoing relevance of fundamental principles that underlie linear control theory justifying their integrated place in defining the innovation of modern engineering solutions (N/A, 2016) (Figure 7.2).

## 7.1.1. Key Concepts in Control Theory

The control theory is based on certain basic concepts that are critical for the understanding of system dynamics of behavior. Stability is one of the most significant among them, meaning the ability of a system to come back to equilibrium after being disturbed. Stability is usually categorized into forms such as asymptotic stability, where trajectories become equally distant from an equilibrium point, and Lyapunov stability, which utilizes Lyapunov's direct method for assessing the stability conditions of nonlinear systems. Such distinctions help engineers create systems that are capable of maintaining performance regardless of variations in input or environmental conditions. The stability of value does extend to theoretical models, crossing over into practical applicability in fields such as robotics and aerospace, where stable performance in a variety of conditions is essential. Feedback is another fundamental concept of control theory, the process by which a system self-regulates. Feedback loops are necessary for the maintenance of levels of desired output in dynamically shifting circumstances. Positive feedback amplifies systems responses to the extent of either explosive growth or instability, while negative feedback seeks to introduce stability into systems by opposing drifts away from goals predetermined. Implications of the feedback mechanism are strongly positioned in linear control systems' study where feedback is mathematically modeled to predict and correct for responses. Modeling not only optimizes performance but also system robustness with regard to uncertainties. Hence, understanding feedback dynamics becomes essential in creating reliable systems that can offer stable operational effectiveness. Control theory also intersects with emerging fields such as signal processing, in the context of prospective technologies such as 5G networks. As the demand for high-speed communication increases, researchers are constantly working on areas involving aspects of sparsity in signal reception and transmission. Development of these techniques has parallels in control theory, with sparse representation perhaps requiring more efficient processing and transmission techniques. By making use of ideas such as compressive sensing, engineers can unlock new potential in system design, particularly for cloud radio access networks and secure communication networks. Since the theory of control keeps developing, its principles remain central not just in the classical engineering field but also to the solution of present-day issues in very fast-evolving technological environments (Boche et al., 2015; Ball et al., 2007).

### 7.1.2. Mathematical Models of Linear Systems

Mathematical linear system models are fundamental tools applied in the design and investigation of feedback control systems. The dynamic behavior of a system is characterized by linear ordinary differential equations, whose use enables the investigation and control strategy design. With the assumption of linearity, the complexity in the dynamic behavior is minimized, and engineers are able to employ such powerful mathematical tools as the Laplace transform and state-space description. Using these models, various features of the system are examined, including stability, controllability, and observability, which are important for the system to perform as anticipated. The application of linear models can also be extended to a number of areas of engineering, including aerospace and mechanical systems, since they are easy to understand and straightforward. By precisely mimicking the characteristic features of complex systems, such mathematical models form the basis of the function of feedback mechanisms that enhance the performance of a system. For feedback control systems, the use of mathematical models facilitates proper adjustment of the response of a system through iterative processes and simulations. One such sophisticated technique of interest is inverse simulation, for which the input trajectories required that produce a given output in a linear system are determined. The inverse simulation procedures applied through traditional techniques can be involved and can lead to difficulties in applications at high speeds. The addition of feedback concepts to the process of inverse simulation completely simplifies this task. This two-stage feedback method not only makes the calculation easier but also enhances system resilience since it effectively eradicates plant uncertainties or disturbances-related variations. The result is a simpler method that is mathematically correct for the model but provides engineering practicality, particularly for complex systems where nonlinear behavior can arise, as evidenced in feedback study literature (Anonymous et al., 2011). The usage of mathematical models in linear systems is not limited to traditional control scenarios; they also have vast applicability in the complex networks prevalent in today's applications. The study of dynamic behaviors like stability and oscillation in these networks requires sophisticated model-building methods. Recent research shows that the dynamics of complex networks, as in biological networks or economic systems, can be modeled mathematically. This is significant for understanding how the inherent structure of a network can influence its dynamic behavior. For instance, in feedback control, how node states and topology relate to each other can have explicit consequences for the systems stability and synchronization ability. As more research is pursued on these complex interactions, the use of mathematical models becomes even more significant in the development of chaos control, stability, and synchronization methods for diverse applications to further underscore the role of feedback mechanisms in linear control systems (Liang et al., 2012) (Figure 7.3).

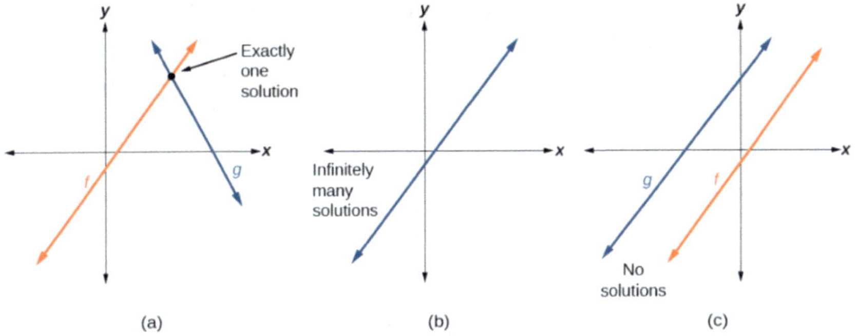


**Figure 7.3**. *Mathematical models of linear systems.*

*Source: https://files.mtstatic.com/site_4425/1142/0?Expires=1756117843&Signature=bsQS6uiXiei4SQtUZJOGfLYec1Yn~VC2usWyhZSPg6poMN0v2fRtovyq-xTNDQXqkqTmUOVJKzdrlYGzA~zl~hrLOQ1LfGQMc1WuP~bj88aIamgnZOwtR-2Urt~5bnffnxCwYf6HZOHaxmZxN7npqZkAivkIbtdtNP34XtKDqXug_&Key-Pair-Id=APKAJ5Y6AV4GI7A555NA.*

## 7.1.3. Stability and Performance Criteria

In the case of linear control systems, stability and performance criteria are the most significant aspects for successful operation. Stability refers to the ability of a system to continue stable even when it undergoes perturbations, while performance criteria include the precise requirements that a system needs to meet in terms of response, accuracy, and robustness. To determine stability, conventional methods such as gain and phase margin analysis are often employed; these methods provide information on control system robustness in varying operating conditions. These concepts, when applied efficiently in combination, can reinforce design objectives, especially in complex systems where multi-faceted performance specifications and stability requirements need to coexist. A fascinating example of the challenges involved here can be seen in the aerospace industry, where flight control systems have strict certification requirements. Under these circumstances, traditional methods of stability analysis tend to fail due to the very high complexity involved in modern launch vehicles. As noted, most adaptive control approaches are incapable of matching classical stability requirements with the dynamic behavior of conditionally stable systems that include flexible-body dynamics. Exactly, models like those used in the NASA X-15 program exhibit unique behavior that calls for sophisticated stability analysis. In such cases, novel methods like a multiplicative forward loop gain adaptive law have been evolved that can lead to stability without compromising on performance under challenging conditions, thereby expressing the true essence of balancing both requirements in high-risk applications (Orr et al., 2012). Also, integrating fuzzy control systems presents a second avenue to enhance performance while ensuring stability. With the use of the Takagi–Sugeno fuzzy model, researchers have developed new control designs to manage multiple performance indices simultaneously. The technique uses linear matrix inequalities to ensure that control solutions not only remain stable but also satisfy optimality and disturbance rejection needs. The simulation output, particularly in complex systems like inverted pendulums, shows the effectiveness of this technique in achieving quadratic optimality while addressing stability problems. This is one illustration of the ongoing advancement in feedback analysis where researchers find ways to implement solutions that pay respect to the fine balancing act between stability and performance needs in linear control systems (Wang et al., 2016).

## 7.2. TYPES OF FEEDBACK CONTROL

Feedback control systems are essentially categorized into various forms that influence their design and application in engineering, particularly in linear control systems. The proportional control system is one of the key feedback control types with a straightforward mechanism to modify the output proportionally relative to the error signal. This technique is straightforward directly, and as such, it is largely utilized in systems where steady-state regulation is critical. However, good as proportional control is in some applications, it cannot always be relied upon to eradicate steady-state errors adequately, so integral control is added. Integral control addresses this limitation by regularly reviewing the accumulated error over time and leading to more accuracy in tracking the setpoint. The effectiveness of such control measures is particularly highlighted by projects like the Pterodactyl project, in which extensive research is conducted on various feedback means to improve performance at vehicle entry, thereby demonstrating the importance of robust feedback mechanisms in practical applications (Barton et al., 2025). In addition to proportional and integral control, derivative control also plays a significant role in feedback control systems. Derivative control aims at pre-empting future errors with the use of the rate of change of the error signal to bring a pre-emptive adjustment to stabilize the system. From the tracing of the error signals, derivative control helps to reduce overshoot and oscillations to enhance the response of the system to abrupt changes. This technique is best applied in adaptive systems where quick adaptation is critical, such as in aerospace applications like the Pterodactyl project's Lifting Nano-ADEPT (LNA) vehicle.

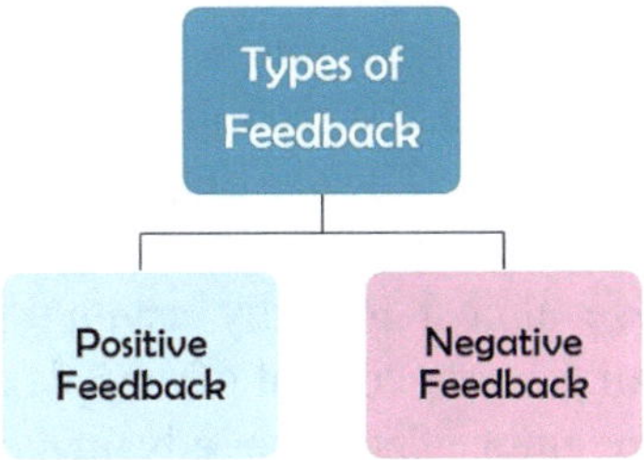


**Figure 7.4**. *Types of feedback.*

*Source: https://electronicscoach.com/wp-content/uploads/2019/12/types-of-feedback-in-closed-loop-control-system.jpg.*

Through the combination of derivative control and other forms of feedback techniques, engineers can create complex adaptive systems capable of maintaining performance under varying conditions. The combination of proportionate, integral, and derivative controls, known as PID control, illustrates the evolution of feedback mechanisms to enhance control systems across applications (Barton et al., 2025). In addition to further broadening the list of feedback control types, modern engineering increasingly utilizes adaptive control systems specifically tailored for variable operating conditions. Adaptive controls flexibly adjust their parameters in real-time based on performance characteristics, thereby enhancing reliability and efficiency in use. This type of feedback control is particularly applied in systems where random external conditions may affect stability and accuracy, such as in aerospace mechanisms during entry and re-entry. Under the Pterodactyl program, adaptive control techniques were pivotal in coping with the uncertainty caused by variable atmospheric conditions and vehicle designs.

Through these sophisticated feedback control systems, engineers can make sure that Deployable Entry Vehicles (DEVs) respond not only to immediate disturbances effectively but also learn from past performance to lead to greater robustness and flexibility in subsequent missions. This advanced analysis of feedback control types reaffirms their intrinsic contribution towards the design of linear control systems as well as the enhancement of the performance of sophisticated technological devises (Figure 7.4).

### 7.2.1. Negative Feedback Versus Positive Feedback

In linear control systems, feedback plays a critical role in maintaining desired system performance and stability. Negative feedback, in specific, ensures error correction through dampening of system response, stability, and robustness against disturbances. It entails comparison between the output response and reference input, which controls adjustments aimed at reducing deviations. Negative feedback systems are very much acclaimed for the achievement of stable equilibrium points, and their ability to satisfy Lyapunov stability conditions stands testament to this. The use of quadratic Lyapunov functions, as stated in Aslam et al. (2025), is one of the means by which negative feedback can be utilized to achieve almost globally asymptotic stability (AGAS), a prominent characteristic of reliable control of complex dynamic systems. This fundamental appreciation emphasizes the advantage of negative feedback to dampen oscillation and provide a reliable mechanism for linear control applications. In contrast, positive feedback amplifies deviations in a system, commonly to the level of runaway response, which generates instability. While such mechanisms seem counterintuitive, they serve applications where change is desirable at a high rate, say, for process start-up or changeovers. Specifically, positive feedback is generally utilized to enhance outputs in specific uses where the aim is to reach a certain level or trigger a state change. For instance, it is used in systems that require prompt action during critical periods regardless of the possibility of overshoot and instability. The research on externally positive systems, as indicated in Sebastiaan et al. (2025), supports providing the right boundaries for input signal properties so that practitioners can harness positive feedback without forgetting the influence of its effect on response dynamics. Last but not least, biopotency of feedback structures underscores the plea for contextual understanding when designing control schemes. The relationship between the negative and positive feedback processes enlightens the fine art of decision-making present in control system design. An advanced design generally involves the use of both types of feedback processes in a complementary manner to achieve the best possible system performance. For instance, in cases of non-linear dynamics, the use of negative feedback to eliminate unwanted oscillations coupled with controlled positive feedback to achieve fast responses can lead to balanced feedback architecture. Such integration not only ensures stability but also enhances responsiveness in complex control environments. Research displays that advanced control techniques employing both feedback types have proved best to respond to the challenges of nonlinear systems (Aslam et al., 2025). Therefore, a delicate understanding of feedback dynamics is necessary for designers and engineers who work on the implementation of effective linear control systems because they manipulate the trade-offs involved in feedback implementations to achieve optimal system performance (Figure 7.5).

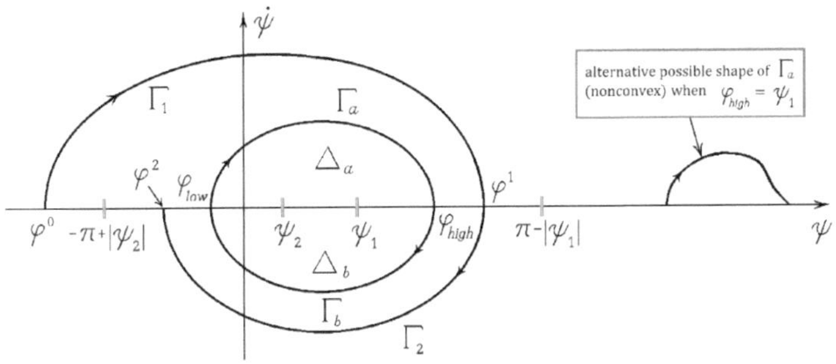


**Figure 7.5**. *Almost global asymptotic stability.*

*Source: https://media.springernature.com/lw1200/springer-static/image/art%3A10.1007%2Fs00498-018-0216-2/MediaObjects/498_2018_216_Fig4_HTML.png.*

## 7.2.2. Proportional, Integral, and Derivative (PID) Control

In the realm of control systems, Proportional, Integral, and Derivative (PID) control stands as a cornerstone due to its simplicity and effectiveness. PID controllers are designed to minimize the error between the desired setpoint and the actual output by adjusting control inputs based on three components: the proportional, integral, and derivative terms. The proportional term addresses the present error, providing an immediate corrective action. Meanwhile, the integral term accumulates past errors, eliminating steady-state discrepancies. Lastly, the derivative term anticipates future errors, offering a predictive capability that can enhance system stability. This multifaceted approach to control makes PID an ideal choice in various applications, from automation processes to dynamical systems, ensuring optimized performance in linear control scenarios. Within specific applications, such as gantry crane operations, the importance of PID control can be exemplified. Gantry cranes are subjected to performance demands characterized by speed, accuracy, and safety. However, they often exhibit unwanted sway, which can compromise operational precision. To address this, hybrid control schemes have emerged that incorporate PID alongside other methods such as linear quadratic regulators and higher order differential feedback controllers. Implementing a proportional integral derivative (PID) control in conjunction with output-based input shaping (OBIS) filters has shown considerable promise in terms of both precise positioning and sway suppression in gantry cranes. Analysis has indicated that this integrated approach fosters enhanced tracking precision, further illustrating the adaptability and robustness of PID control systems in complex operational environments (Alhassan et al., 2020). However, in more unpredictable environments, challenges arise in the implementation of traditional PID controllers. As modern systems evolve, particularly in fields like robotics and online trading, the need for responsive control in real-time has become paramount. The limitations of PID controllers are highlighted when setpoint conditions approach saturation, as seen in CPU utilization control scenarios. While PID can be effective, it struggles under severe overload conditions, leading to prolonged settling times. This has prompted research into alternative strategies such as neuro-fuzzy control, which can outperform traditional PID in heavily loaded situations. By monitoring error feedback and dynamically adjusting control strategies, these advanced approaches promise improvements in response times and overall system performance, indicating a future direction worth exploring in feedback analysis in linear control (Ayav et al., 2002).

### 7.2.3. State-Space Feedback Control

The concept of state-space feedback control forms a fundamental pillar within the framework of modern control theory, allowing for the analysis and regulation of dynamic systems. By representing system states as vectors, this approach provides a comprehensive description of the state dynamics through differential equations. Notably, state-space feedback control facilitates the application of linear control techniques to nonlinear systems, thereby enhancing system performance and stability. For instance, the use of state-space feedback linearization allows for the transformation of nonlinear dynamics into linear forms, thereby enabling the stabilization of systems that are otherwise unstable in open-loop configurations. This methodology is particularly robust, as it not only aids in achieving asymptotic stability but also enhances tracking performance, as illustrated in scenarios involving underwater vehicles utilizing variable ballast for depth positioning (Sumantri et al., 2009). The implications of state-space feedback control extend deeply into synchronization problems among interconnected systems, particularly when communication constraints are present. Recent studies have examined the output feedback controlled synchronization of nonlinear unstable systems, emphasizing the challenges posed by limited communication capacities. Through the application of binary time-varying coder-decoder schemes, researchers have developed theoretical frameworks that address synchronization discrepancies by utilizing a Passification

Theorem-based output feedback control law. The results demonstrate the potential for synchronization errors to exponentially diminish, contingent upon the adequacy of the communication channels capacity. Such insights not only spotlight the versatile applicability of state-space control methodologies but also underscore their significance in addressing real-world problems like controlled synchronization in complex systems (Alexander et al., 2007). In practical applications, the adoption of state-space feedback control has revolutionized various engineering domains, including robotics, aerospace, and process control. By leveraging the ability to systematically analyze and control systems using state-space representations, engineers can significantly enhance system robustness and response characteristics. Furthermore, the versatility of state-space formulations permits adaptation to diverse performance criteria and real-time constraints, making it an invaluable tool in the design of control systems. As research continues to refine these methodologies, new applications in adaptive control and nonlinear system regulation will undoubtedly emerge, solidifying the role of state-space feedback control at the forefront of linear control advancements. Consequently, the ongoing discourse surrounding state-space feedback mechanisms illustrates their pivotal contribution to the evolution of control theory and practical system engineering (Figure 7.6).

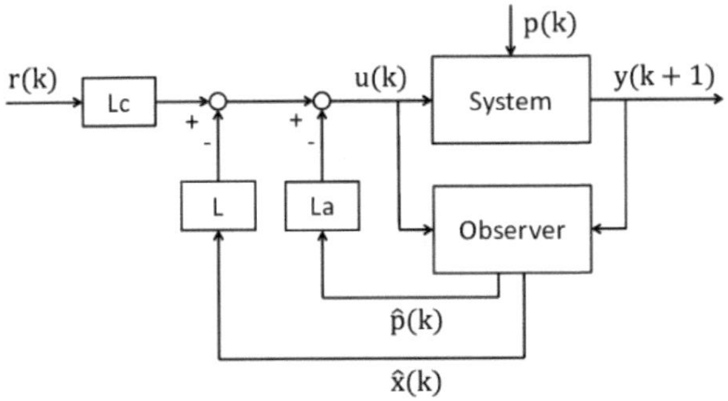


**Figure 7.6**. *State space controller with state feedback.*

*Source: https://www.researchgate.net/profile/Edgar-Ramirez-Laboreo/publication/286470760/figure/fig1/AS:613961888174085@1523391329302/State-space-controller-with-state-feedback-and-reference-and-disturbances-feedforward.png.*

## 7.3. FEEDBACK ANALYSIS TECHNIQUES

The character of modern control systems demands the expert application of feedback analysis techniques to enhance performance and stability. State-space feedback linearization is a leading methodology in such cases, transforming the dynamics of nonlinear systems into an equivalent linear form. This transformation allows the utilization of linear control methods, filling the gap between complex nonlinear dynamics and linear techniques. Particularly in processes such as underwater vehicles with variable ballast for depth control, feedback linearization has been proven to work. Asymptotic stability can be achieved by practitioners through effective error tracking by stabilizing points that were otherwise unstable in open-loop systems. Therefore, this technique not only enhances the performance of control but also enhances systems' functional capacity whose performance has previously been constrained by nonlinear dynamics, making it significant in advanced feedback analysis techniques (Sumantri et al., 2009). Time-delayed feedback control is another critical tactic under feedback analysis that presents a new avenue to the study of dynamical systems' behaviors. By feeding back periodically the outputs with respect to previous data, this technique ensures that the control system is responding to past differences correctly, with stable performance even when there are uncertainties involved. It relies upon the assumption that periodic orbits in the controlled system are essentially parallel to those in the uncontrolled system, so stability analysis is model-independent. Moreover, recent work shows that under some genericity assumptions, time-delayed feedback stabilizes more complex behavior than one would have previously anticipated, and it is therefore a powerful tool to study nonlinear systems. Using intuitions of linear stability analysis, scientists can compute control gains that guarantee successful stabilization, illustrating the method's relevance to modern control theory (Sieber et al., 2016). The comparison of feedback analysis techniques reveals a paradigm shift from traditional linear techniques to more advanced methodologies to accommodate the inherent complexity of nonlinear systems. The coordination of state-space feedback linearization and time-delayed feedback reflects an extensive understanding of system dynamics, with resulting enhancement in stability and performance. Furthermore, the ability to apply linear control techniques to nonlinear systems is indicative of the flexibility and versatility of such methods and enables new developments in numerous fields such as robotics and aeronautics. With increasingly advanced control systems, it will be necessary that ongoing development and refinement of these feedback mechanisms continue to be expanded in the hopes of addressing future challenges effectively, preparing engineers and researchers well to achieve peak performance within increasingly complex environments. Continued research affirms the need for feedback analysis as a foundation to achieving effective control across a broad spectrum of applications (Figure 7.7).

**Figure 7.7**. *Feedback analysis techniques.*

*Source: https://fibery.io/blog/static/076163aec4a13e06a14b7ff8eaf23838/6247c/e30da6dc-e562-4732-ad55-dfc03bfb38a5.png.*

## 7.3.1. Root Locus Method

The Root Locus Method is a classic control system design method, particularly in feedback analysis of linear control systems. It is a visual approach of viewing how the roots of a characteristic equation of a closed-loop system change with varying gain, hence enabling designers and engineers to visualize stability and transient response of dynamic systems. The method allows gain values to be calculated which will provide maximum system performance, allowing an intuitive appreciation of stability margin and response time. The method is especially useful in systems where time-domain specifications, such as settling time and overshoot, are critical. The method is therefore a critical design tool for designers who are responsible for ensuring designs are meeting performance criteria laid down by realistic constraints. Besides simple stability analysis, the application of the Root Locus Method extends to controller synthesis with the aim for desired system behavior. In various engineering applications, engineers use this method for compensator design that alters dynamic characteristics in systems, for example, the magnetic levitation system (MLS) in recent studies. By applying traditional techniques, such as proportional-derivative and phase-lead compensators, system engineers use the Root Locus to regulate the pole-zero configuration of the system. Such regulation is important because it has a direct relationship with improved performance specifications and increased system dynamic control. The application of mathematical and computational software like MATLAB broadens the applicability of the Root Locus such that engineers can apply real-time simulation and validations to their designs, thereby cementing the method's applicability in modern uses. Moreover, the versatility of the Root Locus Method makes it applicable to linear as well as nonlinear dynamic systems. As indicated in recent literature, coupling the Root

Locus with feedback principles enables new solutions in inverse simulation, which is an exemplar of the technique's flexibility in complex cases. This form of flexibility is directly applicable when working with multi-input multi-output systems, which are not dealt with by normal control methods. By integrating feedback principles, engineers can achieve robustness against uncertainties and external disturbances, making the Root Locus Method not only an analysis tool but also a foundation for stability in inverse models. This twin application of the method illustrates its value in helping control solutions and optimize performance across various fields of engineering, so designs are not only theoretically accurate but also feasible (Anonymous et al., 2011; Awelewa et al., 2012) (Figure 7.8).

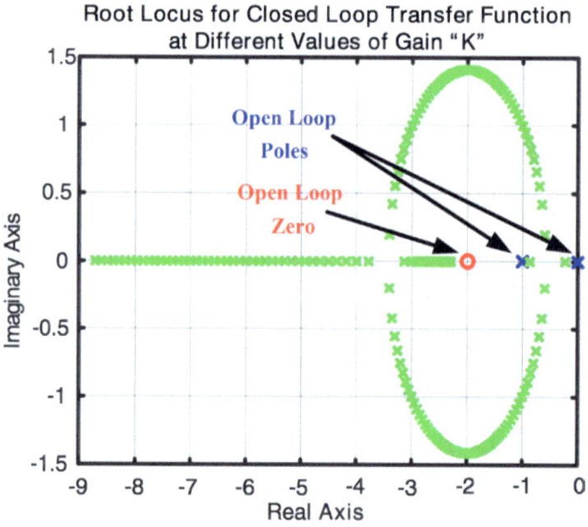


**Figure 7.8**. *Root locus method.*

*Source: https://electricalacademia.com/wp-content/uploads/2017/03/matlab.gif.*

## 7.3.2. Bode Plot Analysis

Bode Plot Analysis is a fundamental analysis in linear control theory's study of feedback systems. By graphically presenting a system's frequency response, Bode plots enable engineers and researchers to determine the stability and performance of control systems with ease. The magnitude and phase plots derived through this analysis facilitate the understanding of how a system would respond to various frequency inputs. Such knowledge is important especially since it informs control design choices towards enhancing the reliability and robustness of control systems. By closely examining the derived gain and phase margins from Bode plots, engineers can predict the systems' performance for different operating conditions, hence laying the basis for successful control design. Bode Plots are also utilized in the analysis of multivariable systems, whose interaction between different variables makes it difficult to design controls. Such an example, as seen in recent studies, for instance, steering control structure designs for 4-wheel vehicles, is helped by using Bode Plot Analysis to a great extent. These systems are assisted by using simplified linear models that allow interdependencies between outputs and inputs to be analyzed. Such plots are also very essential in the design of controllers for systems with multiple outputs and inputs, enabling engineers to investigate potential issues regarding performance as well as stability. Cross-feedback components, based on the results of sophisticated simulations, depict the practical importance of Bode Plot Analysis in modern control system design, particularly where efficiency and safety of operations are of utmost importance, (Leith et al., 2003). Moreover, there are new software packages that have emerged to promote the capability of Bode Plot Analysis in linear control systems. For instance, advanced tools like MATRIXx will automatically set up analysis configurations, hence making it easier for engineers to evaluate. By offering in-built functions consisting of Bode frequency response and associated systems assessments, such platforms reduce the amount of work engineers do during the analysis. The ease of use of interactive decision-making software and various appraisals instantly enables an accelerated and accurate understanding of system dynamics, enabling the interpretation and utilization of Bode plot data with ease. Thus, such technological advancements represent a huge stride towards the functional utilization of Bode Plot Analysis, and control systems can subsequently be designed with more accuracy and efficiency, (Clark et al., 2025).

### 7.3.3. Nyquist Stability Criterion

Nyquist Stability Criterion is a fundamental principle in the study of feedback control systems that provides an important method of ascertaining linear time-invariant (LTI) system stability through their frequency response. By analysis of the open-loop transfer function's frequency response, the criterion allows the assessment of stability in systems by contouring around critical points on the complex plane. Specifically, it emphasizes careful attention to the Nyquist plot—the trace of the open-loop transfer function with frequency going from zero to infinity. The graphical method presented by the Nyquist Stability Criterion permits a direct correspondence to be established between the number of encirclements of the point $(-1, j0)$ and closed-loop system stability, reducing a refined conceptualization of feedback interaction in reaction to changes in system parameters. A distinctive feature of the Nyquist Stability Criterion is its application in the analysis of systems with time delays, a ubiquitous property of actual systems. Time delays produce phase shifts that adversely affect system stability, necessitating accurate analysis. The Nyquist diagram makes it possible for engineers to actualize these delays effects by illustrating how they alter the gain and phase margin of the system. By explanation of the relationship between characteristic roots and encirclements, the designer can use the Nyquist criterion in order to ascertain possible delays-induced instability.

The application is necessary, particularly with systems becoming increasingly more complex as a reaction to technological advancements and convergence of technologies such as renewable energy resources and automated feedforward systems, which have a tendency to introduce unsuspected dynamics into system behavior. In addition, the incorporation of the Nyquist Stability Criterion within the context of large feedback control allows for the possibility of an entire system design framework and robustness analysis. For the evaluation of systems that integrate conventional as well as renewable generation, as proposed by recent studies, establishing stability under varying operating conditions becomes critical.

Decentralization of control test validates the significance of the Nyquist criterion, which enables various subsystems to operate autonomously and contribute toward overall stability without failing to meet established stability tests. This holistic approach to feedback analysis, validated through empirical research and simulations such as those analyzing communication networks, highlights the applicability of the criterion. Such integration allows effective coordination in complex systems that demand reliable performance under different operating conditions, ultimately enhancing the resilience of modern control systems (Mallada et al., 2019; Andrew et al., 2010) (Figure 7.9).

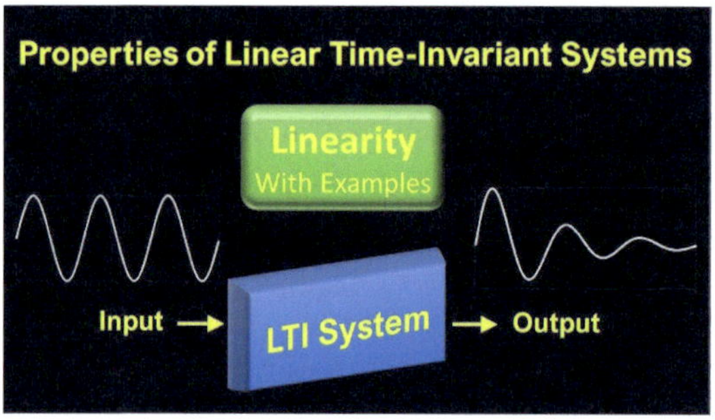


**Figure 7.9**. *Linear time invariant LTI systems.*

*Source: https://i.ytimg.com/vi/QK7qZAvxNAI/hq720.jpg?sqp=-oaymwEhCK-4FEIIDSFryq4qpAxMIARUAAAAAGAElAADIQj0AgKJD&rs=AOn4CLBYqdYI_fYO3yxIKOnKnDCCREXdaQ.*

## 7.4. APPLICATIONS OF FEEDBACK CONTROL

Feedback control applications range widely and cut across numerous fields, most significantly engineering, robotics, and process control. A great example is MRAC application in nonlinear systems, where stability and performance are of primary significance. As discussed in recent work, MRAC is well able to tackle the complexity of nonlinear dynamical behavior, as illustrated by experiments such as control of a Duffing oscillator (Wagg et al., 2003). The ability of MRAC to accommodate changing system dynamics makes it highly valuable in practical applications, ensuring that systems can be predicted even with the presence of perturbations. Such adaptability is of key importance in application fields where operation reliability is a necessity, and hence the significance of feedback control systems in modern engineering applications. Feedback control techniques have also been applied to Markovian jump systems (SMJS) that experience uncertainties in their rate transition matrices. Stability and stabilization processes of the systems involve robust design and analysis for performance regardless of the inherent uncertainties in system behavior. Recent advancements have come to the forefront of the use of Lyapunov theory and linear matrix inequalities (LMIs) as primary tools to design effective control strategies for these systems. This is particularly useful in scenarios where timing and synchronization are most important, and thus it is feasible to control noise and ensure deterministic behaviors (Wang et al., 2014). Feedback control methods enhance the reliability of systems operating under various conditions by allowing engineers to design controllers that are mode-dependent or mode-independent. This enhances the scope of applications for such systems. In addition to these, feedback control applications in adaptive systems have other practical applications in robotics and aerospace engineering. For instance, adaptive feedback linearization approaches are capable of stabilizing systems that are inherently unstable by transforming them into linear systems that are more stable. This method is also highly effective at controlling chaotic systems such as the Lorenz and Chua systems, whose dynamics are hard to control with traditional methods (Wagg et al., 2003). Through the application of these advanced feedback control methods, engineers can achieve improved performance under high-stakes situations where close control of nonlinear dynamics is needed. Hence, the ongoing development of adaptive feedback control systems not only maintains their foundational relevance in engineering but also fuels innovation in other sectors.

### 7.4.1. Industrial Automation

Industrial automation is one of the main contributors to enhancing productivity in the majority of manufacturing sectors. Automation, through the use of sophisticated control systems, reduces human interference to a bare minimum, hence few errors and enhanced efficiency. Automated system design and implementation could involve intricate feedback systems that control and optimize performance. Feedback analysis plays a crucial role in linear control systems where feedback dynamic adjustment of parameters is enabled in real-time. This has vital applications in industries requiring continuous performance under varying conditions because automation systems can readily make adjustments to variations, thus ensuring maximum functionality. The history of automation technology highlights the necessity for a robust feedback control system, not just for process stabilization but also to carry out complex tasks with reliability and precision. The complexity in the design of effective control systems renders iterative tuning inevitable, the most common in proportional-integral-derivative (PID) controllers.

**Figure 7.10**. *The future of industrial automation.*

*Source: https://etimg.etb2bimg.com/photo/116799139.cms.*

While PID controllers are theoretically straightforward, their application in reality is fraught with traps, especially in compromising between contradictory objectives such as rapid response and stability. An effective industrial automation system takes full advantage of sophisticated modeling and simulation techniques to trim down these controllers so that they act as intended over the entire operating range. Research indicates that the majority of PID variants have been developed in order to enhance transient performance, an indication of necessity for standardization against diversity of solutions. Utilization of high-level algorithmic tuning within control modules is perhaps the best example of a giant step towards automating the process, which ultimately enhances productivity and reduces maintenance requirements (Ang et al., 2005). The convergence of intelligent automation with feedback control systems is also changing the industrial production scene. The synergy of machine learning and artificial intelligence allows plug-and-play PID controllers to be developed, which can self-configure to deliver optimal performance with minimal human interaction. This revolution not only streamlines the installation procedure but also reduces system engineers' time taken for tuning and configuring controls by a wide margin, opening up automated systems to greater flexibility and affordability. Employing such advanced technologies leads to a more resilient and efficient industrial system that can easily adjust to fluctuating demand and levels of resources availability. More automation will also see the optimal use of feedback control and smart decision-making, with the balance between them being critical in taking advantage of operational capacity (Aphale et al., 2010) (Figure 7.10).

## 7.4.2. Robotics and Mechatronics

The merger of mechatronics and robotics basically enhances the performance of robotic systems, especially in the context of feedback analysis. Mechatronics combines mechanics, electronics, and computing to generate synergistic systems that can react dynamically to their environment. Feedback systems under the domain of linear control make sure robots can modify their activities from real-time feedbacks. For instance, recent research has indicated the application of an admittance controller that uses passivity theory to provide a natural flow of interaction between robotic systems and human operators. This natural flow of interaction is crucial for scenarios requiring reactive adaptation to unexpected forces, and it illustrates how linear control methods can effectively integrate multiplicity of dimensions of feedback examination and system interaction. The applicability of feedback analysis in mechatronics also involves designing complex control algorithms that maximize robotic performance. In particular, feedback provided by force/torque sensors in robots allows for minute changes that are the essence of human-robot collaboration. The passivity theory empowers robots to respond to the forces exerted by operators or environmental influences without triggering instability. This capability is particularly useful in the use of upper-limbs exoskeleton, where the robotic support has to closely follow human motion and intent. As has been reported in recent works, with increasing control gain, the system performance approaches an idealized model, emphasizing the superiority of feedback loops in terms of maximizing robotic responsiveness and working precision (Choi et al., 2019).

Furthermore, the examination of performance constraints through feedback analysis offers evidence of the working constraints of robotic systems under application conditions. The quantitative agreement between control gain and performance presented in chapters shows how greater control gain can enhance system performance but also provides for the necessity of accurate calibration under operating conditions. This metric is even more significant in situations where the environmental interactions are not observed, thereby necessitating tight control schemes that can compensate for such uncertainties. Such analyses catalyze an even greater level of understanding of how feedback mechanisms can propel the use and design of robotics and mechatronics towards creating the future possibilities of innovations that can lead to greater functionality and usability in different robotic applications (Aphale et al., 2010).

### 7.4.3. Aerospace and Automotive Systems

Aerospace and vehicle systems applications of feedback control systems have become the standard for performance and safety attainment. Feedback analysis, which defines the outcome of a systems inputs, allows engineers to modify control mechanisms in real time, leading to improved efficiency and reliability. Feedback loops, in aerospace applications, for instance, are essential to stabilize flight dynamics and counteract unexpected disturbances. Such systems continuously assess parameters such as altitude, speed, and attitude, adjusting control surfaces in order to maintain optimum flying conditions. These control systems are made necessary by a study detailing aerospace technology improvement, including safety equipment and crew performance, where the role of feedback in preserving operational integrity is pinpointed as being critical (N/A). In automobiles, feedback control systems serve similar important functions but are implemented for resolving automotive problems related to ground vehicles. Various feedback algorithms are employed by current automobiles for controlling engine operation, enhancing fuel efficiency, and facilitating safety features such as anti-lock braking (ABS) and electronic stability control (ESC). All these systems continuously read from sensors, providing real-time information that affects decision-making in a vehicle. The continuous process of system response adjustment to driver commands and road scenarios emphasizes the importance of feedback analysis in linear control, which places emphasis on precision and response. Developments in the practice of manufacturing, as highlighted by current research, further emphasize the role of feedback in system design for meeting rigorous performance specifications in automobile manufacture (Muetze et al., 2014). Furthermore, aerospace and automotive sectors are experiencing increased overlaps of innovative technologies, often enabled through systematic feedback mechanisms. The evolution of electrochemical micromachining (μECM) in production processes is an example of how precise control of micro-scale components leads to the overall system's performance. With the application of fuzzy logic control and adaptive feed rates in μECM, engineers are able to create parts of high quality needs without compromising structural integrity, both critical for both uses. The more sophisticated the equipment gets, the dynamic feedback loops' behavior should be known to engineers who work to innovate and at the same time ensure safety and functionality of their devices. This continuous progress in the feedback control technologies is at the heart of shaping the future of aerospace and automotive systems, promising greater efficiency and greater capability (Muetze et al., 2014) (Figure 7.11).

**Figure 7.11**. *Automotive systems.*

*Source: https://blogs.sw.siemens.com/wp-content/uploads/sites/19/2019/11/Fig-1-Car-ADAS-Driver-futuristic-dashboard-Adobe-206321037-scaled.jpg.*

## 7.5. CHALLENGES IN FEEDBACK CONTROL SYSTEMS

The behavior of feedback control systems is frequently challenged by a number of factors inherent in design and the environment of operation. Difficulty in modeling dynamic systems presents a main challenge. Feedback control relies substantially on mathematical models that over-simplify real systems, thus leading to conflicts between model predictions and system performance. Conflict may cause system instability, especially non-linear systems. In addition, perturbations from outside and variations in system parameters may further aggravate such challenges, rendering adaptive control strategies that are capable of adaptively combating time-varying situations highly valuable, highlighting yet another research area of focus in control theory. Effective feedback control, therefore, is not merely a matter of the robustness of the initial model but also the ability to forecast and anticipate disturbances capable of altering system behavior. Apart from mimicking complexity, feedback control systems usually suffer from issues related to time delays and latency. These can be sensor response times, communication latencies, or computation cycles, all of which can severely detract from system responsiveness.

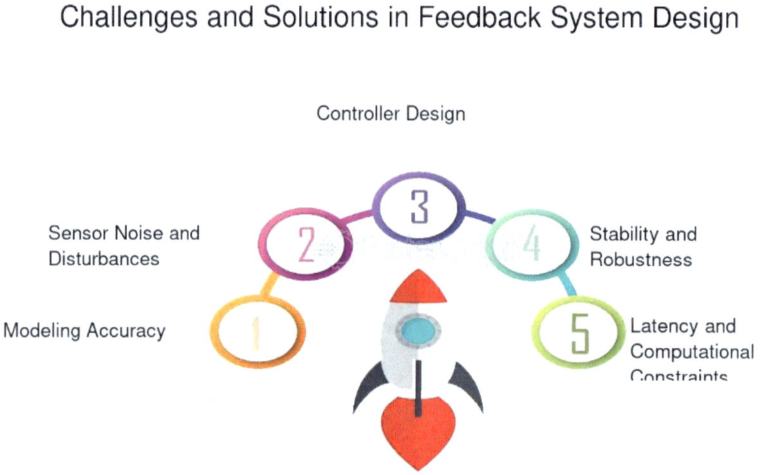


**Figure 7.12**. *Challenges in feedback control systems.*

*Source: https://fastercapital.co/i/Feedback-Control--Circling-Success--Feedback-Control-and-State-Space-Model-Interplay--Challenges-and-Solutions-in-Feedback-System-Design.webp.*

Delays turn such issues into complex ones for the feedback loop, causing it to oscillate or even become unstable if correct control is not exercised across critical thresholds. The impact of these time delays on system behavior must be studied intensively in order to maintain stability; this requires advanced mathematical techniques and simulation methods for the prediction of their effect. Counteracting these difficulties involves the use of control schemes that are sufficiently robust to overcome the effect of such delays without reducing performance. The emphasis on punctual reaction and feedback highlights the delicate balance that must be achieved in well-control systems. Furthermore, integration of emerging new technologies into feedback control systems is promising and poses risks. For instance, the introduction of advanced communication mechanisms, like those anticipated in 5G, poses the question of maximizing the feedback control in those systems. Improved data transmission rates and sparsity of signals present new development opportunities but introduce complexity in controlling multi-dimensional systems. It is at this point that there arises the necessity of rigorous control analysis, as the systems need to accommodate fluctuating power and energy issues (Bajon et al., 2010). Application of multi-physics methods can improve the design and analysis of these systems, but it calls for a good understanding of how various physical phenomena are interacting with one another, thereby indicating an urgent necessity for new solutions that embrace such technological advances well (Figure 7.12).

## 7.5.1. Nonlinearity and Uncertainty

Uncertainty and nonlinearity are the interaction of two basic elements in feedback analysis of linear control systems. Nonlinear behavior has the tendency to bring the type of complexities that make it challenging for typical linear models, particularly with respect to how systems respond to disturbances. Nonlinearity can be quite different in its forms, ranging from varying input-output dynamics to feedback loops that respond uncertainly under certain conditions. Consequently, an understanding of such nonlinear dynamics is essential when developing effective control. By incorporating the knowledge of complexity theory, as exemplified in literature, one is able to realize not only that nonlinearity complicates modeling but also that it demands more adaptive feedback mechanisms (Hornstein et al., 2005). The evolution of these higher-order characteristics in adaptive systems emphasizes the need for robust methodology with capability to manage uncertainty and resulting in improved control performance. While studying uncertainty dynamics with nonlinearity, one needs to understand implications of such factors on system response. Uncertainty is induced by many sources including external sources, model mismatches, and inherent system fluctuation. This lack of certainty causes it hard to build practical feedback loops in linear control systems. For instance, since systems are nonlinear with respect to their responses, small variations in the values of inputs can lead to big changes in output behavior, potentially destabilizing the control process. In addition, the inclusion of adaptive management practices becomes increasingly important in reducing uncertainty, enabling iterative process refinement in system performance (Hornstein et al., 2005). Through emphasis on the need for adaptive and adaptable control mechanisms, scientists and engineers are better able to accommodate the uncertainties invoked by nonlinear dynamics, further increasing the general effectiveness of feedback systems in practical use. Greater advancement of the position of nonlinearity and uncertainty in being contemplated and utilized can be observed with advancements in technology like fluidic valves used for aerodynamic flow control. These systems are a case in point regarding the challenges introduced by nonlinearity, particularly in terms of response time, which is susceptible to fluctuation due to changing operating conditions. Such fluctuation introduces a constraint on their application in the accurate control applications, where predictability of performance is required. New technologies, including closed-loop controls for new piezo-fluidic valves, demonstrate innovative ways to eliminate uncertainty while enhancing operational bandwidth. By applying feedback mechanisms that incorporate nonlinear parameters, engineers can create more resilient systems that adapt to evolving conditions and improve predictability of performance. This convergence of nonlinear dynamics and uncertainty highlights the need for continued evolution and refinement in feedback analysis in linear control systems.

### 7.5.2. Time Delays and Their Effects

In the context of linear control system feedback analysis, the meaning of time delays is crucial as they may have significant effects on system performance and stability. Time delays typically occur as a result of various aspects, such as sensor or actuator processing delay, communication delay, or system dynamics. Time delays may lead to control signal phase lags, which in turn adversely affect the ability of the system to respond against disturbances. When such delays occur, the system can experience oscillations or even instability, which becomes hard to achieve the desired response. Precise modeling and compensation of such delays are thus necessary to ensure robust performance, particularly in applications requiring cautious control, such as in robotics or aerospace systems. Moreover, the effects of time delays are not only restricted to near-instant stability problems; they are also at the center of determining the system responsiveness. Time delays in feedback control systems delay error correction in a timely fashion, resulting in increased steady-state error and potential overshoot or undershoot instances. Eventually, control methods that fail to incorporate time delays result in sub-optimal performance behavior. For instance, collaborative robotics missions usually consist of multiple agents that must cooperate, and optimal task coordination is disrupted by communications delay. In this scenario, Collective Sufficient Richness, which asserts that despite delays in occurrence, good teamwork will result in robotic manipulator parameter convergence, is proposed. This is indicative of integrating time delay into adaptive controller design in a complex situation where various systems interact with each other (Slotine et al., 2018). In networks where stochastic disturbances are dominant, the impact of time delays becomes more dominant, and stronger filtering mechanisms become necessary for reliable state estimation. For example, in genetic regulatory networks with the prevalence of feedback, component interactions possess non-linear dynamics with time delays and extraneous noise involved. Such conditions exist to enhance the necessity to develop advanced filtering mechanisms that are capable of sustaining such uncertainties but remain stable. Current studies have determined that the application of linear matrix inequalities (LMI) can offer sufficient conditions for ensuring filter stability and performance in response to time delays and random disturbances (Fraser et al., 2008). This indicates the real imperative for multidisciplinary approaches to the challenges of time delay complexity in feedback analysis, which results in advances in productive control strategies in a variety of applications.

## 7.5.3. Sensor and Actuator Limitations

Interaction between sensors and actuators is a crucial aspect of the application of feedback control systems but with significant limitations to affect their performance. Sensors, which are responsible for measuring system states, can be limited by resolution and sensitivity. Such limitations can lead to insufficient representation of data, hence affecting the performance of the control algorithms. In design of control systems, most critical to consider are the phenomena of non-linearity in active stabilization, including bounded actuation and robustness of the control parameter. A typical analytical model of the stabilized compressor shows how efficiency and stability are heavily dependent on the choice of sensor-actuator pairs, proving the efficacy of optimal choices made under particular system dynamic conditions. These limitations help engineers create more robust systems that can manage the complexities of real-world use, thereby improving overall control performance (Epstein et al., 2025). Actuators, which convert control signals into physical motion, also possess limitations intrinsic to them that must be taken into consideration in feedback control analysis. One general issue is control bandwidths of actuator response; an actuator that is unable to respond quickly enough to changes can lead to delays or overshoots in system response and erode stability. Control bandwidth versus system performance is at the heart of good actuator design, as shown by research illustrating control bandwidths should be greater than certain values to achieve maximum flow range in compression systems. This finding educates actuators that, in addition to being strong, actuators should be adept at maintaining fast responsiveness to dynamic disturbances. When a person maximizes the system, they ought to appreciate the fact that actuation dynamics regulate the stability and extent of the controlled process, reflecting the capability of well-designed system elements to achieve desirable control outcomes (Epstein et al., 2025). Also, communication constraints may further exacerbate sensor and actuator problems in feedback systems. The constraints arise as power efficiency requirements rise, requiring prudent design of communication channels between the controller and the elements. The channels, if they are poorly designed, can lead to increased external costs, literally slowing feedback accuracy and control effectiveness. This is especially true in the case of memoryless controllers, which, as much as they are valuable for their simplicity, underperform with limited information capacity constraints. Studies have shown that optimizing information rates in a controller can complement external cost constraints while improving operational effectiveness. Balancing the cost-benefit of information transmission needed and saved costs becomes vital to the systems' overall operation, noting the intricate dynamics of actuator and sensor interactions in feedback control systems (Figure 7.13).

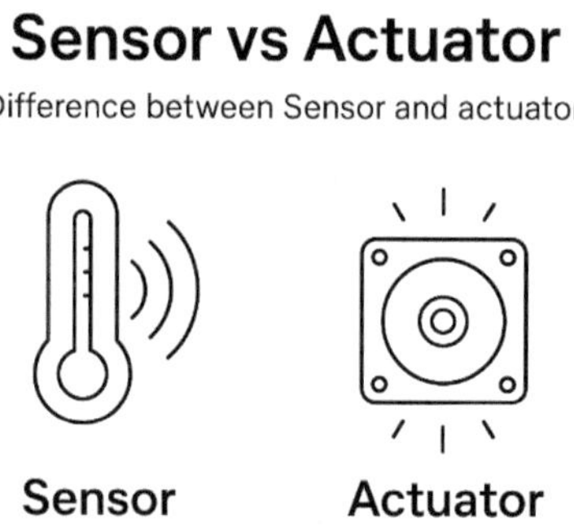


**Figure 7.13**. *Difference between sensor and actuator.*

*Source: https://iotdunia.com/wp-content/uploads/2022/11/sensor_vs_actuator.webp.*

## 7.6. RECENT ADVANCES IN FEEDBACK CONTROL

Budding advances in feedback control have completely transformed the field of control systems, particularly in nonlinear dynamics and uncertain conditions. Researchers have established a strong interest in H∞ control and filtering, which play a significant role in system uncertainty and performance loss. A special issue on reminiscing these developments highlights not only the theoretical contributions but also applications in various fields, such as network-based control and estimation. These developments provide a foundation for tackling modern challenges by intricate systems, showing a significant paradigm shift in control strategies expanding beyond conventional frameworks. The integration of advanced techniques is a step forward toward more robust and adaptable control systems capable of functioning optimally under uncertainty and real-time constraints (Basin et al., 2011). Convex system study is another landmark in control using feedback. By employing models such as Takagi-Sugeno and linear parameter varying (LPV), theoreticians have developed frameworks that successfully mask nonlinearities through a chain of linear approximations. This allows the use of linear matrix inequalities (LMIs) to solve complex nonlinear systems, thus widening the scope for stability and performance optimization. All models' interpolation indeed creates a smoother transition between varying operating states, permitting more refined control methodologies.

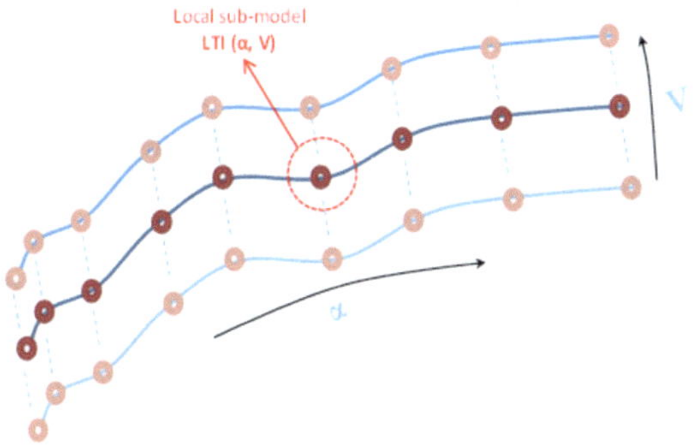


**Figure 7.14**. *Linear parameter-varying models.*

*Source: https://www.mathworks.com/help/control/ug/gainsched2.png.*

In addition, developments around LMI-based methodologies have brought significant enhancement in state estimation, feedback tracking, and model predictive control (MPC). Not only do these developments make control usage efficient but also enhance the safety protocols within these systems by addressing fault diagnosis and fault-tolerant control methodologies (López et al., 2019). New fields of feedback control more and more rely on new methodologies that blend traditional principles and newly developed computational techniques. Incorporation of advanced non-classical methodologies is an indicator that signals a paradigm shift towards higher-level methods in regulating feedback systems. Such methodologies have a tendency to employ newest computational software in order to advance and maximize the control technique, leading to significantly enhanced real-time response as well as error minimization. Besides, emphasis on new applications, particularly networked systems, leads contemporary research into feedback control impacts in distributed systems. As the need for automatic and precise control increases across various industries, progressive advancement in feedback control methods sustains that they remain effective, viable, and adaptable to the complexity of the contemporary technological environment (Basin et al., 2011) (Figure 7.14).

### 7.6.1. Adaptive Control Techniques

Adaptive control techniques are a critical new advancement in modern control theory, particularly in systems with variabilities and uncertainties. The techniques are designed to adjust parameters in real time as a reaction to changing conditions or unforeseen system behavior to enhance overall system performance and stability. In the context of feedback analysis of linear control systems, adaptive control provides real-time adaptation in accordance with performance indices so that more agile and robust operation is facilitated. The development of a variety of adaptive control techniques—ranging from model reference adaptive control to self-tuning regulators—bears witness to the versatility of the tools and to their growing importance in the control of systems with nonlinearities or with time delays, as borne out by the growing demand for methods such as event-triggered optimal adaptive regulation, ensuring asymptotic stability even under uncertainty, as well tested in Wang et al. (2014). One of the key attributes of adaptive control techniques is their connection to singular Markovian jump systems (SMJS).

Their adaptive nature has certain immense benefits in addressing stability problem as well as control when there exists ambiguity in the parameters of the environment. Techniques such as Lyapunov-based methods and linear matrix inequalities (LMIs) are presently one of the central tools of adaptive controller synthesis that not only stabilizes systems of this kind but also robust H-infinity filtering with uncertainty, as explored in Wang et al. (2014). With such advanced techniques, engineers can now stabilize and control SMJSs efficiently even in the case of partial transition rate matrix knowledge or external perturbations. This adaptability not only enhances the performance of the systems but also widens their scope to various industrial settings. Further, the use of optimal adaptive control (OAC) techniques is a key contribution in linear as well as nonlinear control systems, particularly time delays and uncertain dynamics. With increasing pressure on systems to perform optimally, OAC methods based on state and output feedback have gained popularity due to their ability to optimize control responses for an infinite time horizon. By utilizing quadratic cost functions in resolving control issues, the methods ensure that stability and optimal regulation are attainable even with uncertainties. The addition of integral reinforcement learning (IRL) to OAC structures is a revolutionary step towards bringing flexibility into control, as seen in Moghadam et al. (2020). Such decisive synergy provides the direction that can lead to developing adaptive control techniques in the future, making them priceless in the dynamic scenarios of industrial uses and system stability against disturbances.

### 7.6.2. Robust Control Strategies

In control systems, robust control methods have a crucial role to play in ensuring system stability and performance against uncertainties. The methods ensure that system behavior of choice is maintained irrespective of changes in system parameters and disturbances from the external environment. By employing the use of feedback methodologies, robust control attempts to offset the implications of such uncertainties, thereby ensuring a stable operation in real-world applications. The creation of stable control methods would frequently utilize methodologies such as linear matrix inequalities (LMI) and Lyapunov methods. Such mathematical structures allow for the strict synthesis of control laws that not only achieve desired performance criteria but also yield stability even in unfavorable conditions. The combination of these methodologies is especially pertinent in highly complex systems in which modeling mistakes can significantly influence system dynamics and hence prove the actual necessity of robustness in applications to real control systems. Based on the robust control framework, a number of design methods have been formulated to address specific challenges with uncertainties. Amongst these, output feedback robust distributed model predictive control has emerged as leading, especially with parallel system structures. These systems often experience coupling effects that complicate control system design. The employment of distributed buffers and pre-estimators efficiently addresses these interactions while enabling timely communication between system components. The use of moving horizon estimators further enhances estimation accuracy such that control strategies can adapt dynamically to changes in system behavior. The testing of these approaches through simulations demonstrates their real-world efficacy, highlighting how robust control is able to breach the uncertainties of an environment effectively. This strength is essential for maintaining system integrity and performance across diverse operating environments. The deployment of robust control strategies also pushes new boundaries, such as quantum systems, where classical control theory can start to be constrained. In quantum systems closed-loop control, protocols like quantum feedback stabilization and learning control are considered as means to maximize system performance. These strategies exploit the singularity of quantum mechanics while addressing the problem of controllability in the quantum scenario. Quantum systems robust control combines methods that resist uncertainties inherent in quantum processes, just as classical robust control approaches. By using the convergence of classical techniques and quantum concepts, researchers are able to move into new frontiers of control theory to reap the potential of new applications and innovations. These studies identify the evolving context of high-gain control methods, highlighting their applicability not only in classical systems but also in emerging technological arenas, ultimately enriching the feedback analysis discourse in linear control.

### 7.6.3. Machine Learning in Control Systems

The use of machine learning techniques in control systems has gained significant interest for its ability to complement feedback mechanisms, especially in linear control models. Traditional control techniques are liable to depend on well-defined mathematical models and thus are less responsive to the uncertainties and nonlinear dynamics that characterize real-world scenarios. Machine learning offers a way to break these limitations by facilitating systems to learn and adjust from data so as to improve performance and robustness. For instance, semi-parametric models blending physics-based representations with data-driven approaches can accurately model complex system behavior. This multimodal strategy not just enhances the fidelity of trajectory optimization processes but also facilitates uncertainty handling, thus contributing to more trustworthy control performance in applications ranging from robotics to autonomous systems (Alvarez et al., 2017). Besides enhancing the underlying control system models, machine learning's role is otherwise especially beneficial for feedback mechanism optimization in dynamic environments. As systems become increasingly automated and linked, there is a need for such systems to observe and adapt to new situations promptly. Machine learning algorithms enable this adaptation by learning from historical information and instantaneous sensor inputs, thus allowing proactive adjustments. For example, machine learning techniques have been applied by wireless sensor networks to address challenges in the utilization of resources and system resilience, as highlighted by recent literature (Alsheikh et al., 2014). Such a feature is particularly vital in areas where data in real time is constantly generated and processed, allowing systems to respond properly without necessitating excessive redesign. Lastly, the application of machine learning on control systems also raises intriguing questions about the balance between model complexity and computational expense. While better algorithms can be used to achieve better performance, these algorithms can also have a high computational expense which may not be feasible for all applications. The challenge, therefore, lies in selecting suitable models that remain within the boundaries of accuracy and feasibility. Techniques such as Sparse Spectrum Gaussian Processes (SSGP) are indicative of this balance, showing techniques of adaptive online learning that change over time at minimal computational expense. The inclusion of machine learning in the control system feedback loop enables engineers to build robust solutions sensitive to change in the environment to make linear control techniques more accurate and robust (Alvarez et al., 2017).

## SUMMARY

The analysis of feedback mechanisms in linear control systems brings into view the intricacies of sustaining performance and stability. The analysis of control strategies from the traditional linear methods to advanced techniques like backstepping control highlights the evolution of control theory. Specifically, the use of nonlinear feedback in advanced systems, like the ones detailed in Diagne et al. (2016), illustrates the extensive impact that new methods can offer to enhance system behavior. The exponential stability which occurs due to these advanced methods serves to highlight the necessity of solid theoretical frameworks to facilitate practical implementation, enabling efficient control of dynamic systems under varying conditions. Besides, the study of semiactive dampers, as outlined in Sims et al. (2005), is significant in that it provides information regarding the trade-offs between functionality and simplicity in controllers. The contribution stresses the need for developing controllers that not only improve stability but also address the potential limitations of limit cycle behaviors, which can happen in closed-loop situations. The understanding of the nonlinear behavior of smart materials and dampers is critical in formulating strategies that strengthen the system while maintaining the highest level of performance. The integration of theoretical analysis with factual data has a rich bouquet from which advancements in control strategies can be harnessed in the future. Finally, the controversy surrounding feedback analysis in linear control is a reflection of the interactive link between theory and practice. With technology advancing, the only direction is to embrace complexity with models that are able to capture nonlinear behavior to gain powerful control solutions. The future of control systems lies in discovering ways that combine established principles with fresh understanding from new technologies. By developing a unified appreciation of feedback dynamics, academics and practitioners can more appropriately address the needs of contemporary engineering applications, thus providing avenues to enhancing both academic research and industrial practice.

In the context of linear control systems feedback analysis, there are a number of issues that come out strong to highlight the fine balance among stability, controlling mechanisms, and system behavior. At the center of this argument is the problem of stability, a basic characteristic that defines whether a system can return from a disturbance to equilibrium. By analyzing numerous diverse dynamical equations, scientists can define the stability characteristics of the systems and therefore affect their performance. This classification process is primarily based on historical development based on celestial mechanics and control theory, which places the qualitative understanding of system behavior above all else. Additionally, the interplay among nonlinearity, bifurcation, control, and uncertainty constitutes an enormous challenge that brought into focus the need to expound strong feedback mechanisms in engineered solutions as described in the transitions in Ball et al. (2007). The second most important aspect of feedback analysis is the application of these theoretical ideas to real-world use within advanced aerospace systems. One instance is through the use of feedback linearization control techniques, which enables the stabilization of nonlinear dynamics, something particularly valuable when confronted with uncertain orbits in celestial mechanics. The study of displaced periodic orbits for the circular restricted Earth-Moon system is an example, with advanced hybrid mechanisms incorporating solar sails and solar electric propulsion facilitating high-level trajectory control. Applications of this sort demonstrate how feedback systems are capable of providing insight into the tracking and stabilizing of bodies under multiple gravitational forces. As demonstrated in a number of case studies, effective feedback approaches not just enable enhanced control but also enhance system flexibility against the uncertainty that is inherent in space conditions, as advanced in McInnes et al. (2012). Moreover, advances in feedback analysis demonstrate the importance of including current computational approaches and simulation methods in order to deal with real-world complexities inherent in control systems. As research proceeds, research into multi-complex networks and their stability structures illustrates an ongoing quest for solutions that consider nonlinear dynamics and multidependencies of system components. This focus on the expansion of stability theory to multidimensional systems is a critical frontier in determining whether feedback mechanisms can be optimized for emergent behavior. By integrating the feedback analysis advancements with the needs of contemporary engineering, researchers continuously endeavor to create novel control strategies that possess the ability to cope with uncertainty and render systems robust in various applications, upholding the overarching principles introduced throughout the discussions on feedback analysis in linear control.

# REFERENCES


1. Alhassan, A. B., Anene, E. C., Sani, G., et al. (2020). Application of non-model dependent hybrid higher-order differential feedback controller on crane system. https://core.ac.uk/download/480424849.pdf (accessed on 19th September 2025).
2. Alsheikh, M. A., Lin, S., Niyato, D., et al. (2014). Machine learning in wireless sensor networks: Algorithms, strategies, and applications. https://core.ac.uk/download/35456291.pdf (accessed on 19th September 2025).
3. Alvarez, M. A., Gandhi, M., Lázaro-Gredilla, M., Pan, Y., Rahimi, A., Rudin, W., & Wang, J. (2017). Pseudospectral model predictive control under partially learned dynamics. http://arxiv.org/abs/1702.04800 (accessed on 19th September 2025).
4. Andrew, L. L. H., Hjalmarsson, H., Jacobsson, K., et al. (2010). Queue dynamics with window flow control. https://core.ac.uk/download/4886959.pdf (accessed on 19th September 2025).
5. Ang, K. H., Chong, G. C. Y., & Li, Y. (2005). PID control system analysis, design, and technology. https://core.ac.uk/download/1394222.pdf (accessed on 19th September 2025).
6. Aphale, S. S., Feliu, V., Moheimani, S. O. R., et al. (2010). Integral resonant control for vibration damping and precise tip-positioning of a single-link flexible manipulator. https://core.ac.uk/download/370439.pdf (accessed on 19th September 2025).
7. Aslam, F., Khan, H. Z. I., Haydar, M. F., Akhtar, S., & Riaz, J. (2025). Geometrization of higher-order linear control laws for attitude control on SO (3). https://www.semanticscholar.org/paper/3d769803b4f48a60cc39d3b845a1b021486748fc (accessed on 19th September 2025).
8. Awelewa, A. A., Iyiola, S. O., Katende, J., et al. (2012). Root locus-based magnetic levitation system stabilization: An undergraduate control system design approach. https://core.ac.uk/download/19334490.pdf (accessed on 19th September 2025).
9. Ayav, T., & Yılmaz, S. (2002). Neuro-fuzzy controller in real-time feedback schedulers. https://core.ac.uk/download/pdf/324144203.pdf (accessed on 19th September 2025).
10. Bacic, M., & Nicholls, C. (2020). Closed-loop control of a piezo-fluidic amplifier. http://arxiv.org/abs/2005.02968 (accessed on 19th September 2025).
11. Bagiev, M., Brockett, R., Buchholz, J. J., Camacho, E. F., Murray-Smith, D. J., et al. (2011). Feedback methods for inverse simulation of

dynamic models for engineering systems applications. https://core.ac.uk/download/9399528.pdf (accessed on 19th September 2025).

12. Bajon, D., & Wane, S. (2010). Power waves formulation of oscillation conditions: Avoidance of bifurcation modes in cross-coupled VCO architectures. https://core.ac.uk/download/12042374.pdf (accessed on 19th September 2025).
13. Ball, R., & Holmes, P. (2007). Dynamical systems, stability, and chaos. http://arxiv.org/abs/nlin/0702044 (accessed on 19th September 2025).
14. Baños Torrico, A., Costa Castelló, R., & Raveendran Nair, U. (2019). Reset control for DC-DC converters: An experimental application. https://core.ac.uk/download/294829596.pdf (accessed on 19th September 2025).
15. Barton, J. D., D'Souza, S. N., Margolis, B. W., et al. (2025). Pterodactyl: Development and comparison of control architectures for a mechanically deployed entry vehicle. https://core.ac.uk/download/pdf/286392694.pdf (accessed on 19th September 2025).
16. Barton, J., D'Souza, S. N., Margolis, B. W., et al. (2025). Pterodactyl: Control architectures development for integrated control design of a mechanically deployed entry vehicle. https://core.ac.uk/download/pdf/286392660.pdf (accessed on 19th September 2025).
17. Basin, M., & Shi, P. (2011). New advances in H∞ control and filtering for nonlinear systems. https://core.ac.uk/download/10831298.pdf (accessed on 19th September 2025).
18. Belavkin, V., Burgarth, D., Butkovskii, A., Dong, D., et al. (2010). Quantum control theory and applications: A survey. http://arxiv.org/abs/0910.2350 (accessed on 19th September 2025).
19. Boche, H., Jung, P., Strohmer, T., et al. (2015). Sparse signal processing concepts for efficient 5G system design. http://arxiv.org/abs/1411.0435 (accessed on 19th September 2025).
20. Chen, W., Khong, S. Z., Qiu, L., et al. (2019). Phase analysis of MIMO LTI systems. http://arxiv.org/abs/1904.03212 (accessed on 19th September 2025).
21. Choi, I. S., Choi, J. Y., Chung, G., et al. (2019). A passivity-based nonlinear admittance control with application to powered upper-limb control under unknown environmental interactions. https://core.ac.uk/download/275580849.pdf (accessed on 19th September 2025).
22. Clark, D. R., Ostroff, A. J., & Proffitt, M. S. (2025). Integrated tools for control-system analysis. https://core.ac.uk/download/pdf/42829140.pdf (accessed on 19th September 2025).

23. Cremean, L. B., Mabuchi, H., Murray, R. M., et al. (2004). A new approach to teaching feedback. https://core.ac.uk/download/4876294.pdf (accessed on 19th September 2025).

24. Diagne, M., Koga, S., Krstic, M., et al. (2016). Backstepping control of the one-phase Stefan problem. http://arxiv.org/abs/1607.04345 (accessed on 19th September 2025).

25. Doruk, R. O., & Zuglem, I. (2016). Projective DC motor control under disturbance torques. http://arxiv.org/abs/1605.08284 (accessed on 19th September 2025).

26. Epstein, A. H., Greitzer, E. M., Simon, J. S., et al. (2025). Active stabilization to prevent surge in centrifugal compression systems. https://core.ac.uk/download/pdf/42809061.pdf (accessed on 19th September 2025).

27. Fox, R., & Tishby, N. (2017). Minimum-information LQG control – Part I: Memoryless controllers. http://arxiv.org/abs/1606.01946 (accessed on 19th September 2025).

28. Fradkov, A. L., Andrievsky, B., Blekhman, I. I., Andrievsky, B., De Persis, C., Demidovich, B. P., Dudnik, E. N., et al. (2007). Controlled synchronization of one class of nonlinear systems under information constraints. http://arxiv.org/abs/0712.0636 (accessed on 19th September 2025).

29. Fraser, K., Lam, J., Liu, X., Wang, et al. (2008). Filtering for nonlinear genetic regulatory networks with stochastic disturbances. https://core.ac.uk/download/335786.pdf (accessed on 19th September 2025).

30. Hornstein, D. T. (2005). Complexity theory, adaptation, and administrative law. https://core.ac.uk/download/62548681.pdf (accessed on 19th September 2025).

31. Kellman, M. E., & Tyng, V. (2018). Kinetic model of translational autoregulation. http://arxiv.org/abs/1810.01505 (accessed on 19th September 2025).

32. Leith, D. J., Leithead, W. E., & Vilaplana, M. (2003). Control of sideslip and yaw rate in 4-wheel steering car using partial decoupling and individual channel design. https://core.ac.uk/download/9038570.pdf (accessed on 19th September 2025).

33. Liang, J., Liu, Y., & Wang, Z. (2012). Mathematical problems for complex networks. https://core.ac.uk/download/9633413.pdf (accessed on 19th September 2025).

34. López Estrada, F. R., Rotondo, D., & Valencia Palomo, G. (2019). A review of convex approaches for control, observation and safety of linear parameter varying and Takagi-Sugeno systems. https://core.ac.uk/download/288625538.pdf (accessed on 19th September 2025).
35. Mallada, E., & Pates, R. (2019). Robust scale-free synthesis for frequency control in power systems. http://arxiv.org/abs/1804.00729 (accessed on 19th September 2025).
36. McInnes, C., & Simo, J. (2012). Feedback stabilization of displaced periodic orbits: Application to binary asteroid. https://core.ac.uk/download/9037011.pdf (accessed on 19th September 2025).
37. Moghadam, R. (2020). Optimal adaptive control of time-delay dynamical systems with known and uncertain dynamics. https://core.ac.uk/download/524635732.pdf (accessed on 19th September 2025).
38. Muetze, A., Spieser, A., Kamaraj, A. B., Trimmer, A. L., Spieser, A., Ivanov, A., et al. (2014). Design of an electrochemical micromachining machine. https://core.ac.uk/download/29140409.pdf (accessed on 19th September 2025).
39. Oboe, R., Ohnishi, K., & Sariyildiz, E. (2019). Disturbance observer-based robust control and its applications: 35th anniversary overview. http://arxiv.org/abs/1902.09032 (accessed on 19th September 2025).
40. Orr, J. S., & VanZwieten, T. S. (2012). Robust, practical adaptive control for launch vehicles. https://core.ac.uk/download/pdf/10570961.pdf (accessed on 19th September 2025).
41. Sieber, J. (2016). Generic stabilisability for time-delayed feedback control. https://core.ac.uk/download/43098053.pdf (accessed on 19th September 2025).
42. Sims, N. D. (2005). Limit cycle behavior of smart fluid dampers under closed loop control. https://core.ac.uk/download/50555.pdf (accessed on 19th September 2025).
43. Slotine, J. J. E., & Wensing, P. M. (2018). Cooperative adaptive control for cloud-based robotics. http://arxiv.org/abs/1709.07112 (accessed on 19th September 2025).
44. Spurgeon, S. K., Zhang, S., & Zhao, D. (2021). Output feedback robust distributed model predictive control for parallel systems in process networks with competitive characteristics. https://core.ac.uk/download/429123417.pdf (accessed on 19th September 2025).
45. Sumantri, B. (2009). State-space feedback linearization for depth positioning of a spherical URV. https://core.ac.uk/download/12342573.pdf (accessed on 19th September 2025).

46. van den Eijnden, S., Lohéac, J., & Lorenzetti, P. (2025). Externally positive systems of order k and their application to undershoot and overshoot analysis. IEEE Control Systems Letters, 9, 1441–1446. https://www.semanticscholar.org/paper/b2e8544b10619cfeeca1b281d7cdae8037ac6afe (accessed on 19th September 2025).

47. Wagg, D. J. (2003). Adaptive control of nonlinear dynamical systems using a model reference approach. https://core.ac.uk/download/29031384.pdf (accessed on 19th September 2025).

48. Wang, G., Yan, X., & Zhang, Q. (2014). Analysis and design of singular Markovian jump systems. https://core.ac.uk/download/30710107.pdf (accessed on 19th September 2025).

49. Wang, X., & Yaz, E. E. (2016). Robust multi-criteria optimal fuzzy control of continuous-time nonlinear systems. https://core.ac.uk/download/213083946.pdf (accessed on 19th September 2025).

CHAPTER 8

# CONTROLLER DESIGN: PID CONTROL

## Contents



## INTRODUCTION

The Proportional, Integral, and Derivative (PID) controller forms a basic component in control system engineering, providing a reliable method of regulating dynamic systems. A PID controller acts upon three independent parameters: proportional control, which adjusts output in proportion to the current error; integral control, which accumulates past errors to eliminate residual steady-state error; and derivative control, which anticipates future error from its rate of change. This combination enables a successful responsive action of control that is capable of inducing system stability and performance improvement under a wide range of operating conditions. By understanding the interaction between the control parameters, engineers can develop systems with dynamic response to disturbances, thus facilitating optimal control in applications from industrial automation to aerospace engineering. PID controller design and tuning require a fine balance between the three parameters since their tunings directly impact system performance. For example, a highly proportional gain can make the system unstable and generate oscillations, while a low gain can lead to slow response. The integral term will eliminate the steady-state error but, if made too high, will introduce overshoot and oscillations. The derivative term, in return, will enhance the system's stability and response time but, by contrast, will also amplify the noise and result in wild control movements. It is central to learning about such dynamics when designing controller strategies for optimal system reliability and performance.

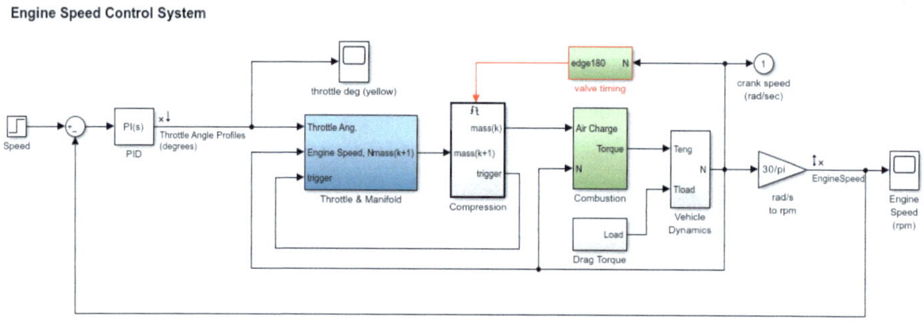


**Figure 8.1**. *Design PID controller using estimated frequency response.*

*Source: https://in.mathworks.com/help/examples/control/win64/DesignPIDControllerUsingEstimatedFrequencyResponseExample_01.png.*

As highlighted in existing designs, using methods such as performance portraits helps visualize and optimize parameter settings for target performance objectives (Mikuláš et al., 2022). New advancement in tuning methods has introduced smart algorithms and heuristic approaches, such as Particle Swarm Optimization (PSO), to further enhance PID controller performance. Employing swarm intelligence, these methods facilitate the identification of optimal PID parameters maximizing control performance in complex systems. Such tactics allow for the adjustment of controller parameters based on current information to enable the achievement of performance targets without much manual tuning. Additionally, the incorporation of intelligent techniques for tuning signals a broader trend towards control design where traditional techniques are complemented by new computational approaches, boosting precision as well as efficiency. This PID control design evolution signals the necessity of drawing from advanced methodologies to keep pace with the ever-increasing sophistication of industrial processes and systems (Ahmed, 2022) (Figure 8.1).

## *Definition of PID Control*

PID control is a general control technique applied in engineering and automation to be used for the control of dynamic systems. It stands for Proportional, Integral, and Derivative, where each characterizes a particular control action. The proportional component reacts to the current error value and provides an output that is proportional to the error, which helps in quick response. The integral term corrects for previous errors by integrating them over time, essentially removing steady-state error remaining. The derivative term predicts future errors by examining the rate of change of the error, which helps in the system's damping response, improving stability. Jointly, the three components form a robust controller that is capable of adjusting the control output so as to achieve desired performance requirements rapidly and precisely, hence making PID control an indispensable part in many applications ranging from robotics to heating and cooling systems. How PID parameters affect system performance is a major aspect of controller design with significant effects on stability and response time. All of the parameters—derivative time, proportional gain, and integral time—have an effect on the behavior of the systems in a distinct manner. For instance, increasing proportional gain tends to increase responsiveness but could lead to overshoot and potential instability if it is too high. Similarly, altering integral time can accelerate the removal of steady-state error but can cause oscillations if it is not well tuned. Derivative time is essential, though, in providing damping to the system, reducing overshooting and enhancing stability, especially for oscillation-prone systems. A clear understanding of how all of these parameters interact is essential so that controller performance is optimized, particularly in systems that have complex nature and require fine control and therefore warrant extensive research into tuning methods and applications. While its performance is better, the best PID control frequently entails fine tuning to treat the special characteristics of the system of interest. PID tuning methods are necessary to give the three parameters' best settings to deliver the best performance. A variety of techniques have been developed, ranging from hand-tuning approaches, such as the Ziegler-Nichols technique, to advanced algorithms involving artificial intelligence and swarm optimization techniques, reflecting a trend toward automation of controller design. These methods allow PID parameter tuning to enhance performance across various operating conditions. From recent studies, successful tuning is seen to cause meaningful system response and stability enhancement, further substantiating the PID controller as an unshakeable foundation in modern control engineering (Ahmed, 2022; Saeed et al., 2021). With increasing industries demanding precision and dependability, research on PID control and PID tuning is further evolving itself, showing its vital place within contemporary technology.

## *Historical Development of PID Controllers*

Evolution of PID controllers has its roots in the early 20th century when industries began taking automation seriously as part of their manufacturing process. Controllers at that time primarily applied proportional control, which provided a straightforward means of control of output variables by varying the input in direct proportion to the error signal. But with the increased demand for more precision and efficiency, the limitation of proportional control became evident. This necessitated the inclusion of integral and derivative actions to enhance system response and stability. The pioneering efforts of engineers such as Elmer Sperry and Harold Stephen Black in the 1920s laid the foundation for PID control, with subsequent sophisticated applications in various fields of engineering. Their efforts placed the significance of feed-back loops in to-the-death drama when it was a question of controlling dynamic systems, radically transforming control engineering terrain. Formalization of the PID controller happened during the mid-20th century with more studies of automatic control systems. With the establishment of the transfer function and the methods of frequency response, engineers were able to systematically analyze systems, and it was understood that with the integration of the proportional, integral, and derivative components, a superior performance could be obtained. The researchers then began examining how adjustment of each component's parameters would result in different responses of the system under diverse conditions. This era also witnessed the emergence of graphical techniques, such as the root locus and the Bode plots, which provided techniques for design and PID controller tuning. Every part of the PID controller was developed in an orderly manner to cater to particular system responses. This increased understanding fueled additional progress in control theory, and the PID controller became a cornerstone of modern automation methods, particularly in aerospace and industrial contexts. With the advancement in technology, efforts were directed towards better PID tuning methods to address complex systems and dynamic environments. The application of computer-aided design and simulation packages in the 1980s also changed the approach to controller design, making prototyping rapid and allowing sequential tuning of PID parameters. Researchers began investigating various tuning algorithms, from heuristic to optimization-based, to automate this process. This development was fueled by the arrival of artificial intelligence techniques, which brought new ways of enhancing PID performance through system identification and predictive control methods. Even more significant, research into hybrid control systems emphasized the potential for combining PID control with advanced AI techniques to address challenging issues in real-time processing and decision-making. Overall, the history of the evolution of PID controllers is not so much a chronology of technological advancements; it is an ongoing dialogue between theory, practice, and innovation within the discipline of control systems (Ahmed, 2022; Zhao et al., 2020).

## *Importance of PID Control in Engineering Applications*

The significance of PID control in engineering cannot be exaggerated, since it is the essence of most modern control systems. In the integration of proportional, integral, and derivative actions, PID controllers provide a general yet robust method for regulating dynamic systems. In particular, the proportional action balances the moment errors by adjusting the control output proportionally, while the integral action eliminates the steady-state errors in the long term by considering previous errors. Moreover, the derivative action looks ahead to future errors as a function of the rate of change, forming a responsive yet stable system. This multi-faceted approach is required by a wide variety of applications for engineering ranging from automobiles to aerospace and heating, ventilation, and air conditioning systems to robots, demonstrating that effective PID tuning has the ability to significantly enhance system performance and dependability. Apart from this, PID controllers are highly adaptable to various fields of engineering and hence meet various operational needs. For instance, in industrial automation, a precision PID controller may propel manufacturing processes towards better efficiency and accuracy and potentially prevent machine breakdown and maximize output production. Similarly, in chemical engineering, PID control plays an important role in regulating reaction conditions to optimal levels and thereby improving product quality and safety. The real-world complexity of these applications calls for precise control strategies, demystifying the importance of PID systems. This versatility is also complemented by evolving tuning methods, allowing for finer-scale tuning that addresses the individual specificity of each application, underscoring the controller's role as an all-purpose tool in engineering design. Emerging trends in intelligent control techniques, such as reinforcement learning, are increasing the relevance of PID control to solutions for new problems confronting engineering applications. Employing intelligent algorithms, engineers can develop adaptive PID control systems that learn from the real world and keep improving. This is particularly demonstrated in fields such as fusion energy, wherein the control of plasma conditions is highly complex with regards to the dynamic dynamics of the system. The new technique involving machine learning coupled with conventional PID methods can lead to radical efficiencies and innovations, such as tokamak magnetic controller design improvements enabling operation over various plasma configurations (Degrave et al., 2022). Thus, as the requirements of engineering evolve, the ongoing relevance of PID control will be demonstrated increasingly and serve as a stage for future innovations in controller design.

## 8.1. COMPONENTS OF PID CONTROLLERS

The Proportional, Integral, and Derivative (PID) controller is a critical control system in control engineering, characterized by its three fundamental elements with individual functions to control the system. The proportional element, represented by "P," primarily manages the reduction of the error between the desired setpoint and the current output. Engineers can boost the systems response by adjusting the proportional gain, Kp; yet high Kp can lead to overshoot and instability. Conversely, low Kp delivers slow system response. The operation of the proportional component, therefore, relies on a delicate balance, which is essential to ensure maximum system performance in various applications, from temperature control to motor speed control, as highlighted in investigations into DC motor control and performance techniques (Zhang, 2025). The integral component, "I," offsets the cumulative error over time by making a correction based on the sum of past errors. This aspect is important in eliminating steady-state errors that could result from system response differences. Incorporating the integral gain, Ki, offers a flexibility to adjust the rate at which the controller responds to accumulated errors, enhancing long-term accuracy. Excessive Ki, however, causes system response oscillations, leading to system instability. Hence, the tuning of the integral component is important in guaranteeing that the system attains and maintains its optimal levels of performance without significant oscillations. The interaction between the integral and proportional components is a significant factor in the tuning of PID controllers for various industrial processes to guarantee an optimum strategy toward system stability and performance enhancement (Botirov et al., 2025). Lastly, the derivative term, denoted as "D," forecasts future system response by determining the error rate of change. The predictive capability serves to dampen the oscillatory effects typically associated with abrupt system output changes. By incorporating the derivative gain, Kd, engineers can make the systems response to transient excitation adjustable, something which is crucial in applications where speedy and accurate responses are necessary, such as in robotics and automated manufacturing procedures. However, too much concern for Kd will increase noise in the signal, thus an unstable system. Proper balancing of the proportional, integral, and derivative terms represents the key to successful PID tuning techniques for optimal operation in different systems as evident through studies of PID parameter effect on DC motor performance and other automated technology (Zhang, 2025; Botirov et al., 2025) (Figure 8.2).

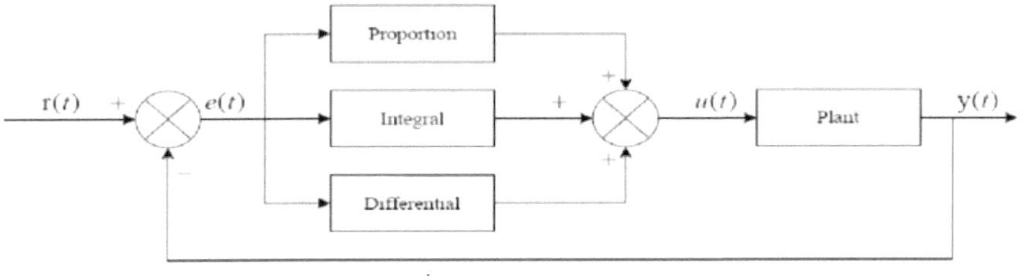


**Figure 8.2**. *Components of PID controllers.*

*Source: https://www.researchgate.net/profile/Mohammed-Desouky-2/publication/331962374/figure/fig6/AS:739304695422978@1553275383242/Figure-2-7-Components-of-PID-controller.png.*

## 8.1.1. Proportional Control

Proportional control is an elementary component in the Proportional-Integral-Derivative (PID) controller used in control systems. Its primary function is output control based on the difference between the desired setpoint and the actual process variable. The strategy controls the output proportionally to the error and can respond quickly to disturbances or setpoint upsets. While proportional control makes the system more responsive, it often creates residual steady-state error, which requires integration and differentiation in the PID configuration. The role of proportional control is particularly important in those applications where instant adjustment is essential, and it justifies its pervasive use in most industrial processes where accuracy and efficiency are required. Moreover, the function of the proportional component depends heavily on parameter tuning, which directly impacts overall system performance. The process of tuning involves careful choice of proportional gain, which determines the controller's aggressiveness in responding to the error. Greater proportional gains ensure quicker responses but may render the system more susceptible to system instability and introduce oscillation or overshoot. Conversely, smaller increases result in stability but may produce bad response times. Careful tuning of the proportional gain is essential to achieve the fine balance between response and stability, whose impact strongly affects the control system performance under various situations. Such close examination in mandating such fine tuning concurrently with research on PID techniques optimized for specific system characteristics and needs is appropriate. Apart from system performance, proportional control has other implications in the broader control system design arena, particularly in decentralized operation for multivariable systems. Industrial processes primarily involve complex interactions within multiple input-output networks, which may make regular proportional control less effective. Decentralized PID tuning methods are thus formulated to minimize unwanted coupling effects and enhance system robustness and stability in these circumstances. By formulating PID parameter tuning as a nonlinear optimization problem, researchers have made approaches that ensure strong performance across interdependent systems. Through this, it not only allows operators the potential to keep levels of performance irrespective of system interaction but also creates avenues for improved industrial application in disparate environments, citing the basic importance of proportional control in modern engineering applications (Thiago et al., 2021).

### 8.1.2. Integral Control

Integral control is the fundamental component of the PID control structure, which plays a critical role in eliminating steady-state error from system operation. Through the summation of the error over time, the integral term continuously corrects the controllers output so that the error is brought down to a minimum, ensuring the system reaches its desired setpoint. This capability ensures integral control is highly beneficial to precision applications, including active magnetic bearings (AMB). In AMB systems, the system may go unstable under bad tuning, but integral control can assist in achieving a stable response with feedback integration over time. Control of this integral parameter successfully can significantly contribute to the efficiency of control systems, bringing vigorous solutions in many engineering problems, particularly for high-precision conditions where operation consistency is required (Zhou et al., 2024). Besides the correction of only steady-state error, integral control impacts the dynamic response of a system. When integrated along with proportional and derivative terms, it together decides system oscillation, overshoot, and settling time. For example, while increasing the integral gain might reduce the steady-state error, it may introduce oscillatory behavior or long settling times if properly not tuned. Fundamentally, the integral term would have to be compensated by means of other PID parameters to avoid destabilizing the control loop. In cases such as power tools' vibration controlling systems, the necessity for tuning techniques such as Crude Approximation and Particle Swarm Optimization becomes evident in order to balance these parameters for maximum performance. This accurate tuning puts an emphasis on the significance of integral control as a critical handle in managing advanced systems (Usamah et al., 2023). The functionality of integral control within PID controllers is essential to understand while developing robust and effective control systems. The integral term not only enhances system accuracy but also demands a very accurate approach to tune and recognize the system. As in the case of advanced applications like AMB systems, using model-free tuning methods can reduce the stabilization process and improve response time. Thus, closely investigating such relationships helps engineers design strategies to counteract potential instabilities while taking advantage of the strengths of each PID component. Finally, mastery over integral control not only aids specific applications but also aids in strengthening the overall organizational discipline of control engineering through better designs that adhere to evolving technological demands (Figure 8.3).

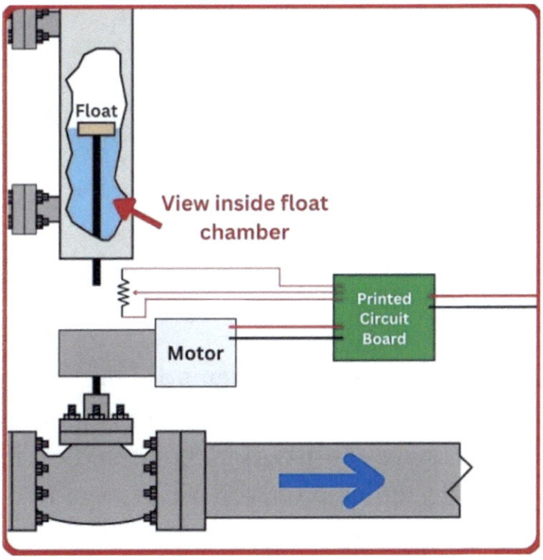


**Figure 8.3**. *Integral (Reset) control.*

*Source: https://control.com/uploads/thumbnails/Thumbnail_32_6.jpg.*

## 8.1.3. Derivative Control

The addition of derivative control to the Proportional-Integral-Derivative (PID) controller design is essential to optimal system performance and stability. Derivative control reacts to the error signal rate of change directly, making provision for predictive actions that decrease response time. The derivative action minimizes overshoot and oscillations in dynamic systems by predicting future trends of errors based on the current rate of change. This is especially true in high settling time and high precision applications where latency can negatively impact system performance. Further, the derivative term enables damping to be achieved, which is important to guarantee stability when the system experiences transient conditions. Therefore, the derivatively controlled response has tremendous impact on the closed-loop performance of systems ranging from industrial mechanical machinery through to advanced robotic utilization and hence its inclusion in a controller ought to be given high priority.

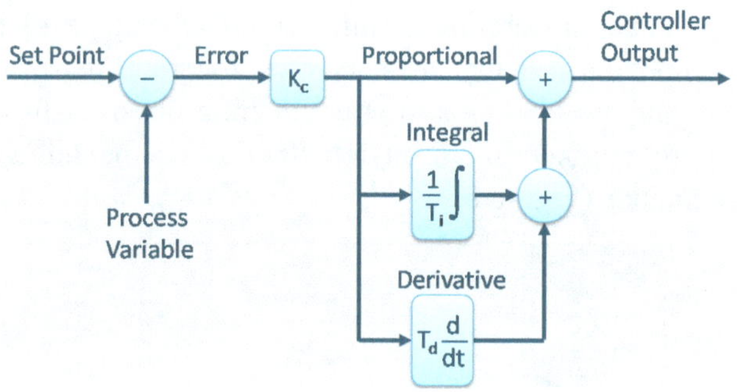


**Figure 8.4**. *Derivative control.*

*Source: https://blog.opticontrols.com/wp-content/uploads/2010/05/pid_controller.png.*

Key to PID parameter tuning, particularly the derivative gain, is an understanding of compromises involved in system response. A derivative gain too high can result in excessive sensitivity to noise and oscillations in the error signal, leading to system destabilization. A derivative gain too low will produce sluggish response, delaying the pre-emptive corrections needed. Optimization methods for such parameters are algorithmic in nature and study how the system behaves under varying conditions. Importantly, the efficiency of derivative control is enhanced through coupling with tight tuning approaches that take into account system dynamics (Thiago et al., 2021). Such methods are shown to be effective in addressing multivariable interactions in closed-loop systems such that measures of performance such as robustness and stability are realized while ensuring responsiveness. In current applications, practical implementation of derivative control needs to consider dynamic system parameter changes, which may be derived from either external or process natural variability. An area of interest of research highlighted in recent publications supports decentralized tuning methods that are efficiently addressing multi-input multi-output (MIMO) system interactions (Thiago et al., 2021). By developing a tuning approach that manages performance and reliability, engineers are able to optimize the derivative element in order to minimize coupling effects across inputs and outputs. In addition, with the complication of industrial systems growing, derivative control not only needs to be effective but also simple to implement in a real-life scenario. With ongoing research and development in PID tuning practice, derivative control remains the focal point in the design of controllers, with scope for improved efficiency and reliability in engineering applications (Figure 8.4).

## 8.2. MATHEMATICAL REPRESENTATION OF PID CONTROL

The mathematical representation of Proportional, Integral, and Derivative (PID) control is fundamental to understanding how these controllers operate in various engineering applications. PID controllers function through the computation of an error value, which is the difference between a desired setpoint and a measured process variable. This error is processed to produce an output signal, typically represented by the equation: $u(t) = K_p e(t) + K_i \int e(t) dt + K_d \frac{de(t)}{dt}$. The proportional term, $K_p e(t)$, responds to the current error, while the integral term, $K_i \int e(t) dt$, accounts for the accumulation of past errors. Lastly, the derivative term, $K_d \frac{de(t)}{dt}$, predicts future errors based on the rate of change, effectively anticipating the systems behavior. Each of these components plays a crucial role in enhancing system stability and performance. Exploring the effects of PID parameters on system performance is essential in the context of controller design. The proportional gain $K_p$ adjusts the immediate response of the system; a high value may lead to overshoot while a low value could result in a sluggish response. The integral gain $K_i$ helps eliminate steady-state error by providing necessary adjustments based on past performance, but excessive integral action can lead to instability and oscillatory behavior. Conversely, the derivative gain $K_d$ serves to dampen system oscillations and improve stability through predictive capabilities. Thus, the careful tuning of these parameters is vital, as improper settings can adversely affect system performance, highlighting the intricate balance required in PID control. Effective tuning methods are pivotal in optimizing these parameters to achieve desired performance characteristics. PID tuning methods are crucial for ensuring the effective implementation of these controllers across various applications. Several techniques exist, such as the Ziegler-Nichols method, which relies on experimental data to set the PID parameters based on the system's reaction to a step input. Another approach is the Cohen-Coon method, which provides guidelines for systems with significant dead time and time delays. These tuning methods facilitate the identification of optimal PID settings that minimize error while maximizing responsiveness.

Moreover, advancements in computational techniques, including modern algorithms and software-based simulations, allow for real-time PID tuning, adapting to changes in system dynamics. As such, these strategies enhance the robustness and efficiency of PID controllers, significantly impacting their effectiveness in diverse fields ranging from robotics to industrial automation (Idrissi et al., 2022; Davies et al., 2021).

### 8.2.1. PID Control Equation

The Proportional-Integral-Derivative (PID) control equation is a simple device in control system design, and it is applied to give a framework to control various dynamic systems. The main function of the PID controller is essentially to minimize the difference between the desired setpoint and the current output of the system with control input adjustment. The equation itself synergistically incorporates three components: the proportional term responds to the current error, the integral term responds to the integration of accumulated errors over time, and the derivative term responds to future errors based upon rates of change. The synergism here represents a complex control strategy that can be adapted to fit varying system dynamics, and it is therefore extensively utilized in various applications across engineering. Moreover, proper tuning of these parameters is of utmost importance since it decides the overall performance of the control system, ensuring stability as well as responsiveness towards disturbances. The success of PID control in real-world applications can largely be decided by an optimal tuning of its parameters. Various tuning methods exist, from basic heuristic practices like the Ziegler-Nichols method to advanced techniques employing optimization algorithms. Proper selection of tuning techniques matters a great deal, particularly in systems where there are nonlinear behaviors and uncertainties involved. For instance, in systems with external disturbances, like tank sloshing modes, the capability to precisely tune parameters will minimize unwanted vibrational responses. Experiments have indicated that the application of a PID controller can yield great improvements in control performance, such as in experiments with tuned controllers applied to advanced systems like functionally graded graphene platelet-reinforced composite materials (Jamalabadi, 2022). This research is further evidence of the importance of applying the appropriate PID parameter values to achieve targeted system responses. New directions in PID control design are also exploring adaptive techniques that adaptively optimize performance under varying operating conditions. One important development is the marriage of fuzzy logic with traditional PID controllers that led to the development of self-tuning controllers that have the capability to adapt parameters in real-time. This method is particularly beneficial in handling the intricacies brought about by nonlinear processes since it enables better control of changes in parameters and noise from sensors. For instance, utilization of an adaptive fuzzy fractional order PID controller has demonstrated improved stability and performance in the control of AUVs in turbulent conditions (Khodayari et al., 2021). Such innovations indicate continued development in PID control methods, with emphasis on fulfilling the needs of sidestepping challenges posed by modern dynamic systems, and achieving optimal compromises between performance, stability, and robustness (Figure 8.5).

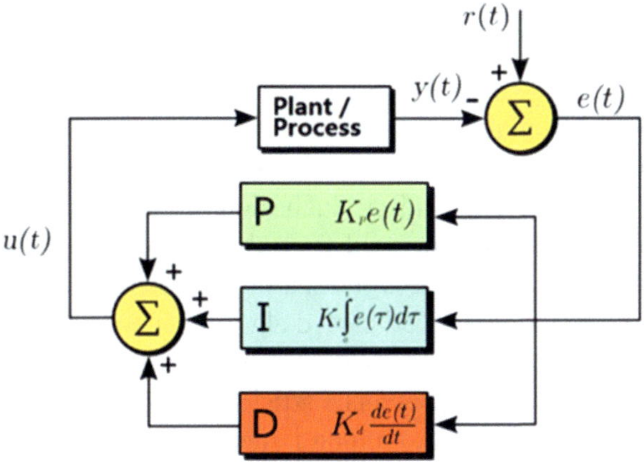


**Figure 8.5**. *PID control algorithm.*

*Source: https://www.veichi.com/d/file/p/e98e5c73fd98eab4592b74920282e07e.jpg.*

## 8.2.2. Transfer Function of PID Controllers

The transfer function of PID controllers plays a crucial role in controller design as it encapsulates the dynamics of the control system. The PID controller comprises three components: proportional, integral, and derivative, each contributing uniquely to system behavior. The transfer function for a standard PID controller is expressed as $G(s) = K_p + \frac{K_i}{s} + K_d s$, where $K_p$, $K_i$, and $K_d$ are the proportional, integral, and derivative gains, respectively. This function indicates how the output responds to changes in the input over the Laplace transform variable $s$. The communication of this relationship is fundamental to the understanding of how modifications to the PID parameters affect the performance of the system as a whole and is therefore of critical importance for use in a range of applications across disciplines like control engineering and automation. A knowledge of the transfer function of the PID controller enables engineers to set up their systems to perform to desired specifications, like stability, response time, and steady-state error. It is conceivable to have a significant impact on the transient and steady-state response of the system by making adjustments to the PID parameters. A case in point is that the proportional gain may be raised to provide a faster response, but possibly destabilize the system. The integral action reduces the steady-state error, but excessive use leads to oscillation. Similarly, the derivative action provides a damping contribution that can minimize overshooting and improve stability. Therefore, carefully exploring the implications of each parameter in the transfer function in the context of the transfer function framework elucidates their combined effect on system performance, opening the way for improved controller design procedures. The implications of the PID controller transfer function extend beyond to theoretical abstractions; they inform applied tuning procedures. Techniques such as the Ziegler-Nichols method and computer-based auto-tuning software use the transfer function to generate optimized PID parameters for a specific application. Such tuning methods have a tendency to utilize the frequency response derived from the transfer function for balancing the trade-offs among responsiveness and stability in an optimal way. Additionally, along with more development in advanced algorithms and artificial intelligence, such as paradigms based on swarm intelligence seen in literature, like those recorded in the research that presents Particle Swarm Optimization, there are chances of innovative solutions to tuning issues emerging (Ahmed, 2022). This integration of classical control theory with contemporary computing techniques provides a rich ground for growth in PID tuning and controller performance, finally enhancing the effectiveness and efficiency of controlled systems (Zhao et al., 2020).

## 8.2.3. Block Diagram Representation

Block diagram notation is a useful tool in control system design and analysis, particularly where Proportional, Integral, and Derivative (PID) controllers are being utilized. The diagrams provide a graphical illustration of the interaction and association among different parts of a control system. Each block in the diagram will typically represent a specific function or action, that is, the controller, the plant, and feedback elements. This block-based approach not only sheds light on the performance of advanced systems but also facilitates the identification of system dynamics and interactions important for effective controller design. With knowledge of how to draw and read these diagrams, engineers are able to make better selections of appropriate PID parameters because they can instantly appreciate how each element of the controller will affect overall system response, thereby guiding tuning processes to realize better response and stability. Of note in block diagram depiction is its application in the PID parameter tuning to achieve maximum system performance. Through the visualization of the control loop as a sequence of operations among the proportional, integral, and derivative components, the engineers can establish how variations in these parameters affect the responsiveness and stability of the systems. For instance, increasing the proportional gain could enhance the speed of response but could also introduce overshooting and oscillation. Conversely, adjustment of the integral component eliminates steady-state errors at the cost of instability if over-adjusted. Block diagrams allow systematic assessment of such trade-offs to enable improved communication between design groups and stakeholders regarding potential impacts of tuning choices. This visualization is vital in sophisticated applications, for example, renewable energy systems, where reliability and efficiency are of utmost concern (Mousavi et al., 2022). Furthermore, the strength of block diagram representation is carried through to several applications, including unmanned aerial vehicles (UAVs), for which high accuracy in control is required. In the design of UAV control, block diagrams represent the relationship between various control methods and the UAV dynamics so that there can be a better understanding of how disturbances, both internal and external, influence flight performance. It helps in making the selection and use of the correct dynamic control methods to combat such disturbances and enhance system reliability (Idrissi et al., 2022). From block diagrams, engineers are able to design iteratively with ease, ensuring that the chosen PID controls will yield the desired performance under diverse work conditions. Overall, block diagram presentation is an effective aid to controller design, bridging the gap between theory and practice in system optimization and PID tuning (Figure 8.6).

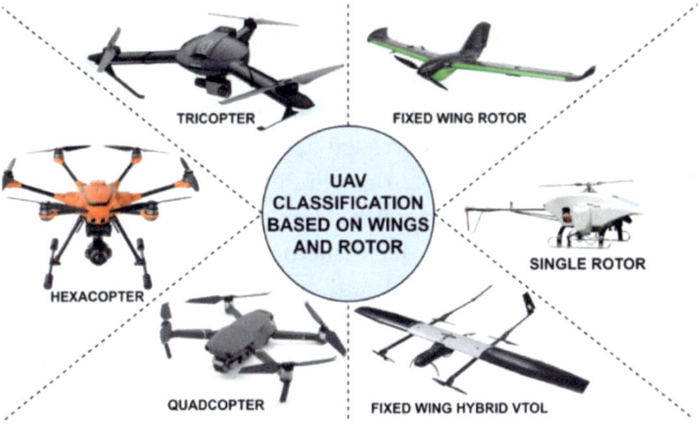


**Figure 8.6**. *UAV technology.*

*Source: https://pigeonis.in/wp-content/uploads/2023/07/Detail-Guide-about-UAV-Technology-Supporting-Aerial-Task-Operations-Google-Docs.jpeg.*

## 8.3. EFFECTS OF PID PARAMETERS ON SYSTEM PERFORMANCE

In control systems, the dynamic response and performance of a system rely heavily on the PID controller parameters. Each component of the PID controller has a different function: the proportional term adjusts the output according to the current error, the integral term adjusts for past errors that have accumulated, and the derivative term adjusts for projected errors based on the rate of change at the moment. The tuning of these parameters has a great influence on system stability, response, and accuracy. Excessive proportion gain, for instance, can provide a fast system but may lead to oscillations and instability if not balanced by correct integral and derivative settings. The problem lies in discovering a compromise among these parameters to provide good performance without exerting adverse influences such as overshoot or excessive settling time. PID parameter tuning is essential to avoiding undesirable system responses, especially in sophisticated applications. Based on research, advanced techniques of tuning, such as the Implementation of the Improved Sparrow Search Algorithm (ISSA), have been proposed to optimize PID controller gains for specific applications. Using optimization techniques that maximize searches for optimal parameters, scientists have been able to reduce overshoot and improve system stability. For example, through the utilization of chaos mapping and dynamic learning factor adaptation, it is possible to enjoy improved controller performance in tasks like overhead cranes, where oscillation and positioning of the payload are required. This approach not only alleviates traditional overshoot problems but also contributes to the robustness and flexibility of the control strategies under various operating conditions, hence highlighting the significance of effective PID parameter tuning for overall system performance (Zhang et al., 2024). Additionally, studies on advanced optimization methods such as Particle Swarm Optimization (PSO) show an increased tendency to utilize nature-inspired methods for PID tuning. Progress of research is further extended into intricate interdependency between the parameters of an algorithm and metrics for system performance with the objective of enhancing precision and effectiveness on various domains, such as industrial and commercial application. Systematic review of PSO methods leads to evidence of widespread applications where parameter optimization is of central importance. Every variation of PSO possesses certain strengths and weaknesses, suggesting that tailored techniques might address some control problems with high efficiency. Future research will certainly illuminate further higher-order tuning techniques, continuing to further the use of PID controllers and solidifying their role in optimal system performance in ever more advanced working conditions (Ahmed, 2022) (Figure 8.7).

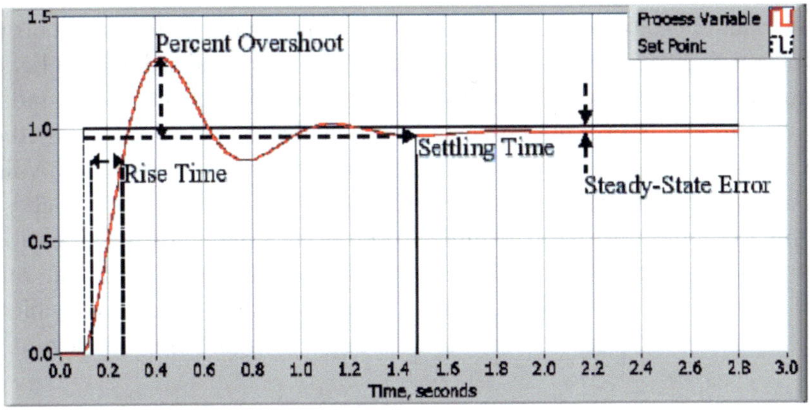


**Figure 8.7**. *The PID controller & theory.*

*Source: https://ni.scene7.com/is/image/ni/12fbdcae1636?scl=1.*

## 8.3.1. Impact of Proportional Gain (Kp)

The proportional gain (Kp) is a critical component in the performance of a Proportional-Integral-Derivative (PID) controller and is a key determiner of system response characteristics. With Kp, the controller adjusts the output proportionally to the current error, and higher gains produce quicker responses for the system. This fast response, however, comes at the expense of stability; excessive Kp can lead to oscillatory response, which converges slowly or diverges totally, especially in systems with heavy inertia or delay. The compromise between responsiveness and stability is the essence of the challenge in designing controllers, with precise tuning of Kp needed to balance performance requirements without compromising system integrity. The interaction of proportional gain with the other PID coefficients is intriguing because Kp is not independent. Along with the integral term (Ki) and the derivative term (Kd), Kp is a deciding factor regarding how the system dynamics will behave as a whole. As an illustration, while Ki addresses steady-state error by eliminating residual differences after a time period, Kp will influence the rate at which the system responds to that error. Conversely, the derivative gain serves to dampen oscillations that might otherwise be amplified through excessive proportional gain. Thus, well-tuned PID control requires experienced knowledge of how Kp depends on Ki and Kd, again highlighting the requirement of well-structured tuning methods considering such dependencies in order to provide optimal performance. Besides, advanced methods of tuning have emerged that further enhance the efficiency and effectiveness of Kp in practical applications. The

Yellow Saddle Goatfish Algorithm (YSGA) is among the methods that brought into focus today's approaches to optimizing not just Kp but also the overall system dynamics based on varying conditions. For example, recent studies indicate that certain new algorithms prove to be better than standard tuning methods, especially when dealing with complex systems such as Automatic Voltage Regulators (AVR) (Micev et al., 2020). Similarly, data-driven approaches are becoming popular for their capability to stabilize systems irrespective of their dynamics (Persis et al., 2019). These technologies represent the evolving reality of PID tuning and how proportional gain tuning is at the center of achieving robust and adaptive control strategies in a range of engineering applications.

### 8.3.2. Influence of Integral gain (Ki)

The Integral gain (Ki) of a Proportional-Integral-Derivative (PID) system is of utmost significance in the elimination of steady-state errors through the summation of the error over time. The dynamics offered by Ki make the system continuously self-correct to minimize the constant deviation from a target setpoint. Upon correct tuning, the integral component of the controller sums up past errors and thus responds to them to push the system output towards the desired point. Nevertheless, if Ki is excessively high, it could introduce overshooting and unwanted oscillations and thus destabilize the system. Conversely, if Ki is excessively low, it could generate a sluggish response and make the system longer to converge towards the setpoint. Thus, comprehension of the impact of Ki is of utmost significance in establishing a balance between stability and responsiveness in control actions. In practical applications, tuning of Ki needs to be considered with care because it has a direct impact on the systems transient and steady-state response. Various techniques exist for the guidance of the engineer in arriving at the optimum value of Ki, including empirical tuning, Ziegler-Nichols methods, and optimization techniques supported by software. Every technique is designed with a combination of strengths and weaknesses in systems behavior and operating conditions. Ziegler-Nichols tuning, for instance, provides a heuristic approach whose outcomes are satisfactory in most cases but fails to take into account complex dynamics occurring within nonlinear systems. Besides, advanced techniques based on Artificial Intelligence, such as Particle Swarm Optimization, have been introduced to provide more control over the influence of Ki on system performance, as discussed in relevant literature (Ahmed, 2022). Such kinds of developments could be most helpful in systems requiring high accuracy and flexibility. Furthermore, the interconnection between Ki and other PID parameters—that is, proportional gain (Kp) and derivative gain (Kd)—is essential to construct overall system control. Integrating these parameters requires a holistic perspective, as their cumulative effects can significantly alter system behavior. For instance, an aggressive Ki may necessitate a corresponding adjustment in Kp to maintain stability and avoid excessive oscillations. Conversely, aligning Kd effectively can also mitigate the overshooting risks spawned by a high Ki. Control systems are increasingly applied in many applications such as power electronics and telecommunications, where precision and reliability are of utmost concern, as mentioned in phase-locked loop studies (Golestan et al., 2016). With the evolution of controller design methodology, the complicated nature of each PID parameter, particularly Ki, will undoubtedly remain a topic of interest for research and application in control theory (Figure 8.8).

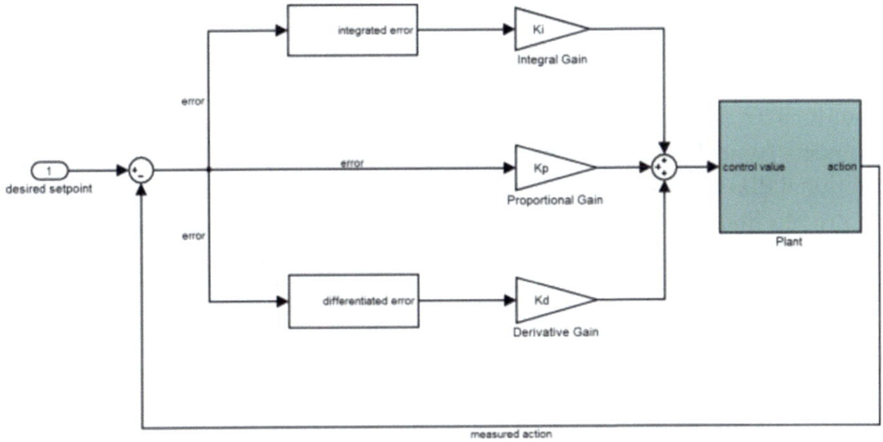


**Figure 8.8**. *Integral gain.*

*Source: https://ars.els-cdn.com/content/image/3-s2.0-B9780080977683000131-f13-05-9780080977683.jpg.*

## 8.3.3. Role of Derivative gain (Kd)

The derivative gain (Kd) is what makes a Proportional-Integral-Derivative (PID) controller perform by looking ahead to future system action from the error signal rate of change. Through its reaction to how fast the error is changing, Kd damps the system response, usually helpful in maintaining the system stable due to disturbances or changes in the setpoint. This forecasting capability enables the control system to respond to upcoming variations, hence reducing overshooting and enhancing transient response. Consequently, where quick fluctuations dominate systems, varying the derivative gain can significantly contribute to the performance of the entire system and suppress the effect of external disturbances. This is particularly evident in industrial processes and robotics, where quick response and precision are crucial to ensure efficient operation. Maximum performance of the control system requires sensitive tuning of Kd. Large values of Kd will make the system too sensitive and, consequently, may generate oscillations with instability potential. Conversely, low derivative gain may cause slow response rates by failing to utilize the anticipatory component of the derivative term. Therefore, the PID tuning is often a case of careful adjustment of Kd in order to balance responsiveness and stability. Techniques such as Ziegler-Nichols and software automatic tuning procedures have been established in order to assist engineers in this tuning process so that they are able to determine the best values of Kd in the presence of system complications. In those nonlinear in nature systems, such as piezoelectric actuators, the need for having effective ways of compensating for Kds interaction with other PID parameters comes about to ensure optimal performance under various operating conditions (Kanchan et al., 2023). Derivative gain is typically coupled in advanced control methods with new approaches to solve for limitations that traditional PID controllers have. Research with fuzzy logic systems, for instance, has proven that combining Kd with fuzzy gain scheduling is feasible to make the controller more responsive to varying system dynamics. It is feasible to tune the controller parameters in real-time, for example, Kd, through sensing system behavior variations, for example, the nonlinear behavior observed in wind energy systems (Hermassi et al., 2023). With this development in their methods, the understanding of derivative gain in addition to other control parameters becomes all the more necessary for implementing high-performance control systems. Ultimately, Kds function is something more than a computation of the rate of error change at the present moment; it has a bearing on the overall effectiveness of the whole control strategy, particularly in controlling complex and time-varying systems.

## 8.4. STABILITY ANALYSIS IN PID CONTROL

PID stability analysis is a critical component of effective controller design, enabling the testing of system performance across various operating conditions. The primary objective in stability analysis is to have the control system act predictably without exhibiting unnecessary oscillation or venturing away from the desired setpoint. Through inspection of the root locus, Nyquist, or Bode plots, engineers can specify stability margins and assess the effect of changes in proportional, integral, and derivative parameters on system performance. Each has a particular purpose: the proportional term controls short-time response, the integral term removes static error, and the derivative term anticipates future system response. Thus, stable analysis with caution ensures that the PID controller can be tuned for maximum performance in a wide range of applications. The PID control parameters not only decide the system response but also decide the stability of the system directly. Variation in the proportional gain, for instance, can lead to faster response but simultaneously can lead to oscillation and overshooting. Conversely, an increased integral gain will enhance steady-state performance but potentially destabilize the system by overcorrecting past error.

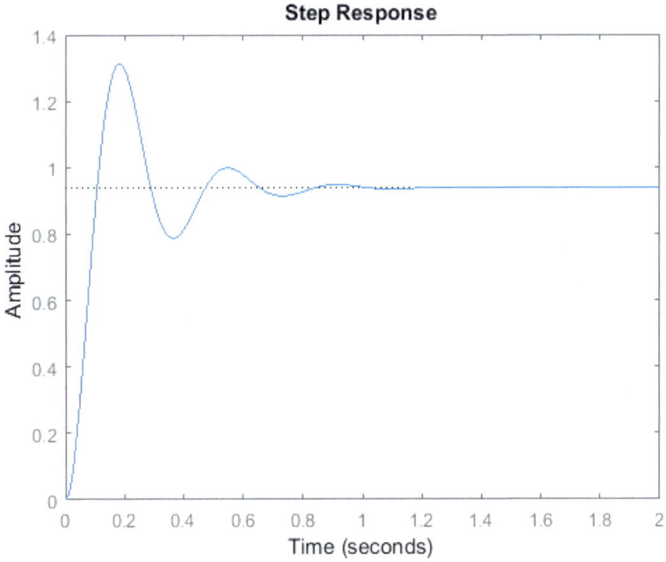


**Figure 8.9**. *Stability analysis in PID control.*

*Source: https://ctms.engin.umich.edu/CTMS/Content/Introduction/Control/PID/html/Introduction_ControlPID_02.png.*

There are tuning methods, such as the use of a Ziegler-Nichols rule or Cohen-Coon method, that provide systematic methods of parameter selection where each of them provides something unique in terms of balancing the trade-off between responsiveness and stability. Furthermore, application of high-level tuning techniques, like optimization procedures and model predictive control, is going to advance the stability analysis with a stronger performance in complex applications, as emphasized in recent studies of the creation of control techniques (Ahmed, 2022) and (Schwenzer et al., 2021). To acquire a well-rounded perspective of stability in PID control systems, it is essential to include simulation tools for software that can predict system behaviors when subjected to varying sets of parameters. One such tool is MATLAB/Simulink, which provides engineers with the ability to graphically model system responses, enabling assessment of stability barriers and potential response improvements before actual realization. Secondly, the evolution of hybrid techniques that blend traditional PID control with optimization techniques opens up opportunities to further extend stability, paving the way for more adaptive and less sensitive-to-changing-dynamics controllers. While the complexity of control systems increases, further research contributes significantly towards developing improved stability analysis methods, keeping this subject relevant in the overall study of controller design (Ahmed, 2022) and (Schwenzer et al., 2021) (Figure 8.9).

## 8.4.1. Definition of System Stability

The concept of system stability is paramount in control system design and implementation, particularly for Proportional, Integral, and Derivative (PID) controllers. Stability refers to the extent that a system can recover to its initial equilibrium after being subjected to a disturbance. In essence, a stable system will not exhibit uncontrollable oscillation or divergence behavior when disturbed. There are a number of methods of studying system stability, including root locus, Bode diagrams, and Nyquist stability criteria, which are all capable of providing information on the effect of changes to PID parameters on overall system behavior. Understanding of this phenomenon is critical in ensuring good PID tuning, enabling engineers to ensure stable operation in a vast majority of industrial processes. Additionally, PID controller tuning is directly associated with the achievement and sustenance of system stability. The parameters—derivative, integral, and proportional—are all utilized for distinct purposes in formulating the dynamic response of a system. A PID controller, optimally tuned, will optimize system performance by minimizing transient response stabilization and steady-state error. Unproper tuning will lead to instability, wherein oscillations and overshoot cause detrimental impacts in a control application. For instance, in decentralized PID control, when more than one input and output are to be optimized, the dependency of these parameters has to be properly tuned to ensure robustness and stability, hence counteracting any coupling effect in the system to the maximum possible extent (Euzébio et al., 2021). Recent advances in controller design have underpinned the requirement to blend artificial intelligence with conventional control methods to further improve the stability of systems. Techniques such as WNN-PID, involving the fusion of wavelet neural networks with predictive PID tuning, have been proposed. These optimize the traditional PID architecture through advanced predictive methods. AI-powered techniques introduce improved control performance through quicker response and reduced overshoot compared to conventional algorithms (Liu et al., 2020). Advanced algorithms employed facilitate real-time adaptation based on predicted output, reflecting the dynamic nature of modern control systems. Lastly, exploring new PID tuning methods bridges the gap between theory and practice, enabling systems to be robust and responsive against increasingly demanding environments (Figure 8.10).

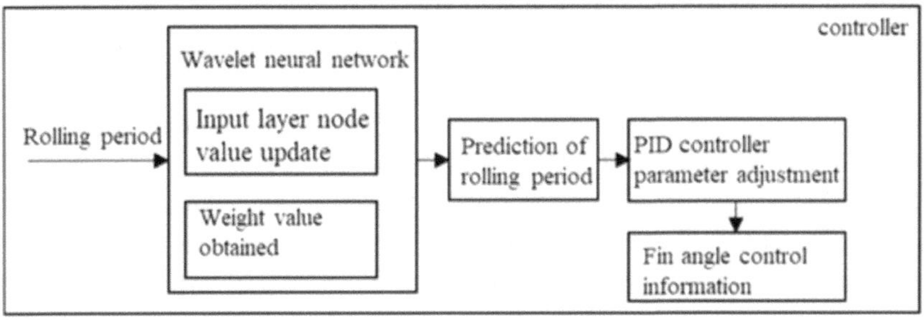


**Figure 8.10**. *Wavelet neural network-based PID (WNN-PID).*

*Source: https://www.mdpi.com/jmse/jmse-11-02205/article_deploy/html/images/jmse-11-02205-g010-550.jpg.*

## 8.4.2. Routh-Hurwitz Criterion

The Routh-Hurwitz criterion is a basic tool of control theory, particularly in robust controller design of dynamic systems. The criterion provides a methodical way of checking the stability of a linear time-invariant system without necessarily having to find the roots of the systems, which is problematic for higher-order polynomials. Stability in PID control is a priority one since system instability during operation will mean there is poor system performance in the form of oscillations and overshooting. Using the Routh-Hurwitz criterion, engineers can analyze the characteristic polynomial from the closed-loop transfer function and thereby find the impact of PID parameters on system dynamics and stability. This design can be utilized directly to choose and adjust PID parameters so that desired dynamic response is achieved along with a stable operating envelope. Use of the Routh-Hurwitz criterion in the design of the PID controller signifies its use in a broad array of engineering applications, ranging from automotive vehicle suspension systems to power system load frequency control. For instance, the research work in question stresses the importance of robust control schemes with disturbance tolerance and operating uncertainty tolerance—a parameter that can be considered through application of the Routh-Hurwitz criterion. Such controllers as the conventional P, PI, and PID controllers have to be critically evaluated in terms of stability margins in diverse configurations as well as under different operating conditions. As observed, perfect PID and internal model controllers have been found to perform better than their conventional counterparts. The Routh-Hurwitz criterion assists in determining such standards of performance by examining the influence of different controller design variations on the stability of the entire system under varying conditions (Provatas et al., 2023). Also, the Routh-Hurwitz criterion has real-world application to other sophisticated PID tuning methods, which would further enhance the performance of control in systems that work under dynamic conditions. Modern approaches to controlling systems, which include optimization techniques and adaptive controls, involve a good understanding of the intrinsic system dynamics, particularly stability. The specification is thus able to enable engineers to design situations where controller gains can be adjusted without loss of stability even in systems with significant disturbance or parameter variations, such as those achieved with electric vehicle fleets during cyberattacks (Raouf et al., 2023). Using the knowledge of the result of the application of the Routh-Hurwitz criterion, control engineers can design PID controllers satisfying performance criteria and guaranteeing stability under a range of operating conditions. Such dual concern about performance and stability is crucial while designing reliable control systems.

### 8.4.3. Nyquist Stability Criterion

Understanding the Nyquist stability criterion is essential in the area of PID control and system performance analysis. The criterion provides a graphical method for assessing the stability of a closed-loop control system from the frequency response of the open-loop transfer function. By constructing the Nyquist plot, one can observe how the open-loop system responds to a range of frequencies and identify encirclements of the critical point ( –1, 0) in the complex plane. The number of such encirclements determines the system's stability, in turn related to the gain and phase margins crucial to robust controller synthesis. This analysis becomes essential while PID parameter tuning as designers have to ensure that modifications do not come at the cost of system stability, bearing testimony to the combined nature of PID control techniques and Nyquist stability assessment. Incorporating the Nyquist stability requirement into PID control design methods allows engineers to make sound decisions. Because the PID controller assigns proportional, integral, and derivative actions weight for the best possible system response, the challenge is how to adjust these coefficients without destabilizing the system. The Nyquist criterion achieves this by offering a tool that allows control engineers to see the stability implication of employing specific PID coefficients. For example, increased proportional gain can increase response rates at the expense of stability. Understanding these trade-offs becomes critical as part of the tuning process, emphasizing the value of a robust design that is capable of accommodating many operating conditions while meeting stability constraints imposed by Nyquist analysis. Further, the use of the Nyquist stability criterion is illustrated in many complex control systems, particularly in modern applications such as microgrids and advanced building controls. For instance, as presented in new literature, model predictive control (MPC) utilization can lead to substantial energy conservation with ensured stability in varying conditions (Ján Drgoňa et al., 2020). This suggests that complete knowledge of stability conditions not only helps in traditional PID tuning but also ensures the optimal performance of more advanced control measures. Similarly, the study of decentralized and coordinated control schemes in dc microgrids illustrates the necessity of implementing stability criteria like Nyquist to ensure robustness for multi-controller systems (Dragi et al., 2015). Stability in these systems is often realized by the inclusion of various control elements with an eye on the dominant forces decided by the Nyquist stability paradigm (Figure 8.11).

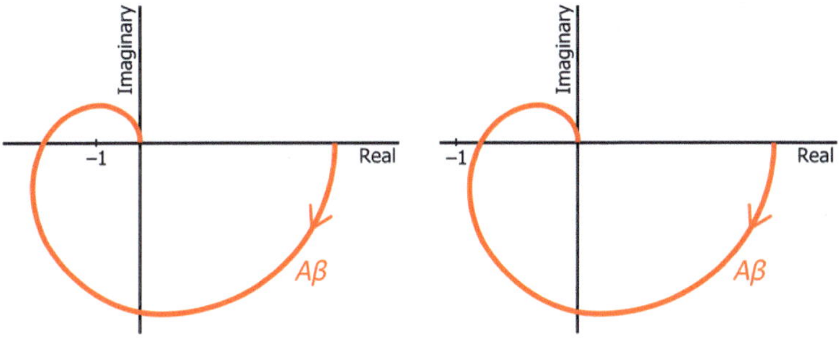


**Figure 8.11**. *Nyquist stability criterion of PID control and system performance analysis.*

*Source: https://www.allaboutcircuits.com/uploads/articles/nyquist_plots_fir_stability_analysis_fig04.png.*

## 8.5. PID TUNING METHODS

PID tuning methods play a crucial role in the optimization of Proportional, Integral, and Derivative (PID) controllers to make sure that control systems respond to disturbances and maintain desired output levels. Among all the different methods of tuning, model-based methods utilize mathematical modeling in simulating system dynamics under different parameters. This technique allows engineers to predict how changes in the PID values influence the behavior of the system and, hence, makes the process of optimization more systematic. Application of such methods can significantly improve system performance by reducing overshoot and settling time, which ultimately results in greater stability and accuracy in controlled systems. As a result, a careful analysis and use of these techniques are made a priority in gaining the reliability of PID controllers in most industrial applications. Another effective technique is the trial-and-error method, which, although being simple, is still widely used because of its simplicity and ease to use. Engineers change the PID parameters in a systematic fashion as they monitor the systems response to inputs and perturbations, hoping to provide acceptable performance. However, this method is often time-consuming and is not always optimum, particularly in more complex systems where there are multiple interactions that complicate the process of tuning.

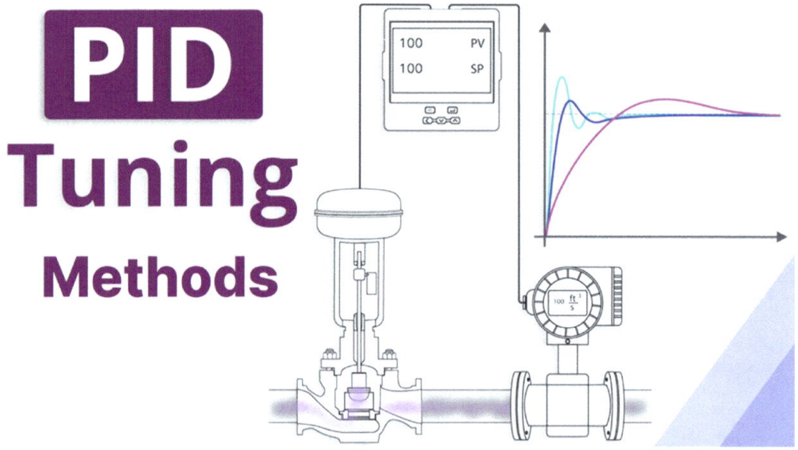


**Figure 8.12**. *PID tuning methods.*

*Source: https://miro.medium.com/v2/1*rc0iv7tRnaQ_Lnc7I1p5lg.png.*

While it has its drawbacks, trial-and-error tuning is advantageous in that it works when there is no model available for the system and allows more manual controller design. Further, the automation of this process has been facilitated by several software tools that are able to simulate and analyze system response, balancing ease with performance in PID tuning. Advanced techniques of tuning, such as the multi-objective PID tuning method, utilize more sophisticated criteria, optimizing performance in multiple directions simultaneously. With the use of a single perturbation strategy, this approach simplifies the process of tuning by analyzing the effect of each parameter to the system in separated time intervals, hence reducing computational complexity. It is not only focused on minimizing overshoot but also on enhancing the systems noise rejection and load disturbance resistance, a major factor in feedback control systems. Research illustrates how the said multi-dimensional tuning techniques significantly improve general system performance, as observed in various applications like power systems and industrial automation (Kalyan et al., 2024; Aafaque et al., 2023). This emerging method captures the evolving dynamics of PID control where efficiency and robustness are all in the wake of varied operating challenges (Figure 8.12).

### 8.5.1. Ziegler-Nichols Method

The Ziegler-Nichols method remains one of the most well-known PID controller tuning methods since it is a formal method of determining the optimal gain parameters. The method employs either open-loop or closed-loop testing techniques in the calculation of key values to be employed in designing good control behavior. The idea behind the Ziegler-Nichols method is that it will be able to compute the final gain (Ku) and the oscillation period (Pu) from which to calculate the proportional (Kp), integral (Ki), and derivative (Kd) gains of a PID controller. With these parameters, the engineers can enhance the system performance considerably, and hence the Ziegler-Nichols method becomes a crucial tool to be applied in control engineering, particularly in sophisticated systems where manual tuning is not sufficient. The Ziegler-Nichols method has also been found to be dependable and consistent in application in motor control systems, among others. For instance, studies on Brushless DC (BLDC) motor speed control placed the advantages of employing the Ziegler-Nichols method in perspective, with excellent improvement in the most critical response parameters such as rise time and settling time. The fault-free calculation of the PID parameters as a result of this method enabled the decimation of high steady-state error, illustrating how effective it is in making control systems responsive and stable. This practical application underscores the importance of the Ziegler-Nichols method as a standard technique used in PID controller tuning, thereby streamlining the process of achieving optimal control in dynamic environments (Fayyed et al., 2025). Moreover, the versatility of the Ziegler-Nichols method to other control systems, such as new applications like the Two Degree of Freedom (2-DOF) PID controller to photovoltaic systems, underscores its extensive use. In such advanced systems, the incorporation of techniques such as the Incremental Conductance method using Ziegler-Nichols tuning provides significant improvement against dynamic disturbances and parameter uncertainty. This merging of techniques demonstrates the efficacy of the Ziegler-Nichols technique even in niche applications where regular control measures become ineffective. Research underscores the fact that tuning of PID parameters by Ziegler-Nichols not only enhances performance but also instills systems with robustness to enable them to operate under variable conditions, testifying to its continued relevance to controller design (Lalthansangi et al., 2024) (Figure 8.13).

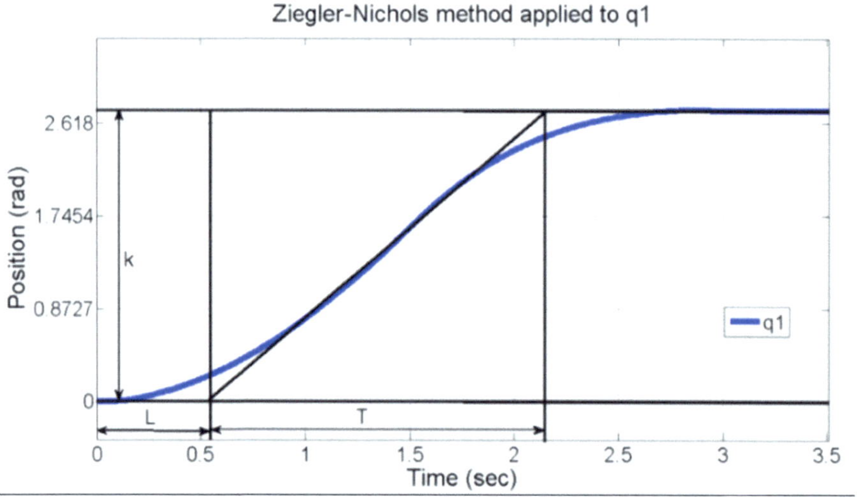


**Figure 8.13**. *Ziegler-Nichols method.*

*Source: https://www.researchgate.net/profile/Abel-Garcia-Barrientos/publication/268436846/figure/fig3/AS:669538521739266@1536641831383/Ziegler-Nichols-method-applied-to-the-transient-response-of.ppm.*

## 8.5.2. Cohen-Coon Method

The Cohen-Coon method is an essential tuning technique of the Proportional-Integral-Derivative (PID) controller, particularly in processes with time delays and non-linearities. The method is particularly useful because it optimizes the response speed and overshooting and hence is ideally suited for industrial applications where system stability is important. With a simple set of equations using process reaction curves, the Cohen-Coon method calculates initial PID parameters with the capability to significantly improve control system dynamics. The calculated parameters ensure that systems respond quickly to changes with little likelihood of over-oscillation, a critical factor in applications that manage sensitive equipment and processes. At the heart of the success of the Cohen-Coon technique is its ability to address processes with an inverse response where the original system response acts opposite to the direction of desired control. This occurs commonly in heat exchangers and other thermal systems and thus poses challenges to traditional tuning methods. As identified in research targeting such systems, tuning techniques like Cohen-Coon are designed to give more stable performance values, for example, reduced overshoot and shorter settling times. In relation to other techniques like Ziegler-Nichols, the Cohen-Coon technique often gives more precise parameters that reverse defects associated with system nonlinearity, hence stabilizing and improving performance under varying operating conditions (Olana et al., 2021). Real-world applications of the Cohen-Coon technique emphasize its flexibility and performance in a variety of industrial situations. For example, investigations have demonstrated that when utilized with empirical or experimental models, Cohen-Coon tuning not only performs to specification but more typically exceeds expectations by eliminating offsets and settling very quickly with little residual steady-state error. In addition, recent studies emphasize the need for incorporating meta-heuristic optimization algorithms with conventional tuning techniques such as Particle Swarm Optimization (PSO) to create hybrid systems. This technique allows PID coefficients to be tuned in Cohen-Coon models, eventually achieving improved system responses with reduced overshoot and shorter settling times in more sophisticated models including second-order process plus delay-time models (Ayas et al., 2019).

## 8.5.3. Hand Tuning Techniques

The manual techniques of tuning are critical in Proportional-Integral-Derivative (PID) controller design, primarily due to the fact that they are easy and accessible to engineers. They all have a step-by-step approach where the parameters of the controllers—derivative, integral, and proportional—are caused to vary based on what is observed to be occurring within the system. Among the common practices is the utilization of the Ziegler-Nichols tuning techniques, which consist of the determination of the ultimate gain and oscillatory period of the system. Although it is highly easy to implement, manual adjustment takes considerable time and, in most cases, leads to less-than-optimal performance, particularly in complex systems where the dynamics may not be known completely. It is important to note that while manual tuning offers immediate feedback, the performance might not be as stable as algorithmic algorithms derived from automated tuning, as shown by research validating the limitations of traditional tuning methods like Ziegler-Nichols against metaheuristic approaches (Hassan et al., 2025). Manual tuning is an important component of PID control, especially in systems where engineers must resort to qualitative measures and their systems dynamics acumen. Successful manual tuning also often amounts to iterative tweaking wherein the engineer observes response characteristics such as overshoot, settling time, and steady-state error after every parameter tweak. This process of experience can be satisfactory in the case of simple applications, but as the complexity in various systems rises, the chances of achieving best parameters decline. The need for precision becomes increasingly evident upon consideration of the performance criteria that stipulate a systems stability and quality of response. For instance, areas such as autonomous robots or flight control need to have optimized parameters to provide stability and responsiveness in dynamic environments, something hard to achieve with manual tuning techniques without the aid of optimization algorithms (Sahin et al., 2024). Using manual tuning techniques can also be an intermediate step for more advanced automated methods, knowing more about PID controllers and the unavoidable challenges. Using manual tuning, the engineers gain deep insights into the effect of every parameter on the overall performance of the system, bridging theory and practice. This basic knowledge is even more valuable as automation finds its way into general control systems, where early manual fine-tuning can direct subsequent optimization procedures. This contention is complemented by research, emphasizing the potential to use manual tuning as an effective entry point prior to using advanced computational methods, that is, those involving the Whale Optimization Algorithm or other metaheuristics, that post superior performance parameters on high-complexity control tasks relative to completely manual approaches (Hassan et al., 2025). The interaction between human and automatic methods thus not only facilitates skill development but also improves overall system performance for PID control synthesis.

## 8.6. APPLICATIONS OF PID CONTROLLERS

In control engineering, Proportional-Integral-Derivative (PID) controllers are ubiquitous due to their general-purpose applicability and effectiveness in various applications. Industrial automation systems greatly value PID controllers since they effectively regulate parameters such as temperature, pressure, and flow rate with high precision. The inherent simplicity of PID algorithms allows them to be easily implemented so that they can respond quickly to system dynamics through robust feedback mechanisms. This adaptability allows PID controllers to be used in a wide range of industries, including manufacturing and process industries, where the optimal performance must be maintained. Through correcting system disturbances efficiently, PID controllers keep processes at required parameters, thus ensuring maximum productivity and minimizing waste. The application of PID controllers also finds its place outside the industrial sector in areas such as robotics and aerospace. In robotic mechanisms, PID controllers are of essential importance to provide smooth and accurate motion control so that robots can perform intricate operations with high reliability. Also, in applications in aerospace, the controllers manage flight dynamics by controlling various aircraft system parameters. Tuning of

PID parameters affects system performance directly in terms of stability and responsiveness. There is proof that addressing tuning issues unique to multi-input multi-output (MIMO) systems, which are often found in aerospace and robotics technologies, is crucial in order to improve controller performance. The development of advanced tuning methods that address robustness and stability problems, as proposed in several studies, underscores the importance of optimal PID parameter tuning in difficult operating conditions (Thiago et al., 2021). Besides industrial and robotics applications, PID controllers also play a significant role in environmental monitoring systems, where they regulate pollution discharges and aid sustainability efforts. PID controllers can be used to regulate air quality systems, having the capability to maintain toxic pollutants to acceptable standards. By controlling control variables from real-time data with high accuracy, PID controllers facilitate compliance with environmental regulations while safeguarding public health. In addition, the ability to operate efficiently in smart city systems highlights the growing significance of PID technology in urban development and management. With societies relying increasingly on intelligent systems in solving environmental issues, the continued improvement and adoption of PID control methods become increasingly significant for green development projects (Ahmed, 2022). Hence, the PID controller applications span a wide range of industries, serving to underscore their fundamental role towards the enhancement of industrial output as well as the greater good (Figure 8.14).

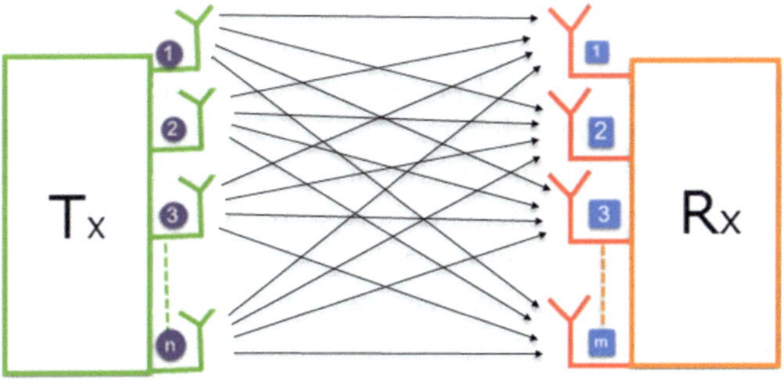


**Figure 8.14**. *Multi-input multi-output (MIMO) systems.*

*Source: https://www.elprocus.com/wp-content/uploads/2017/06/MIMO.jpg.*

## 8.6.1. Industrial Automation

In industrial automation, the integration of sophisticated control systems is imperative in ensuring maximum efficiency in operations. Among the most popular techniques employed is the Proportional-Integral-Derivative (PID) control, which enables precise control over various processes through continuous monitoring and adjustment of the system based on feedback. The advantage of PID controllers lies in their ability to minimize error through dynamic output correction against process variable deviations. In dangerous industrial settings, where relatively small differences can lead to significant inefficiency and loss of money, responsiveness is critical. As more industries turn to automation, demand for sophisticated PID tuning methods increases, with the emphasis on improved performance and reliability in controller design. PID controller design is very much connected with the philosophy of industrial automation, where the system's performance is all dependent upon the optimization of control algorithms. The proportion, integral, and derivative components of the PID controller each play an independent role: the proportional component gives rapid correction based on current error, the integral component monitors cumulative errors to adjust for systemic imbalances, and the derivative actually presages future patterns of error in an attempt to enhance stability. This dynamic is required in an effort to maintain manufacturing processes at their highest levels and reduce variability that can lower product quality. Besides, advanced techniques of controller tuning, such as those involving model-based predictive control (MPC), rely on process models to forecast system performance and facilitate more informed decision-making, further enhancing the extent of automation (Schwenzer et al., 2021). As industries continue to evolve, embracing automation technology is a must. The evolution in PID tuning methods is a sign of a broader technology shift towards intelligent automation, where methods like the Particle Swarm Optimization (PSO) have gained prominence to enhance control parameters. Leverage collective swarm behaviors, researchers are innovatively applying PSO to fine-tune PID controller performance across different applications, from industrial and commercial contexts (Ahmed, 2022). Therefore, the integration of such techniques not only facilitates the process of tuning but also helps in creating more resilient and adaptive systems in case of unforeseen barriers. This syncretism of various methods is a paradigm-shifting power for controller design that sends industrial automation to a new horizon of precision and efficiency.

## 8.6.2. Motion and Robotics Control

The incorporation of robotics into motion control systems inherently improves accuracy in automation operations. The Proportional, Integral, and Derivative (PID) controller is at the center of such systems, coordinating the movement of robotic mechanisms to a desired movement. By determining the difference between a sensed position and a reference position, the PID controller continuously modifies control outputs for maximum performance. Each component of the PID system plays a vital role: the proportional term corrects the present error, the integral term accumulates the past errors to eliminate the steady-state errors, and the derivative term predicts the future errors based on the rate of change. This kind of coordination among the components yields a stable and smooth movement that is called for by the reliability of robot-based applications, especially in those industries with strict demands for precision and efficiency. The performance of PID controllers may be affected significantly by their parameter tuning. Adjusting the tweak to every one of the components—proportional gain, integral time constant, and derivative time constant—may drastically alter the response characteristics of a robotic system. Among one of the classic issues in robotics is rapid response time with reduced overshoot and oscillations. Based on the specific application, for example, for robot arms or UAVs, different tuning methods can be used to achieve optimal performance for specific operating conditions. For instance, while high proportional gain might improve responsiveness, it might lead to instability in some systems. Therefore, understanding the effect of PID parameters on system performance in general is most important, impacting not only responsiveness but also uncertainty and disturbance robustness characteristic of real-world applications. Advances in robotics motion control increasingly incorporate means other than the conventional PID controllers, drawing inspiration from natural systems as well as artificial intelligence. Techniques such as Particle Swarm Optimization (PSO) are emerging ways of controller design to adapt to adaptive and robust control systems that are sensitive to changing environments (Ahmed, 2022). Furthermore, the use of unmanned aerial vehicles (UAVs) is an expression of the importance of optimizing control strategies, particularly under complex flight dynamics and changing environments (Idrissi et al., 2022). Through an examination of these new methods, researchers can address weaknesses found in traditional PID controllers and make robotic systems more efficient and reliable. As the technology continues to evolve, these improvements are likely to provide significantly enhanced motion controlling capability, paving the way for more sophisticated and responsive robotics technologies.

## 8.6.3. Temperature Control Systems

Effective temperature control systems are essential in a wide range of applications, from industrial process to home heating. These systems typically operate on some type of control policy, and one of the most frequently applied are Proportional, Integral, and Derivative (PID) controllers. The efficacy of a temperature control system relies on the accurate adjustment of these PID parameters, which directly influence the response of the systems to a change in temperature. There must be a precise balance between proportional gain, integral action, and derivative response to minimize error and provide stability. Through PID control, temperature control systems can react quickly to variations and achieve a designated temperature within tolerable limits, enhancing system performance and efficiency in energy usage. This flexibility is particularly useful in applications where precision is paramount, such as in the pharmaceutical or food processing industries, where thermal variations can induce massive quality issues. PID parameter tuning is included in optimizing the performance of temperature control systems, and many techniques of tuning can be employed to generate this optimization. Techniques such as Ziegler-Nichols methods allow for systematic calculation of initial PID parameters based on the response of systems. Apart from such traditional approaches, advanced ones such as Model-based Predictive Control (MPC) are being applied more and more in modern applications (Schwenzer et al., 2021). MPC utilizes process models to forecast future behavior of the system, thereby allowing anticipatory tuning of the control strategy. By integrating predictive capability with PID control, temperature control systems are better able to handle constraints and uncertainties in real-world operating conditions. Synergistic integration is a refinement of the overall performance and responsiveness of temperature control systems, aligned with current needs for precision and economy. Finally, selection of the right temperature sensors and actuators is also important to the success of temperature control systems. The sensors should deliver timely and accurate readings, while the actuators should quickly respond to command signals from the PID controller. Interestingly, sensor technology has evolved to enable improved accuracy and lower response times, consequently having a beneficial impact on system performance (Ahmed, 2022). As a result, temperature control systems with better sensors and actuators are more able to minimize errors and maximize desired temperature set points without wasting resources. Continuous research and development in the discipline ensure the successful implementation of PID controllers, leading to improved operation across different applications. Hence, temperature control systems are becoming increasingly sophisticated as it integrates better algorithms and hardened hardware with an aim to meet challenges of modern engineering demands.

## SUMMARY

Research in PID control made up of Proportional, Integral, and Derivative components highlights its important role in modern-day control systems. The controllers are initial devices in most branches of engineering, showing robust performance in various applications. System performance by a PID controller to a large extent relies on the proper tuning of its parameters, which has direct implications on system behavior such as stability, response time, and overshoot. A well-tuned PID controller causes the system to respond quickly to changes without sacrificing precision, hence making it an indispensable tool in automated systems. As discussed, methods at hand for tuning range from manual to sophisticated algorithms, each with unique advantages as well as issues, hence indicating the level of complexity to achieve optimal system performance. PID parameter influence on system performance cannot be overrated. Proportional, integral, and derivative gain tuning offer engineers the means to design systems that satisfy particular performance specifications. For instance, greater proportional gain will lead to quicker response but will induce instability if taken to extremes. Integral gain helps alleviate steady-state error but induces oscillation if not suitably tempered. Derivative gain is a predictive output, dampening oscillations and enhancing stability. Therefore, a fine balance has to be maintained in adjusting these parameters to leverage their complete potential, suggesting the significance of a carefully formulated controller design strategy to ensure reliability in a wide array of applications. The continuous development of control techniques, encompassing improvements in tuning methods, infers that the scenario for PID control will further change. New techniques borrowed from disciplines such as swarm intelligence prove the search for improved control methods, for example, in algorithm studies such as Particle Swarm Optimization (Ahmed, 2022). Furthermore, Model-based Predictive Control (MPC) has also emerged, proving how sophisticated computing power can enhance traditional approaches (Schwenzer et al., 2021). These developments not only provide a hint of future use but also highlight the significance of PID controllers in a world being mechanized day by day. With developing engineering fields, combining new methods along with conventional PID techniques promises better system performance and proper controller design is a field for research and development.

In the context of control systems, PID controllers (Proportional, Integral, Derivative) are a basic means for achieving desired system responses. The operation of PID controllers strongly relies on the parameter tuning: proportional gain, integral time, and derivative time. Every parameter has a specific function: the proportional part corrects current error, the integral part sums up previous errors to cancel steady-state errors, and the derivative part forecasts future trends in errors from current rates of change. Thus, a properly adjusted PID controller can improve system performance considerably by eliminating overshoot, minimizing settling time, and ensuring overall stability, providing a feedback loop that maximizes system response and efficiency. The influence of PID parameter tuning on system response cannot be overestimated as untuned parameters may result in the system becoming unstable and failing to perform well. Overshooting and oscillation can result from too high a proportional gain, while too low a gain results in slow response. Similarly, the integral parameter must also be treated with care to avoid excessive accumulation of past errors, which could further introduce instability. Derivative action, although helpful in cutting down oscillations, causes noise sensitivity that is particularly poor in dynamic conditions. Therefore, understanding interaction of the aforementioned parameters is crucial to obtaining the desired system performance and also in ensuring controllers are operating within specifications designed (Kaufmann et al., 2023). PID tuning methods have been established to assist engineers in realizing optimal values of parameters for their particular applications. These methods range from heuristic methods, such as the Ziegler-Nichols tuning rules, to more sophisticated computational strategies, such as model-based and optimization algorithms. The choice of the tuning strategy might be determined based on the complexity of the system, the nature of the disturbance, and the performance specifications. There has also been recent advancement in tuning methods with the advent of artificial intelligence and machine learning techniques for enhancing adaptability and accuracy in tuning the parameters in order to make PID controllers highly effective in various applications. Technology has continued to advance, with research further enhancing the techniques, unveiling new findings and applications to further develop the PID control systems to remain relevant amid new challenges (Ahmed, 2022).

# REFERENCES


1. Aafaque, M., Boker, A., & Sardahi, Y. (2023). Three-time scale multi-objective optimal PID control with filter design. Asian Journal of Control, 25, 3356–3368. https://doi.org/10.1002/asjc.3060 (accessed on 19th September 2025).
2. Abdollahzadeh Jamalabadi, M. A. (2022). Active sloshing control of a circular cylindrical container with FG GPL-RC. *Journal of Vibration and Control, 29*, 5440–5461. https://www.semanticscholar.org/paper/a1359cff8961b1b3832f9ac6b186408245411125 (accessed on 19th September 2025).
3. Ayas, M. S., & Şahin, E. (2019). Parameter effect analysis of particle swarm optimization algorithm in PID controller design. *An International Journal of Optimization and Control: Theories & Applications (IJOCTA)*. https://www.semanticscholar.org/paper/2911829a7c90a36e49b2dd524b87101833d3be4b (accessed on 19th September 2025).
4. Botirov, T., Sodiqov, B., Jumabaev, E., Salahiddinov, A., & Jurayeva, R. (2025). Design of a controller with an intelligent hybrid structure. E3S Web of Conferences. https://www.semanticscholar.org/paper/348a305208c02ad6b37e1c34e19d2326717b2954 (accessed on 19th September 2025).
5. Davies, M., Wild, A., Orchard, G., Sandamirskaya, Y., Fonseca Guerra, G. A., Joshi, P., Plank, P., et al. (2021). Advancing neuromorphic computing with Loihi: A survey of results and outlook. *Proceedings of the IEEE, 109*, 911–934. https://doi.org/10.1109/jproc.2021.3067593 (accessed on 19th September 2025).
6. De Persis, C., & Tesi, P. (2019). Formulas for data-driven control: Stabilization, optimality, and robustness. *IEEE Transactions on Automatic Control, 65*, 909–924. https://doi.org/10.1109/tac.2019.2959924 (accessed on 19th September 2025).
7. Degrave, J., Felici, F., Buchli, J., Neunert, M., Tracey, B., Carpanese, F., Ewalds, T., et al. (2022). Magnetic control of tokamak plasmas through deep reinforcement learning. *Nature, 602*, 414–419. https://doi.org/10.1038/s41586-021-04301-9 (accessed on 19th September 2025).
8. Dragičević, T., Lu, X., Vásquez, J. C., & Guerrero, J. M. (2015). DC microgrids–Part I: A review of control strategies and stabilization techniques. IEEE Transactions on Power Electronics. https://doi.org/10.1109/tpel.2015.2478859 (accessed on 19th September 2025).
9. Drgoňa, J., Arroyo, J., Cupeiro Figueroa, I., Blum, D., Arendt, K., Kim, D., Perarnau Ollé, E., et al. (2020). All you need to know about model predictive control for buildings. *Annual Reviews in Control, 50*, 190–

232. https://doi.org/10.1016/j.arcontrol.2020.09.001 (accessed on 19th September 2025).

10. Euzébio, T., Silva, M. T. D., & Yamashita, A. S. (2021). Decentralized PID controller tuning based on nonlinear optimization to minimize the disturbance effects in coupled loops. IEEE Access, 9, 156857–156867. https://www.semanticscholar.org/paper/56564df58b8803dd04d3c8690b9aaf28d00b7416 (accessed on 19th September 2025).
11. Gad, A. G. (2022). Particle swarm optimization algorithm and its applications: A systematic review. *Archives of Computational Methods in Engineering, 29*, 2531–2561. https://doi.org/10.1007/s11831-021-09694-4 (accessed on 19th September 2025).
12. Golestan, S., Guerrero, J. M., & Vásquez, J. C. (2016). Three-phase PLLs: A review of recent advances. IEEE Transactions on Power Electronics, 32, 1894–1907. https://doi.org/10.1109/tpel.2016.2565642 (accessed on 19th September 2025).
13. Hassan, A., & Dada, E. (2025). Tuning PID controller for inverted pendulum using Whale Optimization Algorithm. *Nile Journal of Engineering and Applied Science.* https://www.semanticscholar.org/paper/f1c54370c6e9548fb93b73eaa4f2355a3cb93083 (accessed on 19th September 2025).
14. Hermassi, M., Krim, S., Kraiem, Y., Hajjaji, M. A., Alshammari, B. M., Alsaif, H., Alshammari, A. S., et al. (2023). Design of vector control strategies based on fuzzy gain scheduling PID controllers for a grid-connected wind energy conversion system: Hardware FPGA-in-the-loop verification. *Electronics, 12*, 1419–1419. https://doi.org/10.3390/electronics12061419 (accessed on 19th September 2025).
15. Huba, M., & Vrančić, D. (2022). Performance portrait method: An intelligent PID controller design based on a database of relevant systems behaviors. *Sensors, 22*, 3753–3753. https://doi.org/10.3390/s22103753 (accessed on 19th September 2025).
16. Idrissi, M., Salami, M. R., & Annaz, F. Y. (2022). A review of quadrotor unmanned aerial vehicles: Applications, architectural design and control algorithms. Journal of Intelligent & Robotic Systems, 104. https://doi.org/10.1007/s10846-021-01527-7 (accessed on 19th September 2025).
17. Kalyan, Ch. N. S., Rao, S., Bajaj, M., Khan, B., & Zaidan, M. R. (2024). Frequency regulation of geothermal power plant-integrated realistic power system with 3DOFPID controller. *Cogent Engineering, 11*. https://doi.org/10.1080/23311916.2024.2322820 (accessed on 19th September 2025).

18. Kanchan, M., Mohith, S., Bhat, R., & Naik, N. (2023). Application of modeling and control approaches of piezoelectric actuators: A review. Technologies, 11, 155–155. https://doi.org/10.3390/technologies11060155 (accessed on 19th September 2025).

19. Kaufmann, E., Bauersfeld, L., Loquercio, A., Müller, M., Koltun, V., & Scaramuzza, D. (2023). Champion-level drone racing using deep reinforcement learning. *Nature, 620*, 982–987. https://doi.org/10.1038/s41586-023-06419-4 (accessed on 19th September 2025).

20. Khodayari, M., & Balochian, S. (2021). Design of adaptive fuzzy fractional order PID controller for autonomous underwater vehicle (AUV) in heading and depth attitudes. *International Journal of Maritime Engineering*. https://www.semanticscholar.org/paper/46efabe5deef71839ce80c9a29275bbaaa18d7a6 (accessed on 19th September 2025).

21. Lalthansangi, & De, K. (2024). Optimizing solar MPPT boost converter control: Implementation of a two-degree-of-freedom PID controller with Ziegler-Nichols and MATLAB tuning methods. *2024 IEEE Silchar Subsection Conference (SILCON 2024)*. https://www.semanticscholar.org/paper/c3efc6b1c1be0016ff885615c42e7c7f3848e83e (accessed on 19th September 2025).

22. Liu, J., Zhao, M., Li, G., & Chen, J. (2020). Multiple linear regression prediction and wavelet neural network-based intelligent online tuning control method. *IEEE Access, 8*, 160684–160696. https://www.semanticscholar.org/paper/456670edfd3080b6f8883ffaa924da4dd134032d (accessed on 19th September 2025).

23. Micev, M., Ćalasan, M., & Oliva, D. (2020). Fractional order PID controller design for an AVR system using chaotic Yellow Saddle Goatfish Algorithm. *Mathematics, 8*, 1182–1182. https://doi.org/10.3390/math8071182 (accessed on 19th September 2025).

24. Mousavi, Y., Bevan, G., Küçükdemiral, İ. B., & Fekih, A. (2022). Sliding mode control of wind energy conversion systems: Trends and applications. Renewable and Sustainable Energy Reviews, 167, 112734–112734. https://doi.org/10.1016/j.rser.2022.112734 (accessed on 19th September 2025).

25. Olana, F. D., & Abose, T. A. (2021). PID temperature controller design for shell and tube heat exchanger. *International Journal of Engineering and Manufacturing*. https://www.semanticscholar.org/paper/eddd44f661732d10685499e1b93e4562b9f0575e (accessed on 19th September 2025).

26. Provatas, V., & Ipsakis, D. (2023). Design and simulation of a feedback controller for an active suspension system: A simplified approach. Processes, 11, 2715–2715. https://doi.org/10.3390/pr11092715 (accessed on 19th September 2025).

27. Raouf, B., & Mousavian, S. (2023). A robust controller design based on Kharitonov's theorem for frequency control in an interconnected power system. *European Journal of Electrical Engineering and Computer Science, 7*, 1–9. https://doi.org/10.24018/ejece.2023.7.1.482 (accessed on 19th September 2025).

28. Saeed, M. H., Fangzong, W., Kalwar, B. A., & Iqbal, S. (2021). A review on microgrids' challenges & perspectives. IEEE Access, 9, 166502–166517. https://doi.org/10.1109/access.2021.3135083 (accessed on 19th September 2025).

29. Sahin, A. K., Çavdar, B., & Ayas, M. S. (2024). An adaptive fractional controller design for automatic voltage regulator system: Sigmoid-based fractional-order PID controller. *Neural Computing and Applications, 36*, 14409–14431. https://www.semanticscholar.org/paper/3b73949769f840b9b0e1bf3b59e1890c317a90a1 (accessed on 19th September 2025).

30. Schwenzer, M., Ay, M., Bergs, T., & Abel, D. (2021). Review on model predictive control: An engineering perspective. *The International Journal of Advanced Manufacturing Technology, 117*, 1327–1349. https://doi.org/10.1007/s00170-021-07682-3 (accessed on 19th September 2025).

31. Usamah, N. M., Kinn, C., & Ahmad Mazlan, A. Z. (2023). Performance comparison between AFC-GA and AFC-PSO control schemes in attenuating the vibration of suspended handle model. 2023 International Conference on Artificial Intelligence and Power Engineering (AIPE). https://www.semanticscholar.org/paper/05f27b922ba1bd0ddf832f05dfceeb4707d86f9b (accessed on 19th September 2025).

32. Zhang, W. (2025). The impact of PID control parameters on DC motor performance. Applied and Computational Engineering. https://www.semanticscholar.org/paper/4b2ff658591ee787cddc53f6518f19db91662b43 (accessed on 19th September 2025).

33. Zhang, Y., Liu, L., Liang, J., Chen, J., Ke, C., & He, D. (2024). Application of a multi-strategy improved sparrow search algorithm in bridge crane PID control systems. Applied Sciences, 14, 5165–5165. https://doi.org/10.3390/app14125165 (accessed on 19th September 2025).

34. Zhao, S., Blaabjerg, F., & Wang, H. (2020). An overview of artificial intelligence applications for power electronics. IEEE Transactions on Power Electronics, 36, 4633–4658. https://doi.org/10.1109/tpel.2020.3024914 (accessed on 19th September 2025).

CHAPTER 9

# STATE-SPACE REPRESENTATION

## Contents



## INTRODUCTION

The state-space representation is the foundation of control theory and system dynamics, and it provides a framework for model development and analysis in terms of state variables and state equations. It allows for complete description of a systems behavior in terms of its internal state and allows for the modeling of multi-input and multi-output systems. By defining state variables which embody the important characteristics of a system, engineers are able to construct state-space equations that model the dynamics of the system. Such equations model the relationship between input, output, and state variables, and behavior of the system with respect to time can then be forecasted. As systems become increasingly sophisticated, particularly in fields of engineering such as robotics and aerospace, the state-space description offers a structured approach to explaining the complicated interaction involved, paving the way for advanced modeling techniques and control systems enabling high-order system dynamics. State-space modeling also allows for the explanation of linear and nonlinear systems in terms of a uniform framework. This adaptability is necessary in practical applications where systems have a tendency to exhibit dynamic interactions that cannot be well characterized by traditional means, such as transfer functions. In state-space representation, the integration of state variables makes it possible to describe time-varying systems, allowing for greater flexibility to respond to changing inputs and situations. In addition, the incorporation of feedback control within the state-space expression can enhance system stability and performance so that engineers can tune system responses to meet particular operating requirements.

$$\begin{bmatrix} \dot{x}_1 \\ \dot{x}_2 \\ \dot{x}_3 \end{bmatrix} = \begin{bmatrix} 0 & 1 & 0 \\ 0 & 0 & 1 \\ -4 & -3 & -2 \end{bmatrix} \begin{bmatrix} x_1 \\ x_2 \\ x_3 \end{bmatrix} + \begin{bmatrix} 0 \\ 0 \\ 5 \end{bmatrix} v(t)$$

$$y(t) = \begin{bmatrix} 1 & 0 & 0 \end{bmatrix} \begin{bmatrix} x_1 \\ x_2 \\ x_3 \end{bmatrix}$$

**Figure 9.1**. *State space analysis.*

*Source: https://media.geeksforgeeks.org/wp-content/uploads/20231211193928/State-Space-Analysis-in-control-system--3.png.*

The growing reliance on such high-level modeling techniques is a testament to the need for a good understanding of the principles of state-space representation, which is further validated in a broad range of engineering applications ranging from electrical circuits through to mechanical systems. State equation solution, the backbone of state-space formulation, utilizes the knowledge of principles such as the state transition matrix to provide valuable information about system history in time. The state transition matrix not only determines the development of a system from one state to the next but also establishes the impact of initial conditions on system response. Accurate computation of this matrix is essential in system response to external excitations and can be particularly challenging in the case of nonlocal boundary conditions, when conventional approaches will not work. Recent advances in techniques like those presented in Abdullayev (2024) illustrate the pivotal role of auxiliary parameters in identifying conditions necessary for solution existence and uniqueness. Furthermore, new approaches, as depicted by LS4 in Zhou et al. (2022), demonstrate the state-space representation's ability to effectively manage real-world data that have abrupt changes. This broadens the horizon of applications for state-space models, establishing its importance in today's engineering and computational settings (Figure 9.1).

## *Definition of State-Space Representation*

The state-space representation is fundamentally founded on its ability to provide a comprehensive framework for the modeling of dynamic systems. It transforms the standard system of differential equations to a systematic representation that includes both the system dynamics and the state variables. In essence, a state-space representation describes a physical system using a set of variables, or the state variables, that describe the systems current condition. These variables vary with time as per state equations, which are state variable relationships. Not only does this definition facilitate control system analysis and design, but also makes computations more efficient by allowing systematization of multivariable system representations. As evident in some cases, such a representation is important in stability and performance studies, particularly in multidimensional systems such as those whose analysis can be achieved via Partial Integral Equations (PIE) to study coupled dynamic behaviors, as evident in Declan et al. (2025). Additionally, the versatility of state-space representation includes its application in most fields, including engineering, economics, and biological systems. By representing the dynamics of a system in terms of state equations and matrices, it is now possible to represent interactions in systems with numerous inputs and outputs involved. This flexibility supports analysis in time-variant systems where the behavior of the system is determined by external conditions over time. For instance, the Output-Recurrent Gated State Space Model (OR-GSSM) illustrates how state-space models can be architected to represent complex fluid dynamics in multiphase flows. Through gate mechanisms to design transition matrices that develop over varying conditions, systems can obtain great accuracy with computational potency. This demonstrates the immense advantage of applying state-space representations in modeling dynamic systems, as detailed in Chen et al. (2025). Finally, understanding the concept of state-space representation demonstrates its significant contribution in system analysis and design currently. It encompasses not only the clearer identification of state variables and their interdependence but also the derivation of solutions for the state equations through the application of techniques such as eigenvalue analysis and state transition matrices. These matrices encapsulate the response of the systems in terms of time and play a critical role in predicting the long-term behavior of dynamic systems following disturbances. The overall strength of state-space methods lies in the ability to dissect complex system interaction into manageable equations, such that system complexity of behavior is rendered comprehensible and analyzable. With each new wave of technology developing, state-space formulations constitute a toolbox framework that ensures the strength and reliability of system model and control methodologies in all areas of study.

## *Importance in Control Theory and Engineering*

The importance of control theory and engineering arises primarily in the ability to model and analyze dynamic systems through state-space representation, which is interested in system states and their evolution over time. State variables in this context hold the needed information to characterize the system's behavior, and from them, state equations can be formulated to describe how the variables evolve. This approach enables engineers to systematically design controllers that achieve specified performance requirements, that is, stability and responsiveness. With state-space models, engineers have the ability to look into complex interactions in systems, thus enabling system design and performance enhancement. For instance, by applying these principles, engineers can forecast how changes in one state variable will propagate through the entire system and make more informed choices in control systems. Moreover, state-space modeling becomes paramount in the majority of areas of engineering through optimizing control policies, particularly in nonlinear systems where conventional methods fail. For instance, in the advanced controlling techniques such as those employed in the Koopman operator framework. In this method, state transition concepts are used to re-write the nonlinear dynamics in linear forms such that complex system analysis and control are possible. In such applications, the calculated state transition matrices provide the information concerning how the system evolves with time under different inputs that is essential for performance optimization. The use of state-space representations in conjunction with the Koopman approach is an appealing map toward enhancing control methods, especially in new fields like aerospace engineering, where precise control is paramount to mission success, such as in satellite rendezvous missions where linear approximations of polynomial forms are used to render analysis and design simpler. Besides, solutions to state equations provide the foundation for the assessment of system behavior and the application of control strategies. The significance of the solutions results from the ability to describe system responses in terms of initial conditions and input signals. By converting state equations into solvable matrix format, engineers are in a position to apply computational methods to derive system trajectories as well as predict future states with ease. This predictive capability enhances system robustness and adaptability in actual engineering uses—everything from robotics to aerospace systems, where accurate behavior prediction is essential for safety and efficiency. Application of high-order polynomial series expansions, as demonstrated in the case of satellite rendezvous, not only demonstrates the potency of state-space techniques but also testifies to their relevance to tackling real-world challenges. The revolutionary union of state representation with advanced control schemes highlights the central position of control theory in modern engineering practice, guiding the construction of future systems.

## 9.1. STATE VARIABLES

In the realm of state-space representation, state variables serve as the fundamental quantities that characterize the dynamic behavior of a system at any given point in time. These variables encapsulate all necessary information about the system's past and present states, allowing for effective predictions of future states. The selection of appropriate state variables is crucial, as it directly affects the models complexity and its ability to capture the essential dynamics of the system under study. By defining state variables in a coherent manner, engineers and scientists can simplify the analysis and design processes, facilitating the application of control theory and systems analysis. This importance is underscored in various fields, including electrical engineering, mechanical systems, and even emerging domains like graph-based learning systems, where adapting state variables to non-Euclidean data structures presents unique challenges and opportunities (Wu et al., 2020). State variables not only represent the current state of the system but also influence the formulation of state equations, which are vital for state-space modeling. These equations describe the relationships among state variables, inputs, and outputs, thereby providing a comprehensive framework for analyzing system dynamics. By transforming physical systems into mathematical formats, engineers can leverage these equations to simulate behavior, design controllers, and derive insights that guide decision-making. Notably, the structure of the state equations can vary significantly depending on the systems characteristics, such as linearity or non-linearity, and thus requires careful consideration and formulation. Innovations in state-space representation, particularly in the context of quantum simulations and other sophisticated computational models, have broadened the scope for applying these equations across diverse scientific inquiries, from material science to communication networks (Giannozzi et al., 2017). Moreover, the solution of state equations is intricately linked to the concept of state transition matrices, which provide a powerful tool for analyzing time-dependent behavior within state-space models. These matrices elucidate how state variables evolve over time in response to initial conditions and input variations, serving as vital components in predicting future states. Understanding the temporal dynamics captured by state transition matrices is fundamental for both theoretical analysis and practical applications, such as control system design and systems optimization. As researchers continue to explore complex interactions in dynamic systems, particularly through state-space representations, the combination of state variables and their corresponding transition matrices becomes paramount for enhancing system performance and robustness. This convergence illustrates the evolving landscape of systems engineering, emphasizing the need for advanced methodologies that account for intricate interdependencies among variables, as seen in the growing interest in graph neural networks and advanced simulation techniques (Wu et al., 2020) (Figure 9.2).

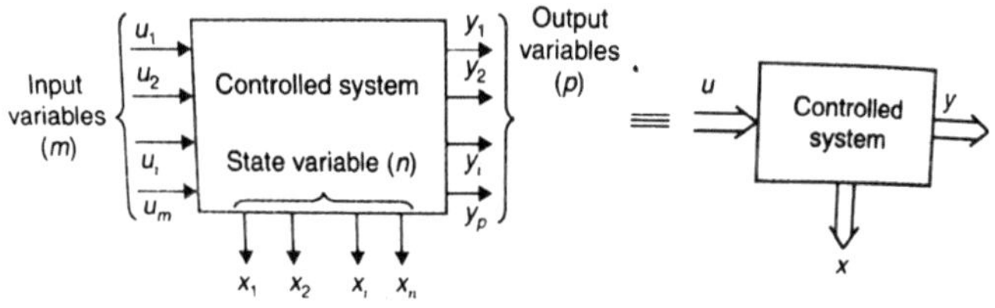


**Figure 9.2**. *State variables.*

*Source: https://lh3.googleusercontent.com/8MVPAkmferO71e1EAvJV3roYa-xwcZO8o7qJhagKHbXMRfCl_00RRP0vGfElALoALVndVl0We4aVkk9-NEu-KURE4ch8pBW0idjAdmMfwH0WaOXqzmEGvdVBOv3pzte7re03F7gBD.*

## 9.1.1. Definition and Role of State Variables

State variables are the fundamental building blocks in the realm of state-space representation and serve as the fundamental descriptors of a system's status at any given moment. State variables, in control theory, encapsulate all the information one requires to forecast future responses based on present situations of the systems. In essence, they provide a mathematical framework for representation of dynamic systems, with the state normally represented by a vector of many variables. This definition is of critical value in modeling interdependencies between various system parameters, detailing how they evolve with time. In the state variable description of a system, engineers are able to translate things to be controlled into understandable mathematical representations, making analysis and control much more feasible. State variables are not only descriptive; they lie at the foundation of formulating and solving state equations. These equations are the dynamics of a system, and major inputs are state variables. State equations can be represented mathematically in a linear or nonlinear format, depending on the system in question. Analysis of these equations allows one to investigate system stability, controllability, and observability. This analytical ability is core in engineering practice, where performance in systems to a large extent depends on their dynamic characteristics. By so doing, state variables not only enable us to prescribe system behavior but also provide a portal through which improvement and adjustment can be made, ultimately ending up converting theoretical abstractions into applied reality. Also, the combined use of state variables leads to the creation of the state transition matrix, a critical tool when evaluating how the states evolve over time as a reaction to inputs. The state transition matrix encapsulates the impact of initial conditions and external factors on the state of a system. In bigger systems, say in communication technology and computational modeling, the concept comes into play in the simplification of complex processes to linear form, making computation and simulation efficient. For instance, recent technologies such as Reconfigurable Intelligent Surfaces (RISs) in wireless communications can apply these concepts to enhance signal transmission and system performance (Renzo et al., 2020). Similarly, software packages like CP2K use state variables for atomistic simulations, demonstrating the utility of state variables to simulate various scientific phenomena (Thomas et al., 2020). In general terms, state variable definition and conception are central to the integration of theory and practice across a lot of fields and providing a solid foundation for state-space representation.

### 9.1.2. Examples of State Variables in Dynamic Systems

State variables in dynamic systems are the key properties that describe the state of the systems at any given time. Common illustrations include position, velocity, and acceleration of a mechanical system, which collectively determine the motion of an object. To illustrate, in simple mass-spring-damper dynamics, position and velocity of the mass can serve as state variables, allowing for the full description of the system dynamics over time. State-space formulation uses these variables in order to define state equations that encompass the system's dynamics within a formal mathematical framework. Essentially, state variables enable one to translate intuitive physical approximations into quantitative analyses so that future states may be forecast from current observations. This approach belongs to the broader philosophy of control and system dynamics, enabling more efficient design and optimization of complex systems. Electrical circuits form an ideal example of another type of systems in which state variables are at the forefront. The inductor current and the voltage across the capacitor, for example, can be used as state variables in RLC circuits. They not only characterize the state of the circuit but also have important influences on how electrical signals change in time.

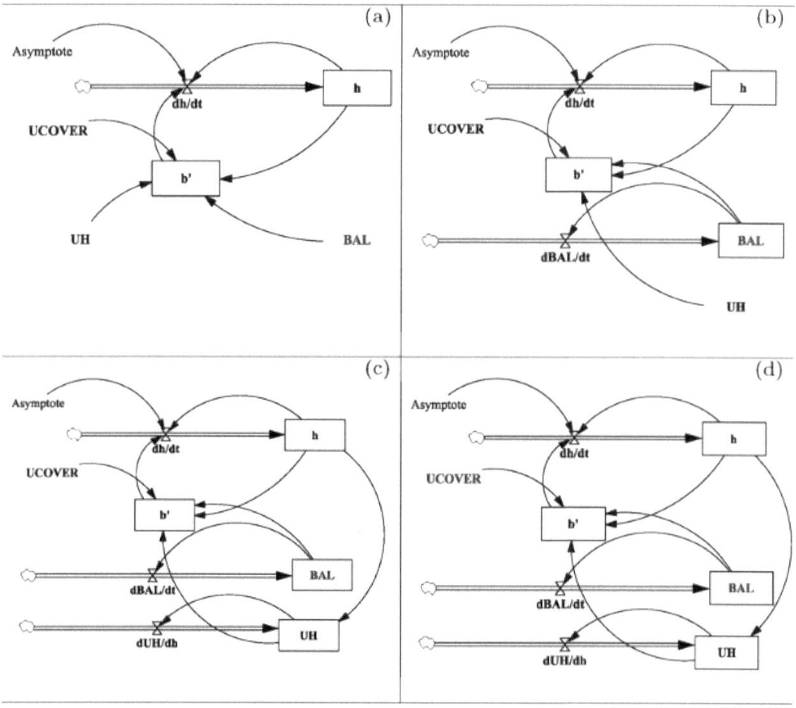


**Figure 9.3**. *State variables in dynamic systems.*

*Source: https://www.researchgate.net/profile/Christian-Salas-Eljatib/publication/43269504/figure/fig3/AS:670025090347018@1536757838230/Dynamic-system-diagrams-State-variables-are-represented-by-boxes-derivatives-by-tick.ppm.*

By using state-space representation, engineers can effectively analyze transient response and steady-state behavior using differential equations formed based on such state variables. As the circuit operates, voltage and current changes may be mathematically modeled, and feedback and control may be implemented within the system. Determining the state variables required for valid modeling not only raises the understanding to the systems response, but it also facilitates designing control methods such that desired performance measures are met in an efficient way. In more advanced applications, state variables enter the realm of fractional-order systems, which involve dynamics with non-integer order derivatives. An example is the viscoelastic material modeling in which response cannot be reasonably characterized by using regular integer-order state variables. Here, the fractional derivatives bring in a new dimension to state-space representations, leading to improved prediction of the material response to loading conditions. The pseudo-state-space representation is a concept that integrates integer and fractional derivatives and thereby facilitates improved description of dynamic system behavior. Using this approach optimal control policies can be formulated, and significantly better system performance can be obtained. In optimal control problems, the presence of fractional-order dynamics is a demonstration of the flexibility and adaptability of state-space modeling, reaffirming its applicability in examining complex systems in various conditions (Biswas et al., 2010) (Figure 9.3).

### 9.1.3. Relationship Between State Variables and System Behavior

State variables and system behavior constitute the elementary fundamentals of control theory as well as system dynamics. The state variables are the primary descriptors that possess the necessary information about a system at any given moment of time, and this in turn influences future behavior. With the change from classical control methods to the state-space formulation, analysis of intricate systems becomes more suitably accessible. This architecture opens up the potential for the incorporation of multi-dimensional interaction between system states into a single model. For instance, differential equation-controlled systems are tersely described by state variables, which both enable analysis and design of dynamic systems to make responses of the systems predictable and ultimately controllable. The state transition matrix also presents itself as a critical aspect in describing development of the system behavior over time. It captures the effects of initial states and determines how they evolve under the dynamics set by state equations. As reasoned from the underlying physical principles, this matrix is a compact summary of how it is that state variables interact and change in an individual system. The dynamics can be modelled in various frameworks, for example, power distribution networks, the behavior of which is governed by variations of nodal powers as given in Wang et al. (2024). This indicates that the state variables' interaction, modeled through a convenient transition matrix, facilitates improved estimation and prediction of system behavior, thereby enabling effective control strategies. In advanced systems like quantum computing, state variable interactions are even more advanced. Qubit dynamics, for example, are characterized by their dynamic probabilities and can be expressed in terms of state variables within a density matrix formalism. By employing observation equations in terms of state variables, operations among qubits can be calibrated accurately, ultimately affecting system behavior in significant ways, such as Oshchepkov (2024). Not only do such mathematical representations capture the dynamics of highly complex quantum systems, but they enable more sophisticated simulations towards predictive modeling. Hence, the state variables' interaction with their own system behavior is no conceptual framework, but an intrinsic tool in designing sophisticated contemporary technologies and analyzing complex systems.

## 9.2. STATE EQUATIONS

In state-space form, state equations provide the mathematical framework that characterizes system behavior. State equations represent the relationship between state variables and their derivatives and constitute a formal structure to study complicated systems. With state equations, engineers and scientists are able to model large types of linear and nonlinear systems in the same manner, facilitating the analysis of their behavior under various circumstances. The state equations' flexibility also allows them to be applied in different fields such as control theory, robotics, and econometrics where the system dynamics are critical for accurate portrayal. With the interconnection of the system components, state equations are equally critical in problem simplification and problem-solving in multi-dimensional space and, thus, paving the way for advanced modeling techniques required in current engineering problems. Solutions to state equations are a key to predicting the future behavior of dynamic systems. Such equations are typically solved by application of techniques such as the Laplace transform or numerical integration algorithms that facilitate more convenient determination of time-domain responses from known initial conditions. While analytical solutions are desirable because they are simple and beautiful, they may not always exist for complex systems and thus become replaced by numerical methods instead. State transition matrix figures as a core element in this process, encapsulating the systems development over time. Using the state transition matrix, engineers practically reconstruct the state of a system at any instance of time, thereby enabling control strategy synthesis and performance evaluation. As such, the resolution of state equations and construction of the state transition matrix are integral building blocks essential to the design and analysis of contemporary engineering systems. The real-world applications of state equations go beyond theoretical applications; they directly influence the design and implementation of real-world systems. In fields such as robotics and automation, state equations are utilized to develop algorithms for maximum control and stability within dynamic systems. An example of this is using state equations to model the dynamics of robotic arms, where the relations between joint angles, velocities, and torques are explained by state equations. Similarly, industries make use of the power of state equations to design network systems, conceptualizing and controlling data flow effectively, which also occurs in machine learning contexts as graph representations become increasingly prevalent. These contexts illustrate the importance of state equations in describing system dynamics as well as shaping engineers toward innovative, educated solutions for a variety of modern technological situations. As research progresses, attempts to harmonize traditional state equations with new paradigms, for example, those present in Quantum ESPRESSO, could release new efficiencies into computational modeling and simulation approaches (Wu et al., 2020; Giannozzi et al., 2017) (Figure 9.4).

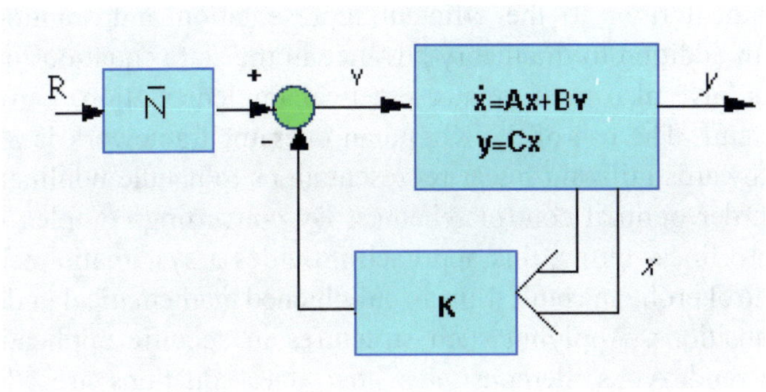


**Figure 9.4**. *State-space representation.*

*Source: https://www3.diism.unisi.it/~control/ctm/examples/motor/Ss4.GIF.*

### 9.2.1. Formulation of State Equations

State equation formulation is a foundation in the development of state-space formulation for the representation of the behavior of dynamical systems across various engineering disciplines. The fundamental idea in the process is the creation of state variables, which provide essential information regarding the current state of the system. Through establishing a relationship among the variables, state equations are formulated based on linear or nonlinear differential equations. This enables one to capture rich interactions within the system and enable rich comprehension of the underlying dynamics. As systems become ever more complex, such as in multiphase flow dynamics or control systems, the need for well-founded state equation formulations takes precedence so that exact modeling and analysis of system behavior under varying conditions are enabled, and hence the importance of well-founded state variable selection and equation formulation. The use of advanced mathematical structures enhances the state equation presentation to allow more sophisticated descriptions of dynamical systems. For instance, semigroup theory and Galerkin projection techniques have been applied to modeling multiphase flows in order to arrive at state-space representations from the complex Navier-Stokes equations. This approach also allows dynamic coupling of numerous fields, such as velocity, pressure, and volume fractions, to be characterized and encourages the development of adaptive computational methods for accuracy and efficiency improvement. Such approaches, such as Output-Recurrent Gated State Space Models (OR-GSSM) and others, play a great role in eliminating errors accumulated over a long period of time, hence improving fidelity of solutions derived from state equations. Continuing advances in such frameworks demonstrate the preeminence of mathematical rigor to the efficient representation and solution of state equations. In addition to the theory advance in the state equation formulation, applications have also seen greater practical implementation, particularly in control systems. The use of the Koopman operator framework is an example of a trend towards utilizing linear representations to handle nonlinear systems with high-order optimal control schemes. By converting complex dynamical behavior into linear terms, this approach provides a systematic technique for optimal control problems built from an enlightened mathematical understanding of state transitions. Applying such structures in specific applications, such as satellite rendezvous, demonstrates how state equations are adaptable in actual applications. The novel feature of such formulations not only enhances computational efficiency but also widens the scope of systems to which efficient control and analysis are feasible, reflecting the relentless progress in the field of state-space modeling and application in engineering.

## 9.2.2. Linear Versus Nonlinear State Equations

The distinction between nonlinear and linear state equations plays a pivotal position in the representation of state space and has consequences for control, analysis, and modeling of dynamic systems. State equations that are linear obey the law of superposition with output being directly proportional to input and easily yielding easy solutions and analysis. These kinds of systems can be effectively modeled with matrices and there is a uniform behavior with varied inputs. For instance, linear system state equations can often be addressed using such tools as the state transition matrix to simplify identification of the response of the system over time. Nonlinear state equations, however, tend to be extremely complicated in dynamics, for example, several equilibria and chaotic behaviors, which are hard to analyze. As systems enter nonlinear regimes, their control strategies may require higher order levels of sophistication to ensure stability and performance. Mathematical structures for linear and nonlinear systems differ widely, with consequences for their representational ability and computational complexities. Linear state equations are generally expressed in canonical forms associated with highly developed theories and methodologies enabling application of linear algebra techniques for solutions to system behavior. For instance, linear system design is usually founded on pole placement methods that can be easily acquired with algebraic manipulations. Alternatively, nonlinear state equations may need numerical solutions, simulations, or approximation techniques in order to find solutions because there are no closed-form solutions. For example, nonlinear dynamics sometimes need phase portraits or bifurcation analysis in order to understand their behavior in a comprehensive manner. Consequently, choice of linear or nonlinear formulations directly influences the complexity of solution procedures involved. Additionally, ramifications of these multiple state equations extend even beyond their mathematical solution; they also have practical implications in a broad array of uses across engineering applications. Linear state equations, in control systems, allow for designs that are firm and stable, well-suited to uses like robotic arms or simple thermal systems. Nonlinear state equations are involved, however, in the modeling of complex interactions typical of biological systems, financial markets, or higher-order aerospace dynamics. The inherent complexities in the nonlinear system require advanced control techniques, such as feedback linearization or sliding mode control, to produce desired performance. As emphasized in current studies, such as the study and control of coupled PDEs and MIMO systems, high-level methods tailored to nonlinear equations are much needed since they more accurately represent the intricate character of most phenomena in the natural world (Declan et al., 2025; Bajodah et al., 2025). Hence, the challenge still remains to manage the richness presented by nonlinear state equations while leveraging the simplicity presented by their linear counterparts (Figure 9.5).

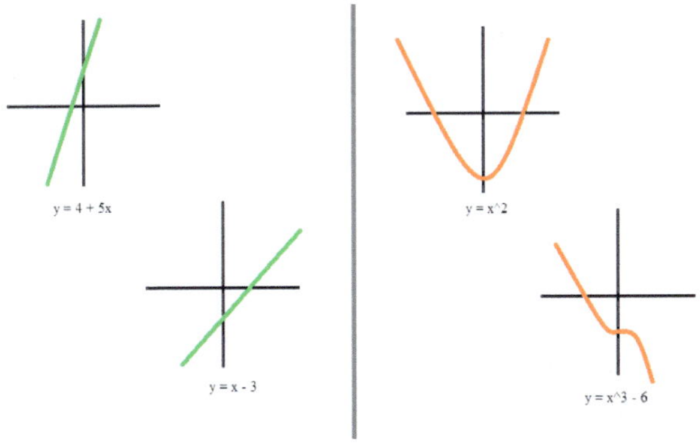


**Figure 9.5**. *Linear versus nonlinear functions.*

*Source: https://study.com/cimages/multimages/16/exponent_functions_linear_vs_nonlinear_functions.png.*

### 9.2.3. Applications of State Equations in System Analysis

In system analysis, state equations play a pivotal role in understanding complex dynamics and behavior of many systems. State-space model building is a robust model building framework for linear and nonlinear systems based on state variables and equations. The method enables researchers and engineers to transform all information of the system into a collection of first-order differential equations, hence simplifying analysis and designing the control. State equations make it simpler to analyze system stability, controllability, and observability, which are essential in ensuring that a system reacts predictably under various conditions. Thus, the application of state equations is an integral part of system analysis that allows for an inclusive approach to comprehending complex dynamic system behavior under numerous applications. Furthermore, the $H\infty$ filter captures the adaptability of state equations for application in practical systems. While the classical Kalman filter is popular, it is often constrained by system modeling uncertainties and noise dynamics. The $H\infty$ filter addresses these limitations by establishing a connection between the Krein space Kalman filter and $H\infty$ state equation-based estimators to achieve more precise estimations under uncertainties. This is particularly critical in the case of state transition matrices (STMs), which play a core role in establishing the system dynamics over time. With the application of state equations and $H\infty$ techniques, enhanced performance in system estimation is facilitated, ultimately leading to more reliable system analysis and implementation in robotics, aerospace, and control systems (Mu et al., 2023). Second, the decomposition of nonlinear differential equations confirms the viability of state equations in analyzing more complex system analyses. By using group-theoretic techniques, researchers are able to decompose the equations into simpler forms, allowing them to better comprehend nonlinear dynamics. As the process demonstrated in Krasnoschechenko (2024) illustrates, the decomposition of linear equations is a fundamental technique that can be adapted for nonlinear applications. This results in the development of simplified models that preserve basic system characteristics, and also enable the implementation of model predictive control schemes. These schemes are an expression of the utilized potential of state equations in the real world, where they play a key role in dynamic system stabilizing and control. Utilization of state equations thus extends beyond academic theoretical analysis, providing practical solutions to contemporary technological challenges in multidisciplinary contexts.

## 9.3. STATE-SPACE MODELING OF SYSTEMS

State-space modeling is a dominant paradigm in control theory and systems engineering that facilitates the complete analysis of dynamic systems. The modeling of a system by state variables and corresponding state equations, the technique summarizes the character of system behavior in an orderly manner. State variables are used to represent the internal state of a system at any instant of time to allow for the total description of its dynamics. To that end, the state equations, typically a set of linear or nonlinear ordinary differential equations, represent the manner in which these state variables evolve with time. This setting not only facilitates the investigation of system stability but also the design of controllers that can affect the behavior of the system to an effective degree. Also, it extends the system's representation to cover multiple inputs and outputs, thereby increasing its applicability in various fields like robotics, space exploration, and electrical engineering. State-space model application is also expanded into numerous fields, with their impact on both the theoretical development and practical applications being vast. Recent advances in generative modeling, particularly sequence data analysis, pinpoint the framework's versatility. For instance, LS4, a new generative model, employs a state-space ordinary differential equation to effectively deal with latent variable evolution. This is considerably enhanced modeling capacity with which numerical challenges intrinsic in optimizing systems with sudden transitions are dealt with. Utilization of convolutional representations in LS4 offers important computational advantages, wherein rapid evaluations can be carried out without the need to compute hidden states explicitly. Consequently, these methods make possible better representation of complex phenomena, enabling greater prediction and operational precision in real-world applications, appropriate to the state-space model flexibility and robust performance across a large variety of applications in the fields of systems analysis. Aside from numerous advantages of state-space modeling, there remain certain problems, like the solving of state equations and computing the state transition matrix. The state transition matrix is what decides how the system state changes with time and is the central focus for the study of dynamics across various intervals of time. State equation solutions are not only useful in terms of system response but also affect the design of controllers and stability analysis. It becomes an issue in attempting to find solutions for nonlinear systems, which generally require advanced techniques such as numerical methods or linearization approximations. Also, recent progress in machine learning has introduced novel approaches that bring forth enhanced optimization processes for state-space models, which bridge the gaps created by intrinsic limitation. With the integration of new approaches, researchers aim to continue advancing state-space methods, thereby enhancing their capacity to capture dynamic system complexities evident in current applications, as noted in Sobhy et al. (2024) and Zhou et al. (2022) (Figure 9.6).

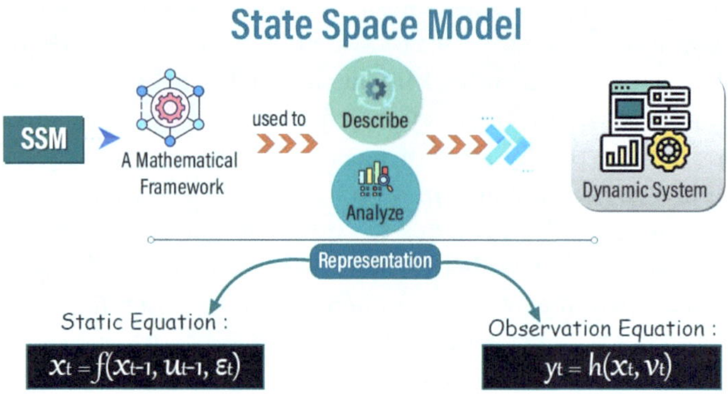


**Figure 9.6**. *State space model.*

*Source: https://cdn.educba.com/academy/wp-content/uploads/2023/07/State-Space-Model-1.jpg.*

## 9.3.1. Process of Developing State-Space Models

State-space model formulation is an important task in control theory and systems engineering, particularly while modeling dynamic systems. It begins with the identification of state variables that characterize the system's behavior in time. State variables, which represent system information at any instant, form an important part of developing state equations that determine the dynamics. By representing physical laws of the system as mathematical equations, engineers can formulate a model that correctly simulates the system's response to various inputs. The state-equation approach enables detailed variable relations to be captured and draws emphasis on interactions which might not be evident in other modeling techniques. Establishing an able proficiency in state variables forms the foundation for achievement in state-space modeling and ensures an all-embracing system representation. Following the definition of the state variables, the next step is to derive the state equations formulated in terms of matrices that reflect the system's dynamics. The matrix form involves the state transition matrix for how the system state evolves over time. The solution of the state equations closely hinges on the stability and controllability of the system, factors that are pivotal in controlling control strategies effectively. With the most recent advancements in state-space modeling, such as those in Output-Recurrent Gated State Space Model (OR-GSSM), comes more accuracy and computational efficiency. Attentive matrix operation structuring reduces errors formulated in terms of long-term forecasting while accurately representing complex dynamics in systems, like gas-liquid relationships in multiphase flows as shown in Chen et al. (2025). This approach optimizes the learning process with the additional assurance of reduced prediction errors. The final step in the state-space model building is fine-tuning and checking of the constructed representation against available real-world data or pre-defined standards. This is necessary in the validation of the stability and observability of the model, as reflecting deep complexities within the system dynamics. Applying methods such as controllability analysis can bring about sweeping enhancements in models since it guarantees matrices of the model, particularly the A matrix, are stable under various conditions. Recent studies show successful applications of enhanced structured state-space models (SSMs) such as Sparse-Mamba that enhance stability and controllability, leading to improved prediction accuracy and computational efficiency (Hamdan et al., 2024). Not only do these advances bolster the theoretical underpinnings of state-space models but also provide practical methods for applying these models in computing and engineering, opening a new frontier for state-space modeling across fields.

## 9.3.2. Advantages of State-Space Modeling Over Transfer Functions

State-space modeling possesses the special benefits over standard transfer functions in that it can incorporate a wider variety of systems. It has a significant benefit in that it can support multi-input and multi-output (MIMO) systems more comfortably. Transfer functions, residing in the frequency domain, struggle with complicated interactions that come from multiple inputs and outputs and are oftentimes addressed separately for each subsystem. In contrast, state-space representation employs state variables that represent the dynamics of the entire system in an integrated manner so that analysis becomes more global. This integrative approach is simple to formulate mathematically, which allows one to readily obtain system behavior under various operating conditions without loss of generality or accuracy. State-space modeling is also better equipped to manage time-varying and nonlinear systems than static transfer function methods. While transfer functions are optimal in linear systems with static parameters, they fall short in dynamic settings where parameters will change over time. State-space models accommodate these alterations more naturally through their structure, maintaining the systems varying dynamics without needing a model redefinition. This capability becomes increasingly important in fields such as robotics and space exploration, where the decision needs to be in real-time. The ability to also represent nonlinear relationships is a value added to make it even more valuable, enabling engineers to construct systems that are robust and sensitive to dynamic shifts in operating conditions. Solution of state equations and application of the state transition matrix in state-space modeling also provide experts with the ability to model system stability and control more precisely. The state transition matrix that governs the temporal evolution of the state of a system permits analysis of system behavior over a set of initial conditions and inputs. This data is vital in the determination of system stability and performance as it permits the analysis of transient behaviors and steady-state responses that cannot be reproduced by transfer functions with any level of efficiency. By taking advantage of such state-space representation characteristics, control engineers are able to design more sophisticated control strategies that reduce system performance, stabilize stability problems, and maximize overall system robustness. Such capability reflects the enhanced analytical ability and flexibility inherent in state-space modeling, thus rendering it much preferred in modern control theory and engineering practice.

### 9.3.3. Case Studies of State-Space Modeling in Engineering

In the discipline of engineering, state-space modeling has been profoundly influential, as evidenced by a number of case studies that explore its application across a range of different systems. One such case study is the dynamics of autonomous systems, which are defined as having no external inputs. It was with the assistance of the novel polynomial expansion procedure detailed in Kanbero Hğlu (2025) that researchers were able to approximate state transition matrices in an efficient manner such that system dynamics could be predicted and analyzed on a real-time basis. Such approximations were compared with traditional analytical solutions, highlighting the utility of state-space representations for modeling complex system behavior. This approach serves to highlight the capacity of state-space modeling for enhancing the accuracy and credibility of engineering simulations carried out automatically, which confirms its status as a crucial paradigm in current engineering practice. Further investigation of non-autonomous systems that dynamically respond to external inputs shows the utility of state-space modeling tools. The complementary methods developed in Kanbero Hğlu (2025) for dealing with input-dependent dynamics indicate a considerable progression in dealing with difficult system interactions. As a demonstration, an important case study involved a system that was created to respond to specific external influences to enable real-time performance to be tested under varying conditions. This application emphasized the broader practical relevance of state-space modeling by showing its versatility for use in diverse and nonlinear engineering applications. These studies illustrate how state-space models can be used to develop perturbation-insensitive control strategies with optimal system performance and meet the requirements of modern engineering applications effectively. Nonlinear differential equation decomposition has also played a significant role in advancing state-space modeling methods. As noted in, group-theoretic decomposition allows for effective analysis of complex nonlinear systems. Through the application of vector field isomorphism and Lie derivatives, researchers have contributed significantly to the decomposition of nonlinear system dynamics from the underlying transformation processes. This helps in the development of complex predictive control techniques, making it important in practice. A case study of this decomposition using a model predictive control architecture showed its applicability in stabilizing nonstationary systems. With the inclusion of these new approaches, state-space modeling continues to evolve, enhancing our capabilities to solve challenging engineering problems effectively in many areas.

## 9.4. SOLUTION OF STATE EQUATIONS

Solution of state equation forms the core of understanding state-space representation dynamics. State equations, represented in matrix notation, describe how system state variables evolve with time based on inputs and initial conditions. State equation solutions allow one to predict the future dynamics of a system from its given state, particularly meaningful in control theory and engineering design. Solution process typically involves finding the state transition matrix that embodies system dynamics to allow construction of analytical or numerical solutions. By means of linear algebra techniques, one can effectively solve the equations to enable system modeling and analysis that is instrumental in robotics, aerospace, and electrical engineering. Mathematical solution of state equations typically invokes differential equations and linear algebra concepts to yield solutions. In linear time-invariant (LTI) systems, the state equations are usually expressed as a first-order linear differential equation. Analytic solutions such as the matrix exponential are generally employed to determine the state transition matrix. This matrix is also instrumental in estimating both the transient and steady-state response of a system. By real-world applications, these calculations can be facilitated by software programs such that system behavior can be simulated under varying conditions. Additionally, these solutions provide the basis for feedback control law design, optimization of system performance, and stability problems. Hence, the complete comprehension of state equation solution is imperative for any scientist or engineer who has to deal with dynamic systems. Moreover, the solution of state equations must be able to address external inputs and disturbances that may affect system behavior. Inputs in state-space models are usually introduced as additional matrices connected to the state variables. The system response combined is finally characterized by both inbuilt dynamics contained within the state equations and by external influences characterized by inputs. Elaborate methods such as Lyapunov's stability theory and controllability theories advance to offer understanding by determining the extent to which a system can respond to input changes. Such theories allow for sound design of systems, including offering robustness and reliability in real-world application. Lastly, becoming proficient in solving state equations not only illuminates theoretical understanding but also holds significant relevance for technical engineering practice in a number of high-tech disciplines, for instance, those documented in Qi et al. (2011) and Giannozzi et al. (2009) (Figure 9.7).

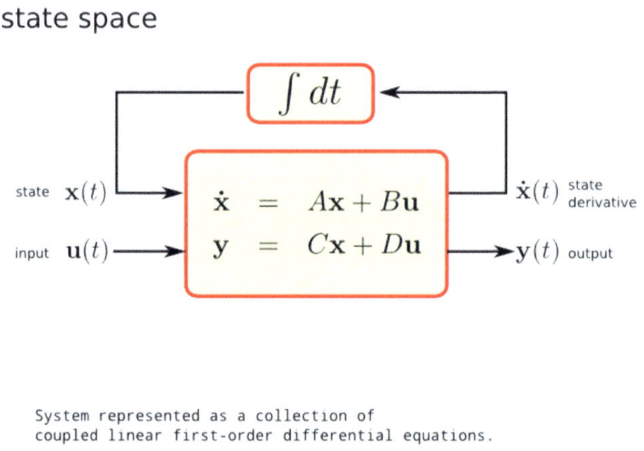


**Figure 9.7**. *Solution of state equations.*

*Source: https://i.sstatic.net/yC4Eu.png.*

## 9.4.1. Methods for Solving State Equations

The methods for solving state equations are pivotal in the field of control theory and systems engineering, particularly when representing dynamic systems through state-space models. One efficient approach is the eigenvalue method, where the systems behavior is examined through the eigenvalues and eigenvectors of the system matrix. By determining the eigenvalues, one can ascertain the stability of the system and the response characteristics to inputs. This method is especially useful for linear time-invariant systems, where the state equations can often be expressed in a form conducive to analysis through characteristic polynomials. The insights gained from this approach allow engineers to design effective controllers and observers, fundamentally crucial in dynamic system management (Ioffe et al., 2019). Another prominent method for solving state equations involves numerical techniques, especially when dealing with more complex, nonlinear systems where analytic solutions are infeasible. Techniques such as the Runge-Kutta methods and finite difference methods allow for stepwise integration of state equations over time, providing approximate solutions that converge to high accuracy. These methods, while computationally intensive, are necessary when working with systems exhibiting dynamics that cannot be captured through simpler mathematical formulations. For instance, in the simulation of tsunami propagation, models like Cliffs utilize such numerical frameworks to compute tsunami propagation and inundation based on the non-linear shallow-water equations, demonstrating the application of numerical methods in tackling real-world problems (Tolkova, 2016). Finally, the Laplace transform stands as a powerful analytical tool in solving state equations, particularly in converting differential equations into algebraic ones. This transformation simplifies the process of finding the system response to various inputs, enabling easier manipulation and extraction of time-domain solutions through inverse transformation. The application of the Laplace transform is essential in control systems analysis, providing insights into system behavior through its transfer function, which relates input to output in the frequency domain. Furthermore, it supports the development of transfer functions for multi-input, multi-output systems, facilitating comprehensive system design and stability analysis. Through such diverse methods, engineers achieve a robust understanding of dynamic behavior essential for effective system control (Ioffe et al., 2019).

## 9.4.2. Importance of Initial Conditions in Solutions

The role of initial conditions in state-space form cannot be overstated, as initial conditions play a pivotal role in specifying the dynamic system's behavior and stability. In state-space modeling, a state-space model of a dynamic system is always specified by a set of state variables describing the systems internal condition at any moment of time. While solving state equations, the selected initial conditions have direct impacts on resulting trajectories of state variables with respect to time. This is particularly critical in nonlinear dynamics-driven systems, whose almost similar initial states could lead to enormously different outcomes. Initial condition definition must thus be accurate to anticipate system performance and effectiveness of control strategies, especially in disciplines such as aerospace engineering, where disturbances must be closely controlled (Rigatos et al., 2025). Also, the initial conditions specify the setting of all of the solution space of the state equations, and they influence how trajectories evolve as time progresses. In control theory, the changes to the initial conditions are often a method of achieving appropriate performance qualities such as stability, responsiveness, and robustness. For instance, in nonlinear control systems, failure to include initial states can lead to suboptimal controller designs and poor system performance when in use. Sensitivity to initial conditions is most evident in complex systems, demonstrating why scientists and engineers need to carefully analyze the ramifications of their choices. Additionally, with the inclusion of interdependent systems, the ripple effect of the initial conditions will affect the whole behavior, hence necessitating the full knowledge of every subsystems origin point (Li et al., 2023). Further, the inclusion of initial conditions in the state transition matrix shows its essential dimension in dynamic system modeling. The transition matrix of states, representing the manner in which the state evolves over time from some initial point, is eventually fixed by the initial state. In practice, what this means is that having an idea of what the conditions are initially can lead to better control and prediction strategies. Good solutions for control problems, such as described in the nonlinear control methods for autonomous reentry space vehicles with system parameters only infrequently constant which are dealt with in a proper treatment of initial conditions. Hence, the initial conditions not only serve as the departure point of analysis but also determine the solutions in general, making their careful choice and representation a cornerstone within dynamic system analysis and state-space modeling (Li et al., 2023; Rigatos et al., 2025) (Figure 9.8).

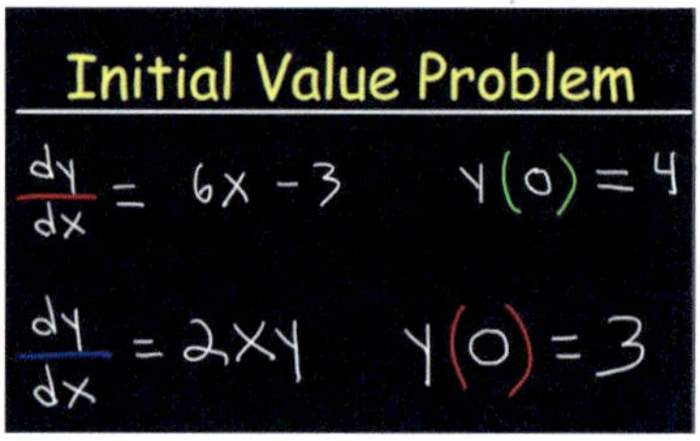


**Figure 9.8**. *Initial value problem.*

*Source: https://i.ytimg.com/vi/kwGukY_2qWQ/sddefault.jpg.*

### 9.4.3. Numerical Techniques for State Equation Solutions

The solution of state equations through numerical methods has an important place in state-space representation research so that analysis and design of complex control systems are possible. One of the principal methods used is the discretization techniques such as using Euler's method or Runge-Kutta techniques for converting the continuous-time state equations into a computationally convenient form. These methods quantify the change of state variables from discrete intervals of time, making it possible to simulate and analyze the dynamics of a system. The choice of numerical method plays a critical role in affecting the accuracy and stability of the solutions, so that it is essential to select an appropriate technique for a specific application. Researchers therefore prefer conducting comparative studies to evaluate the merits and demerits of different methods in a specific context. The success of these numerical approaches hence serves as the foundation for their use within the broader area of systems modeling and control synthesis. Besides the more traditional schemes of Euler and Runge-Kutta, advanced numerical methods have recently been developed addressing specific types of systems, for example, those described by fractional order state equations. For these systems, new methods like the Grünwald-Letnikov approximation have been proposed, which present solutions to fractional differential equations where conventional methods may fail. Numerical methods of this kind give a more accurate description of dynamic behaviors characteristic of fractional systems, eventually expanding the scope of state-space modeling. Furthermore, they are in excellent agreement with present trends in optimizing control strategies since numerical solutions bring profound insight into system performance under various scenarios. Biswas et al. (2010) indicates towards the significance of such advanced methodologies in recasting optimal control problems and hence demonstrating their significance in state-space formulations across a wide range of disciplines. The utilization of numerical methods in solutions to state equations has the prime feature of developing computing tools that ensure maximum ease of access and performance. Libraries such as NumPy and frameworks developed within the scientific Python community have emerged to be necessary tools for engineers and researchers alike. By their means of handling multi-dimensional arrays as well as matrix operations, these libraries make it possible for practitioners to adopt several numerical algorithms in an efficient manner. The possibility of handling and analyzing data in holistic perspectives makes it possible for practitioners to run complex simulations fast and accurately. Besides, with advancing computational tools comes improved interaction between practice and theory in ways that lead to finer methodologies allowing for on-going innovation within state-space representation and control systems. Harris et al. (2020) brings out this trend with the emphasis being on the foundation of such computational tools in present-day scientific research.

## 9.5. STATE TRANSITION MATRIX

The State Transition Matrix, STM, is a key component of state-space modeling since it enables the description and understanding of dynamic systems through analysis. The STM describes how the system state evolves with the elapse of time, compactly summarizing the relationships between various state variables. STM is mathematically a matrix whose elements represent the state variables at t+1 in terms of those at t. Its analytical properties enable it to deliver information on system predictability, stability, and controllability. From state equations, STM is the key to solving differential equations for linear dynamic systems. Its application does not extend only to theoretical frameworks; it is essential to engineering practice, particularly in control systems and simulation, where accurate prediction of system behavior is critically required to guarantee performance specifications. The use of the State Transition Matrix in system analysis allows for the application of sophisticated modeling approaches, particularly in complicated dynamics and stochastic systems. Applying semigroup theory in the formulation of state space, researchers can arrive at more precise descriptions of systems subject to various influences, such as external disturbances or uncertainties. A traditional case in point is the Output-Recurrent Gated State Space Model (OR-GSSM), extending orthodox use of STM through incorporating control inputs alongside tunable timescales. Besides optimizing computational efficiency, the technique also increases physical interpretability of the system behavior (Chen et al., 2025).

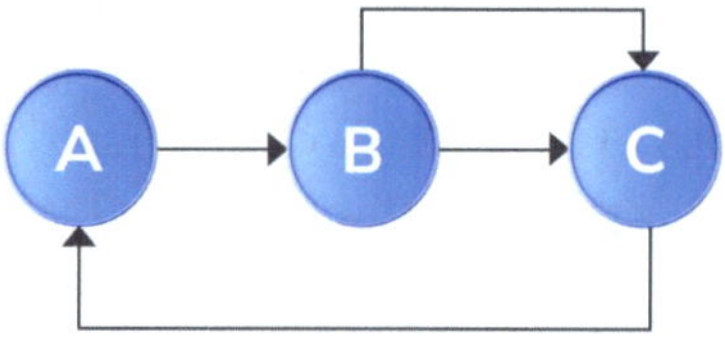


**Figure 9.9**. *State transition matrix and diagram.*

*Source: https://media.geeksforgeeks.org/wp-content/uploads/20240903183740/State-Transition-Matrix-02.webp.*

With the investigation of parameterized transition matrices, the OR-GSSM paradigm addresses the constraints of traditional computational fluid dynamics, and the individuality and ingenuity of the State Transition Matrix are emphasized in next-generation modeling paradigms. In addition, the significance of the State Transition Matrix is emphasized further in multi-dimensional analytical settings, wherein various physical forces must be précised. For instance, the analysis of laminated plates in magnetic fields emphasizes the ability of STMs to identify multi-faceted interactions, such as bending behavior under imposed loads, temperature, and moisture (Sobhy et al., 2024). With the application of the matrix method to derive solutions for governing differential equations, engineers are able to predict the impact of multi-faceted causes on system performance. This theory-state-space hybridization is merely one instance of the STMs ability to merge intricate phenomena into practical frames of analysis.

Ultimately, the State Transition Matrix serves as a valuable bridge between conceptual constructs and practical application that enables greater predictive accuracy and understanding of dynamic system behaviors (Figure 9.9).

## 9.5.1. Definition and Significance of the State Transition Matrix

The state transition matrix is a control theory and systems engineering primitive concept and is a valuable tool utilized to examine linear time-invariant (LTI) systems. Mathematically, the state transition matrix, historically denoted as $\Phi(t, t_0)$, represents how the state of a dynamic system evolves over time from an initial system state at time $t_0$ to some future time state at time t. Such a matrix is inherently related to the systems state equations, normally linear differential equations. The state transition matrix derives its significance in the sense that it summarizes the system dynamics into a compressed form for the full understanding of stability, controllability, and observability without necessarily involving complex computations or simulations whenever a new system state is to be understood. Moreover, the state transition matrix is helpful when solving state equations in the state-space representation. Through the matrix, designers and researchers can compute the system's state at any point in time, provided that its initial conditions are determined. Such a feature is beneficial in simulations and forecasting analysis, where future state computation from existing configurations is important. In addition, it acts as a bridge to combine time-domain analyses and frequency-domain methods, further enriching the toolset available for system stability evaluation. According to various researches, for example, integration into other mathematical structures (Kusaka, 2015), the state transition matrix not only makes state evolution analysis easier but also aids in intuitive comprehension of system behavior across different variables and scenarios. The state transition matrix finds its practical applications outside of theoretical purview; it serves as the foundation for practical systems such as robotics, aerospace systems, and economic models. To illustrate, in control systems, where deployed, the state transition matrix can be used to direct feedback mechanisms so that responses of a system can be dynamically controlled. In economic modeling too, the same matrix can be employed to forecast trends based on current economic indicators, to prove its versatility across domains. It has been said by experts that with the employment of the state transition matrix along with computational techniques, one can achieve robust and consistent results, which reflect real system behavior in the long term (Fernández et al., 1995). The awareness of the definition and significance of the state transition matrix, thus, is not only academic but essential for designing systems that behave in an effective way in response to anticipated as well as unforeseen issues.

### 9.5.2. Calculation of the State Transition Matrix

The computation of state transition matrix is a fundamental component in state-space representation that serves to bridge the systems initial state and its subsequent dynamics. The state transition matrix $\varphi(t)$ encapsulates the dynamics of a linear time-invariant (LTI) system, transforming the state vector at time $t=0$ into the state vector at any arbitrary instant in the future t. This matrix allows for system reaction analysis and prediction to various inputs and is thus vital in control system design.

The calculation is commonly achieved by means such as the matrix exponential, where the transition matrix can be derived from the systems state equations. Specifically, if cast in canonical form, the state transition matrix makes the intricate computations of dynamic system analysis more manageable by allowing practitioners to specifically focus on those crucial parameters that impact the stability and performance of the system. Moreover, the computation of the state transition matrix has a tendency to leverage system matrix A's eigenvalues and eigenvectors. Using eigenvalue calculation, system stability can be analyzed; that is, if all eigenvalues are negative in real, the system is stable. The eigenvectors of these eigenvalues are used to construct a transformation matrix linking states and the system dynamics, ultimately simplifying the state transition matrix to a basic form Korovin et al. (2022). Together with initial conditions, this transition matrix makes it possible to thoroughly examine the system response over time, including the impact of changing inputs and external disturbances.

This eigenvalue-based approach, in addition to making the computation of $\varphi(t)$ easy, provides information about the qualitative system characteristics, which are vital for effective design and control. In addition, advancements in computing devices have ensured that calculations in the state transition matrix are much simpler, particularly for complex systems whose dynamics vary with time. Numerical methods, such as using state-space models in computer-aided design, ensure that it is possible to implement the transition matrix in a practical manner under real-time applications Simikova et al. (2019).

These methods allow transient processes to be modeled with the potential for real-time system feedback-based dynamic adaptation of control strategies. One such example, the introduction of wavelet transformations as the present literature emphasizes, is a demonstration of such a shift towards more sophisticated analysis techniques that capture the nuance of signal change and system state at transition. These innovations are needed for systems like control engineering and power systems, where state variations with time knowledge will provide improved systems resilience and performance optimization (Figure 9.10).

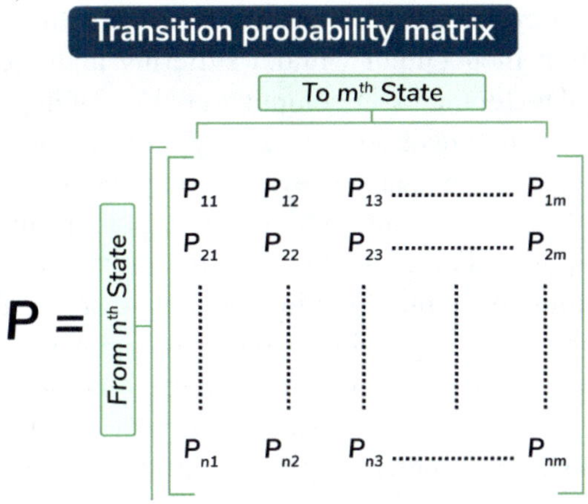


**Figure 9.10**. *Transition probability matrix.*

*Source: https://media.geeksforgeeks.org/wp-content/uploads/20240802153903/Transition-probability-matrix.png.*

### 9.5.3. Applications of the State Transition Matrix in System Dynamics

The state transition matrix is an important system dynamics tool, especially when system dynamics are to be modeled over time. It is a mathematical device that allows one to analyze how a systems state changes based on initial conditions and inputs. From this matrix, system analysts can predict future states of the system by repeatedly multiplying the initial state vector with the transition matrix over time. This allows for both linear and non-linear systems to be modeled and therefore serves a broad range of purposes. The ability to quantify change in state provides valuable information on system stability, response behavior, and control strategies, and thus enhances the potential to design systems with high robustness and potential for meeting specified performance requirements. From engineering to economics, the state transition matrix derives its usefulness in characterizing dynamic responses and system performance. For instance, in control systems, it provides engineers with a paradigm to design feedback structures in order to ensure stability and optimal performance. Secondly, in finance, it assists in modeling the behavior of markets where various factors influence the systems' behavior. The state transition matrix's flexibility is also present in being compatible with simulation software, for example, which can be used for outcome prediction within complex networks or risk estimation for multiple scenarios. The computational efficiency of the matrix allows it to handle high-dimensional data, such that the practitioner is able to gain useful insights from their output without suffering undue computational lag. In addition, new developments in computational techniques further expanded applications of the state transition matrix to system dynamics. New algorithms now have the capacity to analyze systems that contain large numbers of interacting variables, which result in higher accuracy in simulation for a broad spectrum of disciplines. For example, the state transition matrix is applied to quantum simulations to examine the interactions of molecules and properties of materials, exhibiting its applicability to new research areas (Giannozzi et al., 2017). It also finds core application to the exploration of topological insulators, where examination of transitions between different quantum states is crucial for material science development (Qi et al., 2011). These examples not only reflect the maturity and strength of the state transition matrix but also highlight its usefulness in current research, innovation, and real-world applications in various fields.

## 9.6. PRACTICAL APPLICATIONS OF STATE-SPACE FORM REPRESENTATION

Applications of state-space representation to engineering are numerous, most significantly the control of complex systems, such as reentry space vehicle attitude dynamics. State-space models allow engineers to describe these systems in dynamic form, which hold the inter-dependencies and interactions that occur within non-linear systems. One of the most significant methodologies addressed is nonlinear optimal control methods, in which state variables of systems are used to stabilize without the usually cumbersome transformations used in state-space formulations. This has been shown in recent developments where control algorithms make use of approximate linearizations and algebraic Riccati equations to derive feedback control laws such that global stability is achieved through structured approaches such as Lyapunov analysis. These approaches greatly aid aerospace missions by virtue of their capacity to enable precise control at low computational cost, facilitating real-time adjustment necessary for complex maneuvers in unstructured environments (Rigatos et al., 2025). Moreover, state-space representation lies at the foundation of model predictive control (MPC), which utilizes predictive models of the behavior of the systems to optimize control actions over a specified future horizon. By breaking down the nonlinear equations into separately analyzable systems, state-space methods enable modular design of control. Breakdown based on group-theoretic notions such as Lie algebra is usually employed to systematically divide up control questions and input-output relations. The paradigm produced not only simplifies the modeling procedure but enhances the robustness of control strategy employed since it is able to quickly adapt to disturbances and dynamic changes in the system. Such adaptability in handling nonlinearities is essential for high-reliability applications, such as autonomous transport and robot systems (Krasnoschechenko, 2024). Moreover, the inclusion of state-space representation across disciplines in engineering encourages breakthroughs beyond conventional control systems. An example is the linearization of nonlinear systems with the use of state-space that models complex behaviors in processes like chemical reactions, electrical circuits, and mechanical systems. All of these applications gain from state variables encapsulating the present state of the systems to be able to make clearer sense of how the inputs influence the outputs over time. In energy system or environmental modeling tasks, state-space concepts' control of complex interactions becomes increasingly vital. Not only do these approaches maximize controller algorithm design and deployment, but predictive maintenance management as well, enabling systems to respond to failures before they happen. The continuous ongoing evolution of state-space application continues to influence multiple fields, keeping them abreast of modern engineering issues and innovative solutions (Rigatos et al., 2025) (Figure 9.11).

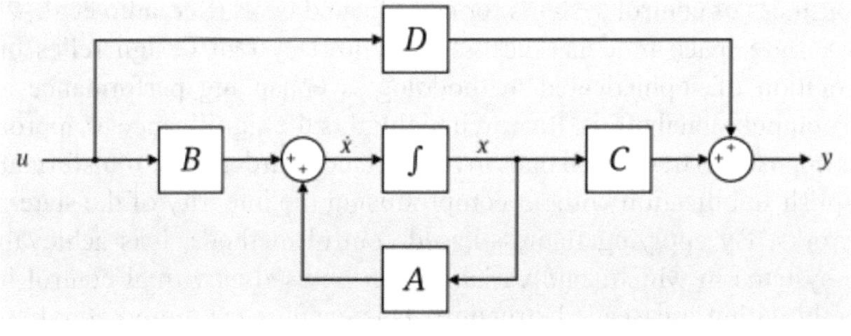


**Figure 9.11**. *State-space representation of linear systems.*

*Source: https://storage.googleapis.com/static.prod.fiveable.me/search-images%2F%22State-space_model_components_for_linear_systems%3A_state_vector_input_vector_output_vector_equations%22-472px-Typical_State_Space_model.svg.png.*

## 9.6.1. Use in Control System Design

The addition of state-space form to control system design has gradually played a significant role in constructing robust and responsive systems, particularly in aerospace engineering. By using state variables and state equations, designers have been able to describe complex systems, such as autonomous reentry space vehicles, where normal procedures fail. The nonlinear dynamic analysis obtained through state-space modeling allows for the design of sophisticated control strategies, for example, sliding-mode control or model predictive control, that enhance the stability and performance of the system. Additionally, the ability to leverage methods without significant state-space model transformations, as highlighted in the research using nonlinear optimal control methods, alleviates problems with singularities and diffeomorphisms, enabling more robust system performance in uncertain conditions (Rigatos et al., 2025). In building control methods, tools such as the Koopman operator framework have been game-changing in optimizing state-space representation. This approach reworks nonlinear dynamics into linear systems of infinite dimensions, hence facilitating easier analysis of complex behavior. Using high-order orthogonal polynomials, the method facilitates the description of state dynamics in linear combinations, hence enabling easier identification of system responses to various inputs. Use of this framework in application, particularly in the case of rendezvous between satellites, proves its utility in formulation of energy-optimal control solutions that ensure system integrity during the process. Consequently, the new approaches not only ease analysis but enhance the robustness of control systems for complicated tasks (Servadio et al., 2023). Besides, state-space models' success in control system design relies on the incorporation of sophisticated methodologies enhancing performance at the cost of computational effort. Research highlights the significance of approaches that, as opposed to traditional ones involving such burdensome transformations, accomplish stabilization without compromising the integrity of the state-space frameworks. By applying flatness-based control methods, it is achievable to devise systems in which state variables can be used as virtual control inputs flawlessly within a cascaded structure. This encourages more natural control design, where engineers can focus on the engineering realities of their designs and not on details of transformations. Following developments in control methodologies, the impact of state-space representation in the control system design paradigm remains important, enhancing reliability and responsiveness in various branches of engineering (Rigatos et al., 2025).

### 9.6.2. Implementation in Real-World Engineering Problems

In engineering, the state-space model is a useful framework for describing complex systems in real applications. These applications often involve dynamic systems whose variables have mutual dependence, such as those in aerospace engineering. Some examples involve the air transport of airplanes, particularly hybrid models such as amphibious airplanes and seaplanes, demonstrating the hydrodynamic-aerodynamic interaction complexities. By using state-space models that model these interactions, engineers are able to predict system behavior while maneuvers such as take-off and landing occur. By using discrete non-linear time-varying dynamics models, extensive simulation of these motions can be conducted including several forces such as gravity and thrust. Success in such simulations demonstrates the feasibility of state-space representation in sound and actionable generation for flight control system development, as stated in the case where the steady-state is taken as a starting point for the dynamic model (Wang et al., 2025). Additionally, application of state-space modeling is also possible outside aerospace engineering to fields like mechanical and civil engineering. For example, in structural dynamics control systems and robotics, the state-space technique comes in handy in modeling and designing controllers that are competent to respond to external stimuli. Specifically, the derivation of the state transition matrix becomes important in establishing how systems evolve over time, particularly in non-linear fields where linear methods prove to be insufficient. An emergent research highlights a time series method to pseudolinear systems and describes how approximations of the state transition matrix may be utilized to enhance the modeling of non-autonomous dynamics within realistic limits. The integration of complementary approaches to treat external inputs further evidences the adaptability of state-space approaches and indicates their application for solving complex engineering problems (Kanbero, 2025). Moreover, the state-space description has significant consequences for fields like robotics, where system precision and flexibility are the most important considerations in terms of control systems. For engineering applications where system response time and adaptability are of critical concern, state-space models enable feedback mechanisms to be integrated, thus enabling closed-loop control schemes that improve performance. PID controller design is a prime example of how state variables are employed in order to derive control laws that enhance the response and stability of systems. Their performance in models, as illustrated in a variety of case studies in satellite dynamics, is evidence of their capacity to model and simulate behaviors within multiple environmental conditions. Lastly, state-space representation not only enhances the engineering design process but also sparks innovation in the creation of new technology meant to meet complex functional specifications under realistic conditions (Wang et al., 2025; Kanbero, 2025).

### 9.6.3. Future Trends in State-Space Representation Applications

While computing capability breakthroughs keep evolving, developing state-space representation application trends more and more tend towards integrating state-space representation with machine learning (ML) and artificial intelligence (AI). Being able to model complex systems efficiently using state-space approach provides solid foundations for AI algorithms to make improved operations. These models enable dynamic phenomena to be described in multi-dimensional spaces, from which AI models can better predict. Future research can be in further developing state-space modeling systems with the inclusion of AI-based optimization techniques to allow real-time state transition matrix tuning. Integration can reshape applications in robotics, autonomous systems, and real-time data analysis, ultimately paving the way to more performance and flexibility in applications. Moreover, structured state space exploration is a focal trend towards developing more efficient state-space representations. With finite-dimensional structured state spaces, researchers are able to respond to increasing complexity in dynamic systems at computational feasibility. Dramatic improvements are possible in areas such as signal processing and nonlinear system identification. As explained in latest research, a move towards structured state-space discrete-time dynamical systems can enhance the accuracy of state-transition laws, which are key to modeling the intricate temporal dynamics of complicated systems. The implementation of classical dynamical systems properties into structured environments provides opportunities for new modeling paradigms, poised to drive engineering and control theory breakthroughs (Fiori, 2025). Further, the employment of state-space models to represent hysteresis behavior is a prime reflection of the current trends in materials and engineering. Utilization of models, such as the Jiles–Atherton (J-A) hysteresis model, within a state space framework supports improved simulation of magnetic material performance attributes. The trend towards state-space representations adds simplicity and convenience in handling hysteresis loops, which is important in developing smart materials and systems. Recent uses, facilitated by techniques like particle swarm optimization for parameter estimation, show the practical uses of these models in fields ranging from electronics to robotics. As industry increasingly calls for increasing control and predictive capability, future development of state-space representations will be central to the development of next-generation technologies (Zhao et al., 2024).

## SUMMARY

The state-space approach is a deep paradigm for modeling and analyzing dynamic systems across diverse disciplines, ranging from engineering and physics to applied mathematics. It not only captures the essential dynamics of systems through the use of state variables and equations, but also enables us to gain a systematic understanding of system behavior over time. With such methods as the state transition matrix, researchers can predict future states from current states, thereby facilitating effective control actions and stability study. The capability of state-space modeling to be applied in linear and nonlinear systems, as well as systems modelled by Partial Differential Equations, has transformed it into a crucial tool in analyzing complex physical phenomena. This adaptability underlines the significance of state-space techniques in modern computational methods and control theory. In addition, the emergence of novel state-space representation techniques, such as those emerging in recent studies, signals rising model complexity. Of particular note, the application of advances like the Output-Recurrent Gated State Space Model (OR-GSSM) signifies a trend toward including understanding from the physical world into machine learning models for enhanced model precision. By discussing dynamic coupling among state variables within state-space descriptions, researchers can manage challenging problems relating to modeling multidimensional systems, such as multiphase flows and vehicle exhaust dynamics. These developments contribute significantly towards computational efficiencies and real-time applicability, emphasizing again the current pertinence of state-space techniques in emerging technology domains. Furthermore, the greater the complexity of systems in interaction, the greater the need for dynamic models that can make reliable predictions while being responsive to uncertainties inherent in actual situations. Enhanced representation and solution of state space have more than academic significance; they translate into viable application in engineering system design for robustness. For instance, improvements in uncertainty quantification and stability analysis enhance state-space model prediction confidence, particularly for safety-critical applications like control of coupled Partial Differential Equations. The application of state-space modeling methods, such as adaptive timescales in transition matrices highlighted in studies employing the OR-GSSM framework, allows engineers to find a compromise between accuracy and utilization of computational resources. This capability is particularly valuable for fields that include real-time decision-making and high-fidelity simulation, thereby opening the door to innovation and optimization. Lastly, embracing these state-space methods generates innovations in fields where system interaction and dynamic behavior are paramount, bringing a new age of smarter, more responsive engineering solutions.

State-space description is a main pillar in modern control theory that allows both analysis and design of dynamic systems. The use of state variables and state equations allows systems to be modeled in terms of a finite number of first-order differential equations that describe the systems' internal state as well as its inputs. This framework facilitates an overt understanding of system dynamics and offers significant information regarding stability, controllability, and observability crucial in effective system management. The transition from classical input-output models to state-space representations is a significant advancement in the modeling of intricate systems as it enables improved handling of time-varying and nonlinear behaviors characteristic of engineering issues. In addition to state equations and state variables, solution of state equations is the foundation of state-space modeling. Methods used can lead to solutions, for example, state transition matrix, which provides a highly effective means of conveying system state evolution with respect to time. Analysis of linear time-invariant systems is easier with the help of the state transition matrix, using initial conditions in terms of state variables at any time. This link establishes the relevance of state transition matrices both theoretically and in engineering practice, including system simulation and control synthesis. Furthermore, application of methods such as the Koopman operator to synthesize optimal control methods speaks volumes about the universality of state-space methodologies across a vast spectrum of engineering fields where state-based analysis yields new solutions to complex control problems (Servadio et al., 2022). The depth of state-space description extends into more sophisticated topics such as nonlinear dynamics and phase-lock effects, again attesting to its generality. An investigation into specific systems, such as the overdamped Josephson junction, illustrates how state-space approaches can provide an accurate description of complex behaviors arising from nonlinearities. The important role of phase-lock regions, identified by their complex dynamics, is an exemplary demonstration of the richness of state-based analysis in revealing the subtleties of system behavior over various parameter regimes. Studies have proved that understanding rotation numbers and their effect on stability becomes critical in this regard, highlighting the synergy between mathematical precision and practical applications of state-space model construction. In the end, all these aspects not only explain the fundamental assumptions behind state-space representation but also reveal its extensive application in both linear and nonlinear systems (Glutsyuk, 2018).

# REFERENCES


1. Abdullayev, V. (2024). To the solution of loaded differential equations with nonlocal conditions. The Bulletin of Irkutsk State University. Series Mathematics. https://www.semanticscholar.org/paper/54570fd8ad89edb25dd36e6484c99a8698dbc749 (accessed on 19th September 2025).
2. Bajodah, A., & Mibar, H. (2025). Static output feedback pole placement control: Restoring state space algebraic roots. *2025 Australian & New Zealand Control Conference (ANZCC)*. https://www.semanticscholar.org/paper/1eec83d63d839979929cf7cfa9707c6bbe93f706 (accessed on 19th September 2025).
3. Biswas, R. K., & Sen, S. (2010). Fractional optimal control problems: A pseudo-state-space approach. Journal of Vibration and Control, 17, 1034–1041. https://doi.org/10.1177/1077546310373618 (accessed on 19th September 2025).
4. Chen, R., Fang, M., & Ma, G. (2025). Output-recurrent gated state space model for multiphase flows modeling and uncertainty quantification of exhaust vehicles. https://www.semanticscholar.org/paper/7d224274eb404035c222d812bce62e943d812a82 (accessed on 19th September 2025).
5. Di Renzo, M., Zappone, A., Debbah, M., Alouini, M. S., Yuen, C., de Rosny, J., & Tretyakov, S. (2020). Smart radio environments empowered by reconfigurable intelligent surfaces: How it works, state of research, and the road ahead. *IEEE Journal on Selected Areas in Communications, 38*, 2450–2525. https://doi.org/10.1109/jsac.2020.3007211 (accessed on 19th September 2025).
6. Fernández, F., & Castro, E. (1995). Algebraic methods in quantum chemistry and physics. https://www.semanticscholar.org/paper/10226ff8a9811c2b620f48c27709a0d9540de166 (accessed on 19th September 2025).
7. Fiori, S. (2025). Discrete-time dynamical systems on structured state spaces: State-transition laws in finite-dimensional Lie algebras. Symmetry. https://www.semanticscholar.org/paper/69b2542204a190c93e58d8c10ae9a8f9b359050b (accessed on 19th September 2025).
8. Giannozzi, P., Andreussi, O., Brumme, T., Bunău, O., Buongiorno Nardelli, M., Calandra, M., Car, R., et al. (2017). Advanced capabilities for materials modelling with Quantum ESPRESSO. Journal of Physics: Condensed Matter, 29, 465901–465901. https://doi.org/10.1088/1361-648x/aa8f79 (accessed on 19th September 2025).
9. Giannozzi, P., Baroni, S., Bonini, N., Calandra, M., Car, R., Cavazzoni, C., Ceresoli, D., et al. (2009). QUANTUM ESPRESSO: A modular and

open-source software project for quantum simulations of materials. Journal of Physics: Condensed Matter, 21, 395502–395502. https://doi.org/10.1088/0953-8984/21/39/395502 (accessed on 19th September 2025).

10. Glutsyuk, A. (2018). On constrictions of phase-lock areas in model of overdamped Josephson effect and transition matrix of the double-confluent Heun equation. *Journal of Dynamical and Control Systems, 25*, 323–349. https://www.semanticscholar.org/paper/82acfb3df9a725267eae1595a497bc1b013b952d (accessed on 19th September 2025).
11. Hamdan, E., Pan, H., & Cetin, A. (2024). Sparse Mamba: Introducing controllability, observability, and stability to structural state space models. https://www.semanticscholar.org/paper/8fa8df190c0ff896b3446d857baddbccb3fee0db (accessed on 19th September 2025).
12. Harris, C. R., Millman, K. J., van der Walt, S., Gommers, R., Virtanen, P., Cournapeau, D., Wieser, E., et al. (2020). Array programming with NumPy. *Nature, 585*, 357–362. https://doi.org/10.1038/s41586-020-2649-2 (accessed on 19th September 2025).
13. Ioffe, D., Kozma, G., & Toninelli, F. (2019). Scaling limits in models of statistical mechanics. *Oberwolfach Reports*. https://www.semanticscholar.org/paper/93a54af8f063dcc3cd80df784f41489931228671 (accessed on 19th September 2025).
14. Jagt, D. S., & Peet, M. (2025). A state-space representation of coupled linear multivariate PDEs and stability analysis using SDP. https://www.semanticscholar.org/paper/2ae78f31ae95611e682969cb6c7b4d196b23aa1a (accessed on 19th September 2025).
15. Kanberoğlu, H. (2025). Time series approach to the response of time-invariant pseudolinear differential equations with application in satellite simulation. *Proceedings of the Institution of Mechanical Engineers, Part G: Journal of Aerospace Engineering*. https://www.semanticscholar.org/paper/6335ffb945b73ed520ff93e8bec16f4cd041545e (accessed on 19th September 2025).
16. Korovin, A., & Aleksandrov, I. V. (2022). Coordinate transformations of three-phase variables using quaternions. *Vestnik Chuvashskogo Universiteta*. https://www.semanticscholar.org/paper/484198a3e4d65a97759be0d100da43d2beb4d7f1 (accessed on 19th September 2025).
17. Krasnoschechenko, V. I. (2024). Decomposition of equations of nonlinear affine control systems and its application to the synthesis of regulators. Mekhatronika, Avtomatizatsiya, Upravlenie. https://www.semanticscholar.org/paper/af164147b21962bda4a3b986d3a14cb3eb94211f (accessed on 19th September 2025).

18. Kühne, T. D., Iannuzzi, M., Del Ben, M., Rybkin, V. V., Seewald, P., Stein, F., Laino, T., et al. (2020). CP2K: An electronic structure and molecular dynamics software package–Quickstep: Efficient and accurate electronic structure calculations. The Journal of Chemical Physics, 152. https://doi.org/10.1063/5.0007045 (accessed on 19th September 2025).

19. Kusaka, I. (2015). *Statistical Mechanics for Engineers*. https://www.semanticscholar.org/paper/f18eaaed89d2ac53858fa262f0a44174ea087731 (accessed on 19th September 2025).

20. Li, Y., & Wang, C. (2023). On a partially synchronizable system for a coupled system of wave equations in one dimension. Communications in Analysis and Mechanics. https://www.semanticscholar.org/paper/034b356450456586455 9b49a4caf416d1c65c474 (accessed on 19th September 2025).

21. Mu, Y., & Zhao, H. (2023). H∞ state and disturbance joint one-step smoothing estimation for a class of linear discrete-time fractional-order systems. Proceedings of SPIE, 12940, 129402N–129402N-12. https://www.semanticscholar.org/paper/3c08bd002a075cbcb173e6111ad830220c545349 (accessed on 19th September 2025).

22. Oshchepkov, A. Y. (2024). Evolutionary equations for studying the dynamics of qubits in variable magnetic fields. *Perm University Herald. Physics*. https://www.semanticscholar.org/paper/cebc7edc720d8838cd3e38d24c4dabf2c886671a (accessed on 19th September 2025).

23. Qi, X. L., & Zhang, S. C. (2011). Topological insulators and superconductors. Reviews of Modern Physics, 83, 1057–1110. https://doi.org/10.1103/revmodphys.83.1057 (accessed on 19th September 2025).

24. Rigatos, G., Abbaszadeh, M., Pomares, J., & Cuccurullo, G. (2025). Nonlinear optimal and multi-loop flatness-based control of the attitude of reentry space vehicles. *Aircraft Engineering and Aerospace Technology*. https://www.semanticscholar.org/paper/ee305dc976a691279cded0ed69ffec6e0bc70c4f (accessed on 19th September 2025).

25. Servadio, S., Armellin, R., & Linares, R. (2022). A Koopman-operator control optimization for relative motion in space. AIAA SCITECH 2023 Forum. https://www.semanticscholar.org/paper/5f785739accd589452ca93d6700a4f064ed8dafa (accessed on 19th September 2025).

26. Servadio, S., Armellin, R., & Linares, R. (2023). Koopman-operator control optimization for relative motion in space. Journal of Guidance, Control, and

Dynamics. https://www.semanticscholar.org/paper/7821c005e07a7c209edf48edc5d7fd9c75f646a2 (accessed on 19th September 2025).

27. Simikova, A., Fedosenkov, B., & Fedosenkov, D. (2019). Modal control of feeding processes in the medium of state space and wavelet transforms. *Vestnik of Astrakhan State Technical University. Series: Management, Computer Science and Informatics*. https://www.semanticscholar.org/paper/e7ff60fb7d3c94b30a8fef5273f9213aad627ea5 (accessed on 19th September 2025).
28. Sobhy, M., & Radwan, A. F. (2024). Levy solution and state-space approach for hygro-thermo-magnetic bending of a cross-ply laminated plate on Kerr foundation under various loadings. Journal of Composite Materials, 59, 3–28. https://www.semanticscholar.org/paper/ac429fd693dc49eec61efe89030f024c5e1bb846 (accessed on 19th September 2025).
29. Tolkova, E. (2016). Cliffs benchmarking. https://www.semanticscholar.org/paper/84db6d0dc1e4973c8498627982dd74d567fcc03c (accessed on 19th September 2025).
30. Tzeng, E., Hoffman, J., Saenko, K., & Darrell, T. (2017). Adversarial discriminative domain adaptation. https://doi.org/10.1109/cvpr.2017.316 (accessed on 19th September 2025).
31. Wang, M., Wu, G., Lv, H., Liu, S., Huang, L., & He, N. (2025). Nonlinear time-varying modeling and simulation methods for hydrodynamic–aerodynamic coupling near-surface flight scenarios. *Aerospace*. https://www.semanticscholar.org/paper/05eac03aec1e2ba311e91fc8d84c477c10f0e6ba (accessed on 19th September 2025).
32. Wang, Z., Xu, Z., Qi, D., Yan, Y., & Zhang, J. (2024). Approximate power flow solutions-based forecasting-aided state estimation for power distribution networks. IET Generation, Transmission & Distribution. https://www.semanticscholar.org/paper/506769e6a92b5ba391b8b063eba8778d870c8de0 (accessed on 19th September 2025).
33. Wu, Z., Pan, S., Chen, F., Long, G., Zhang, C., & Yu, P. S. (2020). A comprehensive survey on graph neural networks. IEEE Transactions on Neural Networks and Learning Systems, 32, 4–24. https://doi.org/10.1109/tnnls.2020.2978386 (accessed on 19th September 2025).
34. Zhao, J., Zhou, J., Zhang, L., & Sun, J. (2024). State space representation of Jiles–Atherton hysteresis model and application for closed-loop control. *Materials, 17*. https://www.semanticscholar.org/paper/e947e6cf32f5958399abfbb275b54c29c3184ce0 (accessed on 19th September 2025).

35. Zhou, L., Poli, M., Xu, W., Massaroli, S., & Ermon, S. (2022). Deep latent state space models for time-series generation. https://www.semanticscholar.org/paper/31b27d90c2d5732386dd3a46a7d8336c0555c6fe (accessed on 19th September 2025).
36. Zhu, J. Y., Park, T., Isola, P., & Efros, A. A. (2017). Unpaired image-to-image translation using cycle-consistent adversarial networks. https://doi.org/10.1109/iccv.2017.244 (accessed on 19th September 2025).

# CHAPTER 10

# ROBUSTNESS AND PERFORMANCE IN LINEAR CONTROL

## Contents



## INTRODUCTION

In the context of control systems, linear control, both concepts performance and robustness are key to controller design and operation for effective design and operation. Robustness describes a system's ability to handle variability in system parameters or external disturbances without loss of performance, while performance is concerned with how a system achieves its desired targets, say stability and responsiveness. The relationship between these two attributes becomes of specific concern in accuracy-demanding applications, such as robotics and aerospace engineering. Thus, an introduction to these topics is critical to understanding their applicability. By studying methodologies and paradigms that optimize robustness without compromising performance, researchers and engineers can create systems able to maintain functionality despite unforeseen operating conditions. Current advances in control techniques prove the ongoing pursuit of improved robustness and performance in linear control systems.

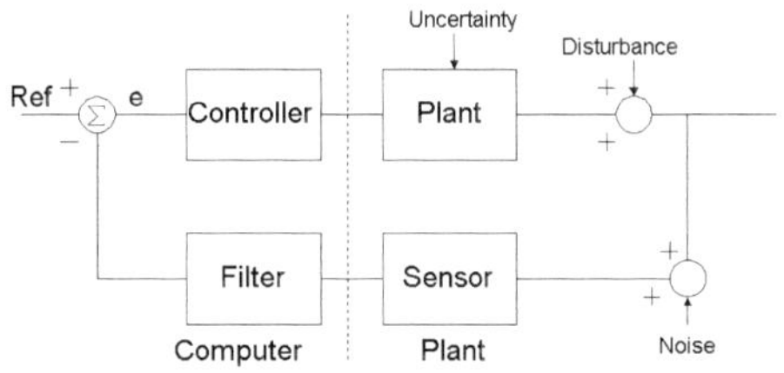


**Figure 10.1**. *Robust control theory.*

*Source: https://users.ece.cmu.edu/~koopman/des_s99/control_theory/plant.gif.*

For instance, the advent of iterative learning control (ILC) techniques using linear extended state observers (LESO) has transformed the management of disturbances and uncertainties. The techniques enhance the ability for trajectory tracking, especially in dynamic systems like six-degree-of-freedom (6-DOF) micro-motion stages (Li et al., 2024). Such innovations not only support better initial error compensation but also ensure faster rates of convergence, with control systems remaining stronger despite disturbances. With the integration of linear control systems with intelligent algorithms, the very roots of robustness have to shift in order to address the nonlinear nature that will subsequently enhance system dependability and efficiency. Besides, the use of complex mathematical frameworks, such as strong control theory, has opened up new possibilities for analysis and synthesis of systems that employ neural networks. The use of recurrent neural networks (RNNs) for control systems highlights the necessity of robust performance measures that are insensitive to input perturbations (Pauli et al., 2022). Incorporation of linear matrix inequality constraints into the learning process of such networks promotes enhanced robustness. With these approaches, engineers intend to bridge the gap between theoretical stability and real-world performance, improving control systems' stability in the real world. This intersection of theory and practice encompasses the importance of designing robust linear control techniques that not only meet performance specifications but also maximize stability under varying conditions, setting the stage for further advancements in the field (Figure 10.1).

## *Definition of Linear Control Systems*

Linear control systems, the hallmark of control theory, dictate dynamism of variables in an estimated system exemplified by linearity. Linear control systems are founded on the principle of superposition, whereby multiple inputs can be aggregated to precisely predict outputs. Linear control systems are typically defined by mathematical models formulated in terms of differential equations, transfer functions, or state-space representations. These theories allow for the evaluation of system performance and dynamic characteristics like stability, transient response, and steady-state accuracy. These insights form the basis for engineers to develop systems capable of responding in a dependable and good manner to disturbances and changes. The theory of linear control systems forms a fundamental foundation upon which strong control methodologies can be built to make sure that systems operate under uncertain environments and variable conditions. Linear control system robustness is not an ancillary problem but a core part of the design and behavior of such systems in real situations. Robustness refers to the ability of a system to preserve its performance even in the presence of uncertainties, for example, model inaccuracies or external disturbance. Robust control methods aim to develop controllers that can deal with such variations without compromise of system stability or performance. For instance, coordinated control strategies employing proportional-integral-derivative (PID) techniques have been found to be capable of stabilizing low-frequency oscillations (LFOs) under operation uncertainties with varying power system conditions in wind power systems. This adaptability points to the necessity of creating control systems not merely optimal in idealized circumstances but resilient in actual working conditions, thereby making robustness a prerequisite of linear control systems (Sadiq et al., 2024). Moreover, the performance of linear control systems can be quantified by a number of parameters like response time, overshoot, and settling time. Evaluations of performance allow engineers to tailor control strategies that cater to specific operating conditions with assured stability and reliability. The presence of sophisticated simulation tools, such as LTSpice, and optimization algorithms also promotes linear control system design based on the capability to test systems rigorously under multiple conditions. Advanced methodologies integrate simulation results to achieve optimized performance parameters with adherence to design specifications. For example, the use of particle swarm optimization along with dynamic simulation has been shown to produce exceptional improvements in the operational efficiency of operational amplifiers, which are such key components (Thanmay et al., 2024). Integration of performance indices with robust design will, in turn, result in innovation within control systems, allowing for different applications to improve.

## Significance of Performance and Robustness

The robustness-performance relationship is relevant in linear control systems, where performance and stability need to be maintained regardless of uncertainties. Robustness is concerned with the insensitivity of the systems to variations in parameters and external interference; performance is concerned with the effectiveness and rapidity of the control method in achieving desired effects. In practical applications, for example, in microalgae culture systems for the removal of nutrients, it happens that robustness is not an add-on characteristic but rather a crucial one. As mentioned in Carecci et al. (2025), the complex interplay between elements in such systems calls for a design philosophy that puts resilience at the forefront of its agenda in order to be able to guarantee that even under varying operation conditions the system can successfully provide its desired effects. Thus, the union of performance and robustness is paramount in the successful implementation of control techniques in real scenarios. In dealing with issues resulting from delays in control systems, the focus on robust design principles increases even more. Delays in time, which are quite prevalent in industrial processes, significantly weaken control performance by destabilizing the system and causing phase lagging. The use of contemporary control techniques such as root locus and linear quadratic regulators is hampered by the presence of such delays, and therefore robust techniques become very important. The development of the Padé approximation, as discussed in Cağfer et al. (2024), is a key tool to handle these delays adequately by enabling more accurate simulations and controller design. By representing time delays in terms of rational polynomials, system designers are able to enhance the reliability of their models and robustness of their control systems, which in turn culminates in improved stability and performance in systems with time delays as an inherent challenge. The interaction between robustness and performance not only enhances the stability of control systems but also optimizes their overall performance. A good control system is characterized by how it can adapt to dynamic conditions and maintain performance metrics regardless of uncertainties. This sensitivity is particularly important for scenarios involving complex processes such as in bioprocessing systems. By making careful considerations of the technical-economic feasibility of biorefinery setups, as noted in Carecci et al. (2025), engineers can employ techniques incorporating robustness into performance optimization goals. This ensures that control systems not only react but are also capable of sustaining high performance at varying operating parameters. In essence, the continuous enhancement of robustness and performance supports control systems that are not only feasible technically but also economical, underpinning the need for sophisticated design methodologies in modern engineering disciplines.

## 10.1. FUNDAMENTALS OF LINEAR CONTROL THEORY

The control theory of linear systems is a basic background for understanding how systems can be controlled to achieve desired outputs. Linear control theory is essentially about describing and analyzing dynamic systems by linear differential equations. Based on this mathematical model, engineers can build controllers capable of controlling system behavior even when there are system parameter variations or disturbances from the outside world. Through the action of feedback mechanisms, linear control systems can enhance robustness and maintain performance even in the presence of uncertainty. Predictability and the management of system dynamics are essential across a wide variety of applications, including manufacturing and aerospace engineering, to facilitate more efficient and more reliable performance under realistic operating conditions. By establishing a solid foundation in linear control, engineers can create systems that balance best between response and stability. In addition, the use of linear control theory in real-world problems requires a deep understanding of system stability and performance criteria. The control designs often require the integration of transfer functions, state-space formulations, and frequency response analysis, all of which collectively define system behavior under various conditions. Stability in a control system directly relates to its capability of minimizing the impact of disturbances and uncertainties. It is here that the importance of employing advanced design techniques, such as H-infinity control and robust controller synthesis, in expanding the margins of system performance gains prominence. Technology advancements, particularly in technologies like telecommunications and automated systems, have generated an ever-growing need for reliable control mechanisms. This chronic demand fuels research toward optimizing linear control designs to meet stringent performance standards without loss of robustness. Even considering the prospective of emerging technologies, linear control theory application crosses intriguingly with machine learning as well as multi-user communication systems. The blend of control theory and machine learning concepts carries the promise of creating control algorithms that learn and adapt through system performance as well as over time. Specifically, linearity control principles can improve the efficiency of multi-user MIMO systems' performance, ensuring robust performance under a wide range of operating conditions (Erik et al., 2014). Such surplus-antenna based systems can achieve high throughput and energy efficiencies without losing adaptability to user demands. This synthesis of traditional control theory and modern data-centric techniques is the dynamic evolution of linear control theory, proving its relevance and feasibility in solving present problems in engineering and technology (Michael et al., 2015) (Figure 10.2).

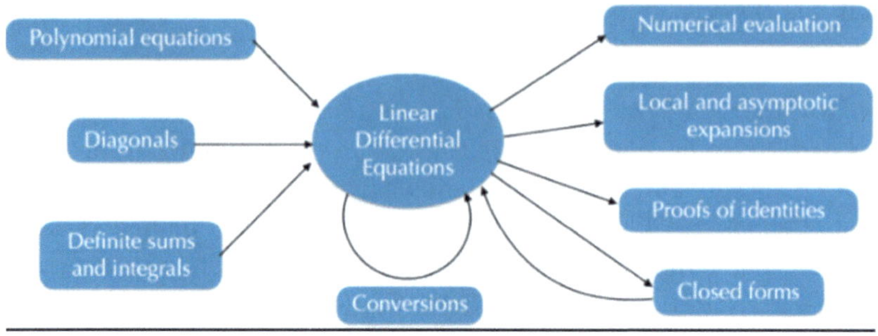


**Figure 10.2**. *Linear differential equations.*

*Source: https://media.springernature.com/lw685/springer-static/image/art%3A10.1007%2Fs10208-018-09411-x/MediaObjects/10208_2018_9411_Fig1_HTML.png.*

## 10.1.1. Key Concepts in Control Theory

Control theory has several fundamental concepts that form the foundation of the design of behavior and stability of systems. The notion of feedback is one of the central themes, and feedback is a property that identifies control systems. Feedback can be positive or negative, but the negative form is typically highlighted due to its stabilizing effect. In linear control systems, the output-input relationship is critical in maintaining desired performance irrespective of disturbances from the outside world or uncertainty in the model. Understanding how feedback influences system dynamics is crucial while designing robust controllers that have the ability to adapt under dynamic conditions so that desired performance goals are realized with effectiveness. Through the study of feedback mechanisms, the engineers can make informed decisions regarding controller design and stability demands, which consequently achieves system robustness and reliability. Robustness within control theory is naturally associated with performance metrics that measure how well a system can resist a variety of uncertainties and disturbances. A robust control system can still work adequately even if the system is struck by parameter fluctuations or random external disturbances. This feature is most prominent when operating in the domain of linear control, when linearity simplifies much of the complexity that is associated with system behavior. Practical applications, however, have constraints that are short of the idealized cases; thus, knowing the compromises between performance and robustness is vital. Techniques like H-infinity synthesis or μ-synthesis provide systems of controller design that can properly balance these competing requirements for maximizing performance without sacrificing system stability. These methods place heavy stress on uncertainty management in design, thus resulting in more reliable control systems in diverse applications. The combination of computational algorithms and tools, in exploring new paradigms in control theory, has transformed the ability to examine as well as develop robust control systems. Utilization of software platforms such as CP2K signifies the achievement of sophisticated computational methodology in the ability to carry out high-performance simulations of complex systems. These devices are made to incorporate the challenges of large-scale systems and changing environmental conditions so accurate modeling is done under both deterministic and stochastic forces. With increasing capability to model and simulate complex behavior, engineers can leverage these advancements to refine their control techniques to become more nuanced and better determine system robustness. Continual advances in computing power and algorithms support the search for improved control techniques, and additionally provide comprehensive views of performance metrics assessment over a variety of applications to guarantee that systems are operating and effective across a broad array of circumstances (Matthew et al., 2021; Thomas et al., 2020).

### 10.1.2. Mathematical Models of Linear Systems

In linear control systems, mathematical models constitute a fundamental paradigm for understanding system performance and designing control approaches. Mathematical models in linear control systems apply linearity and time-invariance as simplifying approximations of advanced system behavior to analyze and design systems. With the application of differential equations and state-space models, engineers can gain information regarding the systems stability, controllability, and observability. However, one must be able to understand that most actual systems exhibit nonlinear behavior or time-varying characteristics. Hence, the excessive reliance on mathematical models is a double-edged sword. While they offer a formalized process of problem-solving, the inherent assumptions will lead to errors and necessitate scrutiny in terms of robustness and performance in linear control systems. The management of uncertainty in such mathematical models of linear control systems is a significant characteristic. Modeling uncertainties or unexpected perturbations complicate the control design task. Successful control strategies must sacrifice robustness—maintaining performance in such uncertainties—to peak performance under ideal conditions. Linear control design models inherently model real systems; therefore, understanding the character of such modeling is vital. As characterized in recent literature, accurate modeling uncertainty representations are significant toward achieving high performance. Specifically, a controller designed for an inappropriately modeled plant can lead to instability or suboptimal performance, reiterating the importance of modeling structured uncertainty in ensuring dependable control systems (Balas et al., 1994). Moreover, new developments in robust control theory emphasize the use of structured uncertainty descriptions of linear systems. These descriptions define the performance versus robustness trade-off that is necessary for any control design. When applied, a controller that has been designed assuming well-defined uncertainties will deliver significant performance enhancement, proving the synergy among model precision and control performance. In an interesting turn of events, experiments have shown that surprisingly resilient performance can be achieved for most uncertainty values, leading designers to pay less attention to determining precise uncertainty weights and more to committing reasonable structured models (Balas et al., 1994). These comments mirror the evolving face of mathematical models in linear control systems with a plea for creating control strategies that are tolerant of unavoidable modeling complexity in the name of high performance and robustness under real-world conditions.

### 10.1.3. Linear Control Stability Analysis

Stability is the very basis of linear control systems, and its impact directly affects performance and robustness. Stability analysis essentially involves knowing if the system will behave in a stable manner under very different inputs over time, with solutions not running out of control. Different methods are employed to identify stability, some being the Routh-Hurwitz criterion, which provides necessary and sufficient conditions for stability of linear systems. Other common procedures include root locus analysis and Nyquist plots, which give a graphical view of the relationship between system parameters and stability. Such procedures allow engineers to design good controllers while considering parameters like disturbances and parameter variations. Ultimately, stability determination ensures that the system can maintain a satisfactory level of performance under a wide variety of operating conditions, thereby enhancing the overall utility of linear control methods. Based on the fundamental principles of stability, robustness is an equally important facet of the performance of linear control systems. Robust stability extends the concept of stability to include the possibility of uncertainties, for example, changes in system dynamics or external disturbances. This feature is crucial particularly in practical applications where conditions are subject to change unpredictably. Techniques like H-infinity control and mu-synthesis are generally employed during the design of controllers to maintain stability in the presence of such uncertainties. Quantifying worst-case scenarios and incorporating them into the control design, engineers can create systems that not only operate well under nominal conditions but can also survive perturbations. This concomitant interest in robustness and performance underlines the requirement for large-scale stability analysis, whereby systems can maintain operational integrity and perform as required in spite of varying environments. Finally, the relationship between stability analysis and performance analysis plays a critical role in the design of effective linear control systems. Whereas stability ensures that a system will behave as anticipated, performance parameters such as settling time, overshoot, and steady-state error determine how effectively a system carries out its function. Optimizing techniques may be applied to optimize stability and performance simultaneously, which is usually hard due to the conflicting requirements involved. For instance, more robustness may be achieved at certain times by increasing the response time, so the system designer has to make a compromise in building the system. Together with ongoing development in the area, advancements in computational algorithm and simulation software provide further aid in performing fine-grained stability analyses, enabling designers to tackle complex control issues more effectively. Integrating these principles, engineers can confidently navigate the complexity of linear control, ensuring assurance of stable and adaptable system performance across varying conditions (Figure 10.3).

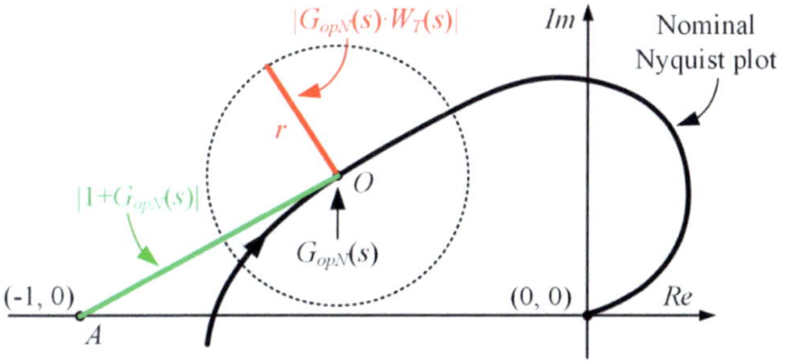


**Figure 10.3**. *Geometric representation of the robust stability.*

*Source: https://www.researchgate.net/publication/361994965/figure/fig2/AS:1143 1281092879927@1667017424274/Geometric-representation-of-the-robust-stability-criterion-for-H-infinity-controller.jpg.*

## 10.2. UNDERSTANDING ROBUSTNESS

Robustness in the context of linear control systems refers to the ability of a system to maintain performance despite variations in parameters or operating conditions. This property is essential, especially as systems encounter uncertainties, be they due to modeling inaccuracies, external disturbances, or variations in the systems environment. The concept of robustness not only emphasizes stability; it also ensures that performance metrics such as tracking and disturbance rejection remain satisfactory. Therefore, robust control design aims to create control laws that can perform effectively across a defined range of uncertainties. Such an approach addresses one of the critical challenges in modern control systems: balancing system performance with the inherent uncertainties that come with real-world applications. The development of advanced techniques, such as μ-synthesis-based adaptive robust control strategies, highlights the progress in achieving robustness in linear control.

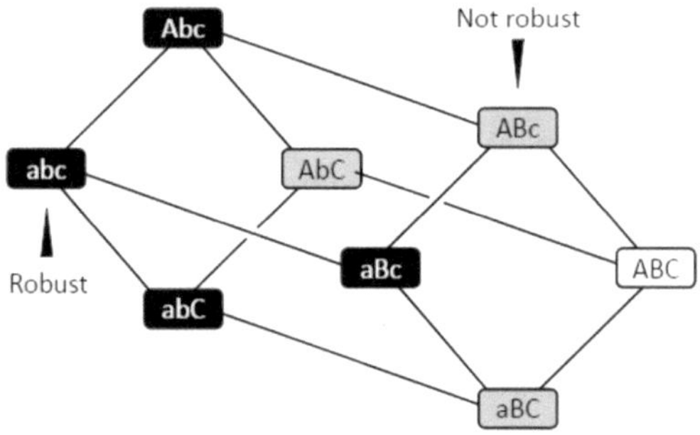


**Figure 10.4**. *Robustness.*

*Source: https://upload.wikimedia.org/wikipedia/commons/0/0b/Robustness_cube_annotated.svg.*

These strategies rely on adaptive model compensation, where accurate online parameter estimation counters various nonlinearities that could otherwise compromise system performance. When applied to linear motor-driven systems, for example, these techniques enhance the robustness associated with high-frequency dynamics—often overlooked in traditional control approaches. The approach detailed in Chen et al. (2014) demonstrates how integrating physical modeling with dynamic analysis can lead to significant improvements in tracking performance, particularly as it relates to achieving higher closed-loop bandwidth. This integration showcases that robust control is not merely a theoretical construct; it is a pragmatic response to real challenges encountered in control engineering. Understanding robustness also intersects with performance evaluation methodologies prevalent in proteomics and other fields. The introduction of label-free quantification techniques, as seen in Jürgen et al. (2014), offers a parallel perspective on how robust methodologies can produce reliable outcomes even under variable conditions. In this context, robustness signifies the capacity to achieve precise measurements without the complications inherent in isotopic labeling, thus ensuring data stability across varying experimental setups. Either in the context of control systems or proteomics, the overarching theme is that robustness is integral to achieving dependable results. Such an understanding fosters the continuous improvement of techniques supporting both theoretical and practical advancements, culminating in enhanced reliability across disciplines (Figure 10.4).

## 10.2.1. Definition and Importance of Robustness

Robustness for linear control systems refers to how well a system maintains its performance in variance with system parameters or external disturbances. It is a very essential feature since real-world applications involve the form of uncontrollable environmental conditions and parameter variations that cause system stability and performance to degrade. Effective robustness provides a control system with the capability to repeatedly deliver desired levels of performance, thereby lessening the likelihood of failure or no optimization. Because measures of performance become more advanced due to evolving dynamics, robustness is all the more significant to be defined. The necessity of precise definitions of performance measures puts the relationship between control system design and robustness into perspective, allowing engineers to design configurations with the ability to withstand uncertainties while preserving desired behaviors (Djari et al., 2024). Robustness has more value than stable performance; it contributes to operational efficiency and safety in industrial applications. For instance, in aerospace engineering, a robust control system is critical for guaranteeing aircraft stability under atmospheric turbulence and mechanical oscillations. In manufacturing processes, the capability to correct and compensate for disturbances can significantly increase productivity while reducing waste. With complex systems, conventional performance metrics like overshoot and settling time must be supplemented with richness of detail from robustness. By articulating these issues, engineers are able to design systems that not only perform well under ideal operating conditions but also react to a greater range of operating conditions, thus ending up with increased reliability and user confidence (Somers et al., 2024). Also, the relationship between verification methods and robustness is relevant in the field of modern control systems. Traditional approaches, such as Monte Carlo simulations, do not succeed in capturing rare but important failure modes, which could jeopardize the integrity of a system. In light of this limitation, deterministic techniques based on worst-case analysis have been developed to provide an all-encompassing picture as far as robust performance and stability are concerned. These techniques, however, do not provide any insight into the occurrence probability of such extreme sets of parameters. Probabilistic μ-analysis development is a step ahead in filling this gap, which seeks to find possible risks that might compromise system reliability. By the integration of robust analysis with probabilistic techniques, control systems can be designed to not only be guaranteed to meet performance specifications but also robust to uncertainty and unlikely adverse situations (Somers et al., 2024).

### 10.2.2. Factors Affecting Robustness in Control Systems

Robustness in control systems means the ability of a system to provide performance insensitivity to uncertainties and variations in its operating conditions. A number of factors can significantly affect this property, such as environmental disturbances and system parameter changes. For instance, external condition fluctuations—temperature, humidity, and electromagnetic interference—destabilize control system performance at the cost of their stability and reliability. Robustness design principles must be accorded special priority as systems able to accept these variations will ensure the prevention of performance deterioration and provide a consistent response. Engineers can set up parameters that improve system reliability by knowing and determining these environmental factors. Proper control algorithms, capable of responding to unforeseen changes, and their application are responsible for fostering robustness as a significant element. The relationship between system feedback loops and model representation accuracy is critical in guaranteeing performance levels of the required quality. In some instances, such as myoelectric prosthetic control systems, must read the intent of a user accurately regardless of aspects such as the position of the limb and shifts in the electrodes (Wang et al., 2024). These may lead to differences in the accuracy of classification, which directly influence the ability of the prosthesis to function intuitively and effectively. Therefore, it is important to perform ongoing research for the purpose of enhancing algorithms to accommodate such variability, in an attempt to enhance the robustness of control methods for complicated operations. Furthermore, constraints provided by actuator dynamics naturally influence system robustness. Motor saturation and range of motion limitations could lead to control failure if considered insufficiently in system design. For example, in a robot system like a helicopter, the hybrid controller synthesis that combines proportional-derivative control with optimized techniques can remarkably increase robustness against operational stresses (Ghiloubi et al., 2024). This approach is used to show how new control strategies are able to counteract basic physical limitations with high levels of performance. Dependable operation under diverse conditions is ultimately a function of control law design and predictive modeling with focus to ensure adequate responsiveness, showcasing the fine balance necessary to derive robust performance in complex control systems (Figure 10.5).

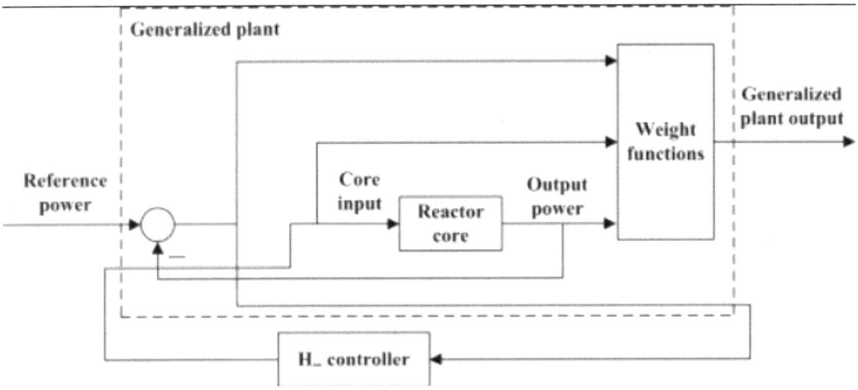


**Figure 10.5**. *Robust control system.*

*Source: https://ars.els-cdn.com/content/image/1-s2.0-S1364032116001465-gr8.jpg.*

## 10.2.3. Methods for Assessing Robustness

Robustness assessment is highly important in linear control systems, where it is necessary to verify that a system is able to preserve its performance even in the presence of uncertainties and external disturbances. Deterministic model-based worst-case analysis is an effective tool for robustness evaluation. This approach examines the space of all possible cases of uncertain parameters so that scholars can achieve mathematically ensured bounds for robust stability margins and worst-case levels of performance. Through identification of potentially dangerous situations, engineers can anticipate beforehand and focus on managing weaknesses in control strategy, thus enhancing designs in terms of reliability and efficiency. Unlike simulation-based methods that risk missing rare critical events, deterministic methods provide a comprehensive study that appears to be pivotal in today's system design and verification practice. Another attractive approach to the assessment of robustness of control systems is to exploit the power of probabilistic analysis. Traditional Monte Carlo simulations, while useful for sizing frequent events, fall short to capture rare but significant combinations of uncertainties. To surpass this limitation, researchers have turned increasingly to probabilistic μ-analysis, an integration of the merits of deterministic and simulation-based methods. This method is interested in the analysis of the probability of occurrence of rare events with the potential to compromise system performance. By estimating the likelihood of exceeding certain stability margins or performance requirements, engineers are able to learn more about the vulnerabilities of their control systems. This type of analysis helps design guarding mechanisms against unlikely circumstances, rendering the control system effective and functional in a variety of contexts. Incorporation of neural network techniques has also turned out to be a crucial methodology to enhancing robustness in linear control systems. The integration of radial basis function (RBF) neural networks with contemporary control philosophies, including fast nonsingular terminal sliding mode control, is one such integration. This new approach improves system performance through increased rates of convergence and effective handling of uncertainties and disturbances. It also greatly reduces chattering phenomena that seek to destabilize the smoothness and effectiveness of control outputs. Stability analysis based on a Lyapunov-based structure guarantees the global finite-time convergence of the system, thereby establishing the excellence of this hybrid approach over traditional sliding mode controllers. Thus, the incorporation of neural networks not only augments strength but also positions such systems for high-precision applications in industrial automation, thus making it an important step forward in control engineering as highlighted in Le et al. (2025).

## 10.3. PERFORMANCE CRITERIA IN CONTROL SYSTEMS

Measurement of performance indices in control systems is crucial in the estimation of robustness and efficiency in linear control design. Performance indices such as settling time, overshoot, and steady-state error provide quantitative measures that enable comparison of different control techniques. Settling time is the rate at which a system can approach its final value, while overshoot is the extent to which a response exceeds its final value. The control system with the least settling time and overshoot is usually the most preferred since this means efficiency and stability. Furthermore, steady-state error compares the desired output to the actual output when equilibrium is reached and gives a measure of the accuracy of the control system. The careful selection of such parameters therefore forms an excellent platform on which effective control measures can be derived and optimized. In order to have effective control systems, it becomes important to implement a number of performance measures that acknowledge the uncertainty of actual systems. Robustness is defined as the capacity of a system to act irrespective of uncertainties or variations in system parameters. Significant, performance measures must be implemented in a manner that maximizes robustness along with achieving some attractive response characteristics. For instance, the inclusion of integral control will enhance steady-state performance by eliminating offsetting errors but potentially decrease overshoot if not properly tuned. The compromises needed imply that there has to be some compromise, often envisioned through root locus and frequency response methods, which graphically represent how systems behave under varying conditions. With these analysis tools, engineers can better address dynamical systems complexities when they work to preserve robustness in the presence of performance requirements. Lastly, improvements in computational methods solidify analysis and application of performance measures in control systems, which delivers gigantic advances in model accuracy. Algorithm application, as seen from recent research in protein structure prediction, shows how sophisticated computational operations can lead to correct modeling outcomes regardless of the availability of data (Jumper et al., 2021). Similarly, application of adversarial machine learning methods has revolutionized image-to-image translation tasks where the ability to perform mapping learning without examples in pairs is indicative of difficulties encountered in control system application (Zhu et al., 2017). These emerging approaches broaden the scope of conventional performance metric analysis so that modelers can design more robust and adaptable control systems. The interfacing of these sophisticated computing techniques with conventional performance metrics ultimately results in the development of robust linear control strategies that can deal effectively with dynamic fluctuations and uncertainties in engineering practice (Figure 10.6).

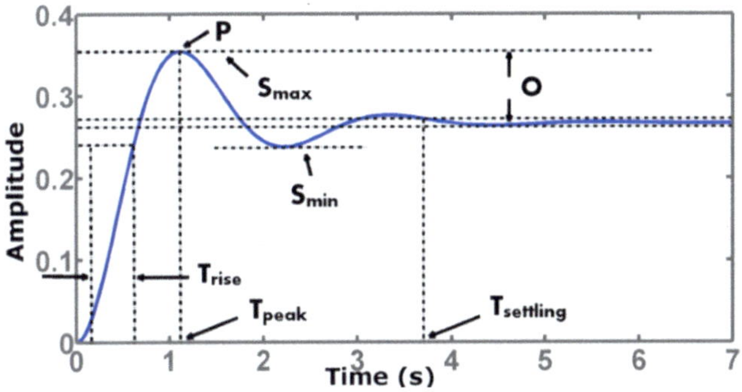


**Figure 10.6**. *Control performance metrics marked in a step response example.*

*Source: https://www.researchgate.net/profile/Junyang-Shi-2/publication/333851280/figure/fig7/AS:818623480348672@1572186455074/Control-performance-metrics-marked-in-a-step-response-example.ppm.*

## 10.3.1. Common Performance Metrics

Performance indicators in linear control systems are the focal points of measuring system behavior and performance in the world. Performance indicators like Root Mean Square Error (RMSE), Mean Absolute Error (MAE), and Mean

Square Error (MSE) quantify the accuracy of forecasts in various environments. Particularly in applications such as supply chain finance, where system dynamics are governed by advanced interactions, good performance indicators become extremely important. The study on a 4D supply chain finance system highlighted the accuracy of predicted outcomes through the implementation of an Artificial Neural Network (ANN) model, which achieved RMSE values of up to 0.00011 on the first future value. These measures not only give an indication of the performance of control actions but also indicate the systems resilience under different operating conditions, making them highly important in linear control system robustness and stability analysis (Zehra et al., 2025). The incorporation of performance measures in the design and analysis of control systems leads to a more structured process of achieving desired outcomes. Other techniques, such as the Second-order Linear Active Disturbance Rejection Control (SLADRC) tuning technique, emphasize the relationship between phase margin and gain crossover frequencies. The technique offers systematic tuning by mapping an acceptable area onto the Nyquist plot that directs design specification selection.

**Figure 10.7**. *Performance metrics.*

*Source: https://cdn4.avada.io/media/shopify/owXgRJu.jpg.*

The use of performance measures like the integral of time-weighted absolute error (ITAE) in conjunction with graphical analysis enhances determination of optimal parameter values. In practice, experimental confirmation via actual real-time control tasks demonstrates the validity of tuning techniques, especially within time-delayed and unstable dynamics systems, validating the fact that performance parameters are crucial to achieving robust control outcomes (Gunasekaran et al., 2025). Lastly, the combination of control methods and performance parameters embraces the aspect of developing effective linear control systems. Good performance does not just depend on statistical measures but also on the capacity of the control mechanisms to respond to varying conditions. With the help of measures such as MAE and RMSE, engineers can refine their control techniques to be responsive to stringent performance requirements. As observed through both theoretical studies and empirical evidences, indicators of performance drive the development of control frameworks that react to uncertainties and disturbances within the system. The integration of the findings shows how common measures of performance form the foundation for achieving utmost robustness and performance excellence in linear control systems, paving way for engineering practice innovations (Zehra et al., 2025; Gunasekaran et al., 2025) (Figure 10.7).

## 10.3.2. Trade-Offs Between Robustness and Performance

For linear control systems, robustness and performance often entail crucial trade-offs. Robustness is the capability of a system to maintain its performance against uncertainties, disturbances, and variation in system parameters. Performance, however, normally entails speed, accuracy, and efficacy in general in achieving control objectives. To reach an optimal balance requires judicious design decisions, especially in compound systems where extrinsic events like communication failure or cyber-attack pose a threat to functionality. In those instances, ensuring resilience in the face of potential interruptions typically entails giving up some of the performance metrics, illustrating the intrinsic trade-offs associated with the design of fault-tolerant control systems. The concept of robustness comes into play, because systems that offer maximum performance are prone to failure when robustness is below par under actual conditions. One fascinating approach to managing these trade-offs is the use of distributed control structures that allow increased robustness by means of decentralized decision-making. Such structures adequately address the problem of reliable control in large-scale interconnected systems without complexity in design. Through the Linear Matrix Inequalities, designers can construct distributed controllers that are resilient towards communicative as well as geometric uncertainties. As highlighted in Massenio et al. (2024), the approach guarantees exponential stability even when the failures of subsystem interconnections occur. Structuring control gains strategically to emulate the nominal communication structure enables systems to restore functionality in case of disturbances. But then, this reliance on a fixed network configuration demands constant monitoring and flexibility, further credence for the necessity of balancing performance needs and robustness strategies. Also, utilization of advanced control techniques such as Active Disturbance Rejection Control (ADRC) further emphasizes how greater capability makes possible traversing the trade-offs between robustness and performance. With the inclusion of nonlinear control strategies and the use of system characteristics such as flatness, they are effectively able to mitigate uncertainties and disturbances found in dynamic systems. Improved Extended State Observer (IESO) is a valuable tool in this context, which is established in Lamraoui et al. (2024), enabling better unmodeled dynamics estimation along with better control performance. This capability is a clear example of a good counterbalance, and it enables systems to achieve high performance under conditions where unexpected complications are also accommodated. Last but not least, the creation of control approaches like ADRC also represents a paradigm shift toward robustness-oriented designs without a considerable penalty in performance, therefore positively contributing to trade-off debates in linear control systems.

### 10.3.3. Criteria for Evaluating Performance

Within linear control systems, the establishment of criteria for the measurement of performance is required to guarantee robustness and efficiency under various working conditions. Performance criteria typically entail measures like stability, response, and precision, which are crucial in defining the system's ability to maintain desirable outputs in the occurrence of disturbances. Stability is the degree to which the control system will settle back to its rest state following perturbations, while responsiveness quantifies how quickly the system attains its desired value. Further, accuracy describes the level of precision with which the system tracks reference inputs. Because system designers are working toward optimum performance, these specifications collectively define a framework in which the designer can assess the probable effectiveness of their control measures. The incorporation of more advanced methodologies will also enhance the robustness of performance analysis of linear control systems. Techniques such as H∞ control have emerged as rich tools for optimization of system response in terms of coping with external disturbances and environmental uncertainties. Robustness is highlighted through the ability of such controllers to exhibit performance under a range of working conditions, as evidenced in a number of research works. For instance, intelligent optimization techniques make the possibility of optimizing controller parameters to achieve minimum sensitivity to disturbances, leading to improved performance measures. Control methods therefore do not just need to be designed with stability and responsiveness in mind but should also incorporate designing around the inherent variability within real systems so that performance is up to standards under standard and worst-case conditions. The integration of predictive models and simulation program code is essential in determining performance in linear control systems, providing more insight into system performance. Predicting performance with artificial neural networks (ANNs) can lead to significant improvements in understanding system dynamics and improving control strategies. For example, ANNs' ability to forecast future states accurately improves the efficacy of the control system on the basis of evidence from research findings showing low values in predictive analysis for root mean square error (RMSE). Moreover, numerical simulation-based performance assessment such as Bode plot analyses allows engineers to visualize the response nature of complicated systems and identify possible weaknesses in controller design. This multi-faceted approach ensures not just that performance measures reflect theoretical abstractions but also empirically try them out through practical applications, yielding robust linear control systems with the capacity to address varied operation problems.

## 10.4. TECHNIQUES FOR ENHANCING ROBUSTNESS

In linear control system domains, robustness has emerged as the top priority, and investigations of various methods of improving performance despite uncertainty continue. A successful method among them is the use of a Direct Torque Control (DTC) method. A special characteristic of DTC is optimizing control over other motors while minimizing speed overshoots and torque ripples—two problems detrimental to system robustness. Traditional control methods, such as Proportional Integrator Derivative (PID) controllers, offer considerable advantages in linear systems but are less effective in non-linear systems. To balance this limitation, optimization techniques can be employed for real-time PID gain adaptation as a response to internal and external disturbances, thereby making the control system more resilient. This adaptability is necessary to maintain healthy performance amidst evolving circumstances, as with recent advancements in PID regulation optimized by new algorithms. Of note in the pursuit of robustness is the incorporation of intelligent optimization techniques, such as the Ant Colony Optimization (ACO) algorithm. The combination of ACO with DTC allows for parameter tuning of the PID controller gains through fine adjustment, significantly enhancing robustness and system performance. By applying a cost function like the Integral Square Error (ISE), this hybrid approach allows systematic parameter space exploration, thus yielding optimum control solutions for specific operating conditions. Experimental confirmation of ACO-DTC method has demonstrated its potential to manage the complexities associated with Doubly Fed Induction Motor (DFIM) effectively. In practical applications, simulation studies conducted on platforms such as Matlab/Simulink have confirmed ACO-DTC's reliability, exhibiting considerable gains in performance indices, particularly under system nonlinearity. Moreover, the emphasis on computational approaches is an epochal advance in control system robustness improvement. Improved advances in machine learning, particularly paradigmatic models such as AlphaFold, have indicated the capabilities of applying biological knowledge and sophisticated algorithms to propel precision in structure predictions. Although they initially were used in protein structures, the fundamental principles in predictive modeling and optimization have far-reaching implications for linear control systems. Through the application of comparable computational techniques, control systems can potentially become more predictive, enabling the adaptation to changes in system dynamics at a higher speed. Control theory integration with machine learning holds a promise for a new future of enhancing robustness, and as such, research in the future should continue aiming at bridging the two disciplines in a bid to find new methodologies sensitive to the intricacies of real-world applications and thus ensuring long-term performance in engineering processes (Figure 10.8).

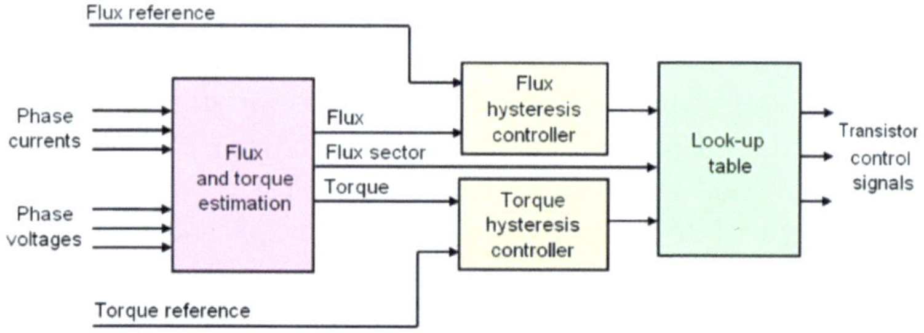


**Figure 10.8**. *Direct torque control.*

*Source: https://upload.wikimedia.org/wikipedia/commons/thumb/9/97/DTC_block_diagram.JPG/960px-DTC_block_diagram.JPG.*

## 10.4.1. Feedback Control Strategies

Efficient methods of feedback control are essential in enhancing the performance and stability of linear control systems. These techniques exploit real-time output data of systems to control inputs correspondingly while guaranteeing desired levels of performance despite disturbances or variations. For instance, the nonlinear feedback linearizing controller (NFLC) as well as its integral version (INFLC) use the approach of exact linearization to cancel out the nonlinearities inherent in systems such as single-machine infinite-bus (SMIB) systems. By allowing for decentralized, linear, and optimal controllers, these approaches offer greater stability and reliability, especially in situations of fluctuations in operations. The real-time adaptability enables not only improved transient response but also more accurate steady-state performance, vital for maintaining system integrity against harsher environments (Vernekar, 2025). Furthermore, the inclusion of sophisticated techniques such as linear-quadratic-Gaussian/loop transfer recovery (LQG/LTR) control significantly broadens the horizon for effective feedback schemes. With the aid of a higher-order Kalman filter, this technique, following a formal frequency-domain loop-shaping approach, offers a compromise between optimal control performance and robustness with respect to noise and disturbance. While traditional linear control methods are likely to fail under the effects of unpredictable changes in system dynamics, LQG/LTR presents a framework that can accommodate various performance specifications like stability margins and rejection of noise. Such a capability makes the performance of the feedback loop more resilient overall, thus enabling the control systems to ride through unknown operating conditions and thus maintain good functionality within real applications (Vernekar, 2025). Within the field of financial systems, feedback control strategies have even been used successfully, demonstrating their applicability outside the classical engineering challenges. Research on a 4D supply chain finance system based on fractal fractional derivatives is a good example of how feedback mechanisms can be utilized in economics and logistics. Fractional-order differential equations render a more refined implementation of system behavior, while stability analysis based on the local and global level conditions the control schemes further. Incorporating artificial neural networks (ANNs) into this, as demonstrated in recent literature, allows accurate predictions of future states to be generated and hence facilitates decision-making in the face of uncertainty. These novel applications of feedback control illustrate its central significance in ensuring resilience and versatility, not only in engineering systems but also in complex financial systems, reaffirming the necessity of robust feedback control techniques in today's complex world (Figure 10.9).

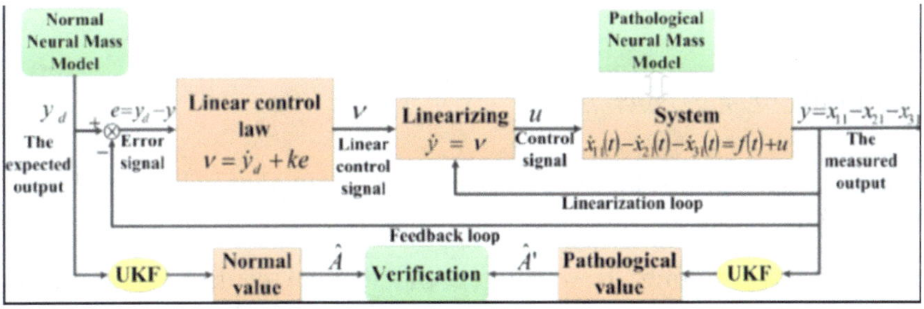


**Figure 10.9**. *Feedback linearization control system.*

*Source: https://www.researchgate.net/profile/Fei-Su-9/publication/283240355/figure/fig2/AS:308213959413761@1450495342283/Block-diagram-of-the-feedback-linearization-control-system-It-consists-of-two-loops.png.*

## 10.4.2. Adaptive Control Methods

The quest for adaptive control methods is significant in optimizing robustness and performance in linear control systems, particularly in dynamic and sophisticated systems. The methods enable the ability of controllers to adjust parameters online to accommodate variations and uncertainties in system dynamics. A good example includes the application of adaptive control in Pneumatic Artificial Muscle (PAM) systems. This is a fine example of how adaptive control techniques successfully handle nonlinearities and parameter uncertainties. In the PAM system, an adaptive control policy may excel in tracking performance through online parameter estimation and hence ensuring a stable system that will react variably (Sun et al., 2019). Theoretically, this speaks to its potential for asymptotic stability without linear approximation of systems, a feature that speaks of its capability to manage complex dynamics in real-world uses.

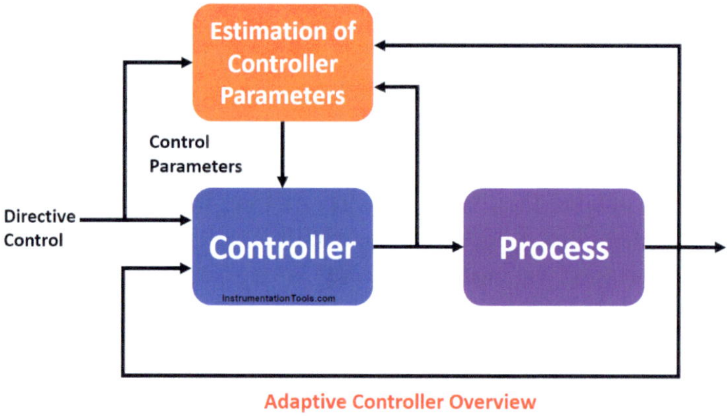


**Figure 10.10**. *Adaptive control.*

*Source: https://instrumentationtools.com/wp-content/uploads/2021/12/Adaptive-Controller-Overview.png.*

Again demonstrating the resilience of adaptive control methods, ship-mounted crane systems offer unique control difficulties due to their operating environment and varying parameters. For those cases, even model-based normal controllers would not be effective due to uncertain variables like changing payload dynamics and external disturbances, which play a key role in performance. The development of a neural network-based adaptive control method indicates a futuristic solution to all these issues. By direct control of the nonlinear dynamics of the crane system, the method alleviates the impact of structure and parameter uncertainties such that it is able to control rope and boom positions accurately. The method also achieves effective damping of payload swing, which is a common source of maritime safety problems, and hence enhances overall robustness and control quality (Yang et al., 2019).

The application of adaptive control methodologies not only improves the performance of linear control systems but also points to real-time adaptability. The application of sophisticated methodologies such as those applicable in PAM and ship-mounted cranes shows the potential of adaptive systems to be stable and fight back against changing operational demands. As technology continues to advance, the development of these methods will be crucial in delivering consistent performance across different applications, from robotics to industrial control. With the employment of real-time data and advanced algorithms, adaptive control offers a pathway to acquiring greater resistance to disturbance and uncertainties, subsequently delivering optimized system performance in linear control environments. The promise of adaptive control methods is to redefine the standard for control system robustness and thereby initiate innovations across a wide range of technologies (Figure 10.10).

## 10.4.3. Robust Control Design Approaches

In control systems, robust control design methods have emerged as fundamental strategies aimed at ensuring system stability and performance in the presence of uncertainties and disturbances. They are characterized by their ability to deliver sought-after performance over a wide range of operating regimes, particularly when the model of the system is susceptible to variabilities. A common approach makes use of disturbance observers which estimate the external disturbances on the system. For instance, Apte et al. (2019) shows this concept by synthesizing a robust speed controller for a permanent magnet synchronous motor (PMSM). Here, the disturbance observer is utilized to estimate load torques that influence system performance to allow the robust controller to be optimally tuned based on feedback about the disturbances as well as enhance overall robustness without compromising performance. Another widely applied method of robust control design is the utilization of sliding mode control, which is particularly useful with its capacity to sustain insensitivity to parameter variations and external disturbances. Using a nonsingular terminal sliding mode method in conjunction with active disturbance rejection control (ADRC), control practitioners are able to simplify complex control problems into straightforward ones. For instance, as illustrated in Wu et al. (2019), the robust control strategy for an autonomous ground vehicle (AGV) employs this combination to solve the complex dynamics of path following. The sliding mode structure not only serves to reduce the control objectives but also significantly enhances the overall vehicle robustness with respect to unexpected changes in the driving conditions. The combination of the robust control methodologies ensures optimum performance and stability during dynamic conditions, highlighting the engineering realities of the methods when applied in real-time. Lastly, the adaptation of robust control methodologies into most engineering applications speaks to the versatility of the approaches. These designs are equally robust and perform better, as the studies above demonstrate with the theoretical and experimental proofs presented. The interaction between linear and nonlinear control techniques provides an understanding of how systems may be designed for different scenarios. For example, the solution in Apte et al. (2019) illustrates how a sliding-mode observer can be cascaded using disturbance observers to facilitate sensorless operation in PMSM, clearly illustrating robust control's cross-domain versatility. Lastly, the common theme running through various applications is a preoccupation with performance sustainability combined with resilience in the face of uncertainty, placing robust control approaches at the center of the evolution of contemporary control theory and practice (Figure 10.11).

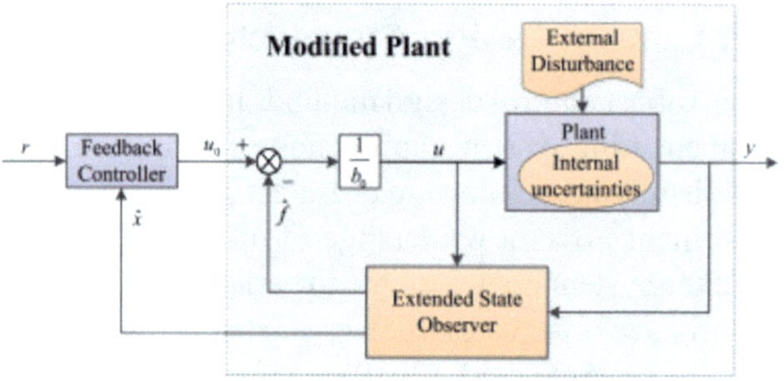


**Figure 10.11**. *Tuning of active disturbance rejection control with application to power plant furnace regulation.*

*Source: https://ars.els-cdn.com/content/image/1-s2.0-S0967066119301339-gr1.jpg.*

## 10.5. PERFORMANCE OPTIMIZATION STRATEGIES

The pursuit of enhanced performance in linear control systems often necessitates the implementation of performance optimization strategies that are both robust and adaptable. Traditional control systems, like PID controllers, may falter under varying operational conditions and parameter uncertainties. To address these limitations, modern strategies employ advanced techniques such as adaptive control and optimization algorithms tailored to the specific dynamics of the system. For instance, the integration of the Artificial Bee Colony (ABC) algorithm allows for the dynamic tuning of controller parameters, ensuring minimized tracking errors and optimized control signal outputs. This method significantly enhances the controllers ability to maintain stability and performance across a range of conditions, illustrating the profound impact of optimization techniques on the robustness of linear control systems (Henry et al., 2024). In addition to adaptive control, robustness in the face of external disturbances is a crucial consideration for performance optimization strategies. As networked control systems (NCSs) become more prevalent, they are increasingly susceptible to threats such as jamming attacks, which can compromise stability and performance. A strategic approach that incorporates stochastic game theory models offers a robust framework for defending against such attacks.

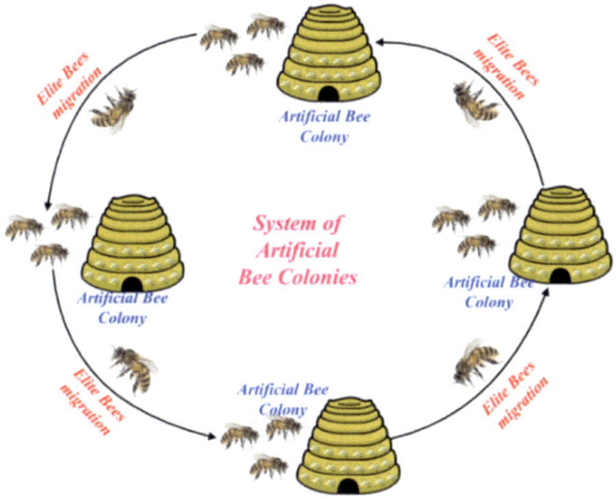


**Figure 10.12**. *Artificial bee colony (ABC) algorithm.*

*Source: https://i0.wp.com/transpireonline.blog/wp-content/uploads/2019/08/Fig-1.png?resize=640%2C506&ssl=1.*

By analyzing the interactions between defenders and potential disruptors, this methodology optimizes mixed strategies to bolster system resilience. The continuous updating of value functions based on adaptive strategies allows for effective mitigation of disturbances, ensuring that control effectiveness is not only maintained but enhanced, even under duress. Thus, the application of game theory in optimizing performance highlights the evolving landscape of robust control systems (Mejdi et al., 2024). Moreover, the synthesis of performance optimization strategies with robust control methodologies leads to significant advances in the design and functionality of linear control systems. Through rigorous application of Lyapunov-based stability criteria, designers can create controllers that ensure bounded errors and stability under parameter fluctuations. This kind of stability guarantees that the system can withstand uncertainties and maintain desired performance levels over time. As demonstrated in various simulation studies, including those comparing sliding mode control with backstepping controllers, the integration of these performance optimization strategies results in smoother control signals with minimal chattering, thereby enhancing overall operational efficiency. The resulting advancements not only solidify the theoretical foundations of linear control but also pave the way for practical applications across numerous engineering domains (Henry et al., 2024; Mejdi et al., 2024) (Figure 10.12).

### 10.5.1. Model Predictive Control

Model Predictive Control (MPC) has emerged as a vital method in the case of robust linear control systems. By implementing a system dynamics model, MPC predicts future behavior over a time horizon and optimizes control actions at each time step as a function of predictions regarding future behavior. The predictive capability renders MPC particularly effective in dealing with multi-variable control scenarios and constraints that must be monitored. Unlike traditional control approaches that generally rely on real-time information, MPC formally considers the impact of future control actions and can perform more strategic control effort tuning. MPC robustness has been demonstrated by various extensions designed to cope with non-linearity and disturbance, as well as through its versatility in real-world implementation. Therefore, it is an important asset in the designer's toolkit for delivering acceptable performance while meeting performance specifications. Another important innovation in MPC is its use of advanced state estimation techniques, which enhance the controller's robustness. Most traditional MPC applications have required advanced calculations to achieve optimal projections; however, recent advances have come up with less intricate techniques that are still efficient. These new formulations make use of transmuted step response models that effectively incorporate double-integrated or integrated disturbances into systems without unnecessarily complicating numerical computations. As introduced in the new literature, these advances offer practitioners a method for obtaining the optimal prediction vector with better efficiency without the need for time-consuming Riccati equations. These improvements not only lead to improved predictive accuracy but also enable on-line tuning parameters, ultimately leading to enhanced control performance as referenced in Garcia et al. (1992). Also, the integration of Disturbance Observer methods within the MPC framework is a giant step towards robust control. Disturbance Observers have long been capable of detecting and estimating disturbances coming into the control system from the outside and thus enabling pre-emptive measures to be taken. With the inclusion of such observers within an MPC framework, control systems are also made capable of responding dynamically to unforeseen perturbations, making overall robustness possible. Recent surveys reveal that one such framework alone achieves efficient control for disturbances and bridges the gap between theory-oriented control methodologies and real-world engineering design. Its ability to derive robust control strategies with model uncertainties as well as external disturbances is reflective of a highly sophisticated development from traditional MPC methods and emphasizes its usability in present control engineering research, as presented in the findings in Oboe et al. (2019).

## 10.5.2. Optimal Control Techniques

Optimal control techniques play a crucial role in enhancing the performance and robustness of linear control systems. The techniques aim to minimize or maximize a given cost function under the system dynamics as well as constraints. Robust control is mainly concerned with uncertainties and external disturbances, which are critical in maintaining system stability. With LPV models, the controllers can be made to change with varying system parameters, thus it is possible to attain optimal performance under different operating conditions. Control with emphasis on robustness is particularly applicable in applications such as cable-driven parallel robots with redundancy, where the disturbances can be highly influential on performance. Here, robust vibration compensator is designed to reverse undesirable oscillations and provide accurate trajectory tracking, indicating practical feasibility of these control schemes for real systems (Jamshidifar et al., 2018). Incorporation of advanced techniques in optimal control systems provides dynamic systems with increased flexibility. Techniques such as Lagrange interpolating polynomials allow modeling of non-linear dynamics with enhanced accuracy, which can become a determining criterion while handling disturbances in system performance. Also, methods like LPV-H∞ control design improve the control methods' robustness by optimizing, in a systematic way, the effect of external disturbances on system performance. These methods ensure suppressive action of vibrations and oscillations is taken care of when designing, with improvement in stability even under heavy perturbations. For instance, applying strong control techniques to redundant cable-driven parallel robots is a prime example of how advanced methodologies can be employed to solve authentic issues in order to solidify the concept of flexibility in modern engineering solutions (James et al., 2011). Furthermore, optimal control techniques' combination of robustness and performance drives continuous research and development in control engineering. A test of the efficacy of various control structures, particularly in uncertain dynamic conditions, has led to the establishment of new and innovative methods that are on the periphery of mainstream control theory. The research cycle of enhancing control schemes—such as variable structure control and robust adaptive control—also focuses on the intricate interaction between theory and practice. From the point of view, engineers can identify new methods of optimizing linear control systems such that they are responsive and stable in other conditions of operation. Lastly, incorporating robustness into optimal control not only enhances system performance but also maintains a level of excellence required for future control technology innovation (Jamshidifar et al., 2018) (Figure 10.13).

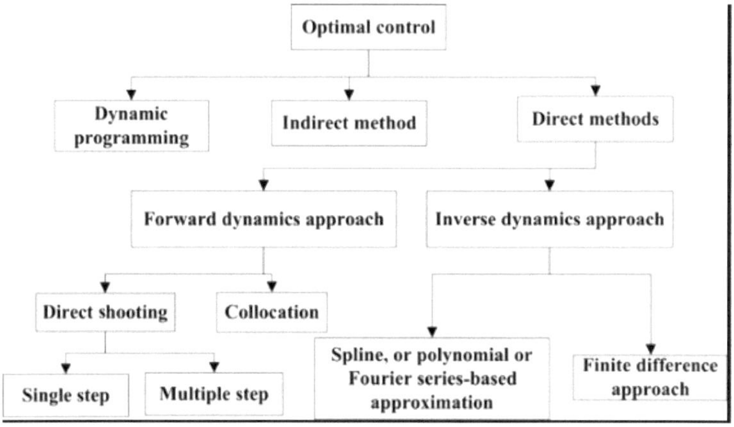


**Figure 10.13**. *Classification of optimal control methods.*

*Source: https://www.researchgate.net/profile/Hayder-Al-Shuka/publication/265412748/figure/fig8/AS:672496747560961@1537347127116/Classification-of-optimal-control-methods.ppm.*

## 10.5.3. Performance Tuning Methods

Performance tuning methods play a critical role in enhancing the reliability and efficiency of control systems, especially for linear control processes. They are aimed at optimizing system parameters to meet anticipated performance requirements such as stability, speed, and accuracy. A critical need of these methods is the balance between robustness and performance because the tuning parameters are optimized too aggressively, leading to instability or decreased responsiveness. Historically derived tuning methods can be trial-and-error-based and inefficient, neither being optimal. New methods as alternatives apply computer-based methodology whereby data-driven models make predictions of system behavior and vary parameters accordingly. This integration provides for expanded possibilities of real-time tuning and designing systems that meet not only the performance specifications but also compensate for varying operating conditions. One of the new performance tuning methods is based on using machine learning algorithms, which significantly enhance parameter tuning accuracy. A case in point is the use of Bayesian optimization for cascade control systems, where parameters are simultaneously tuned to ensure optimal performance across several control loops. This approach guarantees convergence to the global optimality, a step forward from conventional tuning methods that were not able to take into account the interactive contribution of more than one parameter. In addition to that, this method enables data efficiency with fewer iterations needed for tuning and therefore accelerating the process of achieving good performance. These advancements illustrate the dynamic evolution of performance tuning methods in setting the boundaries of control system performance and providing greater flexibility in systems under varying conditions (Khosravi et al., 2021). The robustness of performance tuning methods is also emphasized through their ability to maintain the performance of systems under disturbances. Experimental confirmation of the procedures has proved that automated tuning has been capable of yielding better tracking performance, with robustness and adaptability against external disturbances. Such an aspect is essential in real-world problems since control systems must deal with uncertain changes in their working environments. Methodological advances, particularly those concerning physical and biological aspects, increasingly highlight the great value of robust performance tuning in addressing complex control problems. These integrated techniques do not only ease the tuning process but also improve the stability in operation of the systems, which makes them increasingly important in linear control applications (Jumper et al., 2021).

## 10.6. ROBUSTNESS AND PERFORMANCE CASE STUDIES

Research into performance and resilience of linear control systems is essential for ensuring stable performance in uncertain settings. Case studies are an in-context method of evaluating various control schemes, revealing their merits in application. For instance, use of a distributionally robust data-enabled predictive control (DeePC) algorithm illustrates how uncertain stochastic linear time-invariant (LTI) systems can efficiently be controlled. This algorithm employs noise-corrupted data to predict future system behavior under the output chance constraint. The case study of aerial robotics shows the closed-loop performance of the DeePC, including how it achieves robust out-of-sample performance as well as the satisfaction of strict probabilistic guarantees, thereby demonstrating the theory of robustness in the control design process (Coulson et al., 2021). Through such evaluations, we have knowledge regarding the implementation of theoretical advances into practical efficacy. Of particular importance to performance and robustness is also found in load frequency control (LFC) strategies for power systems under uncertainty for different operational scenarios. Based on a recent case study of strongly resilient delay-dependent PI-based LFC schemes, this model not only deals with strong parameters such as transmission delays, parameter uncertainties, and load fluctuations, but does so effectively. Based on Lyapunov theory and linear matrix inequalities, this method not only constructs the entire LFC model but also formulates a stability criterion suitable for sampled-data control systems. The incorporation of an exponential decay rate (EDR) offers a robust framework to measure performance under uncertainties, hence providing optimum control even in unfavorable conditions.

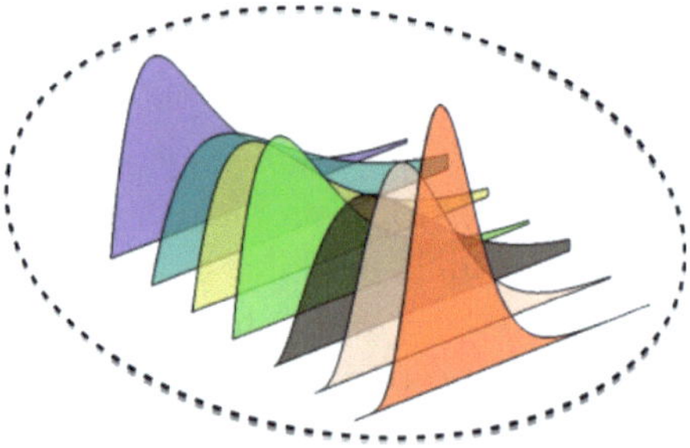

**Figure 10.14.** *Data enabled predictive control.*

*Source: https://control.ee.ethz.ch/research/theory/data-enabled-predictive-control/_jcr_content/par/fullwidthimage_1073292541/image.imageformat.1286.1934051402.png.*

The simulation results also validate the findings, revealing how the introduced LFC scheme can withstand parameter changes and uncertainties in renewable energy sources without compromising system performance (Xing-Shangguan et al., 2020). Moreover, empirical research's interplay with theoretical models refines our understanding of performance and robustness in linear control systems. Not only do these case studies highlight the effectiveness of control mechanisms like DeePC and robust LFC but also their responsiveness under changing conditions full of uncertainties. As can be seen in the case of aerial robots, DeePCs ability to foresee and counter disturbances is a feature of innovative ideas required by modern control systems. Similarly, power system LFC schemes' tolerance to variability in diverse constraints maintains the demand for innovative designs in response to the challenges of the new era. By integrating theoretical methods with practical case studies, researchers and practitioners can create effective control solutions that enable both intellectual research and functional practice in linear control systems, enhancing even more significantly the key role of robustness and performance in a broad variety of applications (Figure 10.14).

### 10.6.1. Real-World Applications of Linear Control

The engineering discipline has numerous practical applications of linear control that prove its efficacy and effectiveness in numerous fields. One of the most publicized ones is the space industry, where autopilot systems rely on linear control techniques to stabilize an airplane while in flight. These systems utilize pre-computed mathematical models to regulate complex variables, for example, height and velocity, such that optimal performance is maintained even in the presence of external disturbances such as turbulence. The stability of linear control to this end is critical in that it helps to produce reliable responses to changes in flight conditions with the final aim of developing safer and more efficient flight. The successful application of these principles illustrates how linear control not only maintains the technical demands of modern aviation but also lies at the basis of the industry's commitment to innovation and reliability. Apart from aerospace, linear control methods are also an essential part of manufacturing environments, namely robotics and automation. Robots used on assembly lines are usually regulated with linear control algorithms that allow precise movement and coordination during operations such as welding, painting, or packaging. Linear control algorithms offer precise real-time adjustment of robot behavior to account for production process variations to facilitate quality control and optimize output. The ability of linear control systems to accommodate such dynamic changes is a testament to their resilience since they are able to accommodate both anticipated and unexpected fluctuations in the production process. Therefore, the integration of linear control is crucial to maintaining operational efficiency and keeping manufacturing costs at a minimum, hence affirming its significance in contemporary industrial operations. Further, linear control is utilized in the field of economic policy modeling where its principles can be employed to study the impact of policy uncertainty on economic indicators. Linear control can be employed to create forecasting models that assess how variations in economic policy uncertainty—for example, those documented in recent research—affect investment, output, and employment levels across various industries. For instance, in linear control models, with the help of tests, economists are able to experiment with policy changes inducing reactions in market activity and thus facilitate helpful inferences of the effects of fiscal measures. These uses underscore the versatility of linear control in the technical and social sciences and the broad utility to analyzing complex system dynamics as well as guiding policy-making in response to economic crises, as championed by work on the effect of economic policy uncertainty on investment behavior and aggregate business performance (Baker et al., 2016).

### 10.6.2. Comparative Analysis of Different Control Strategies

Comparative assessment of multiple control strategies is important for identifying their performance and robustness in any context. Among the prevalent strategies, Model Predictive Control (MPC) stands out since it is capable of delivering optimum control while obeying constraints. Recent applications of MPC in controlling Unmanned Aerial Vehicles (UAVs) demonstrate that MPC is effective in handling uncertainties and exterior disturbances. The design approach centers on the stability and resilience of the systems, particularly when exposed to disparate operating conditions. Through simulation-supported investigation of diverse flight modes, the MPC technique has been contrasted with traditional methods such as the Linear Quadratic Tracker with Integrator (LQTI). This broad comparison not just captures the merits of MPC in complex environmental settings but also gives a point of reference to measure performance across different control mechanisms, thus offering insights about their actual uses in real-world settings (Manzoor et al., 2020). In addition to MPC, there is also another control approach that is exceptional and this is the use of Feature Pyramid Networks (FPNs) in object detection systems. Such networks reveal the superiority of hierarchical structure in the management of multi-scale features, to be analogized with linear control methods, whose selection tends to involve various system dynamics. FPNs leverage deep convolutional networks to conduct feature extraction at multiple scales, resulting in increased robustness toward different input data. Similarly, as robust linear control methods accommodate perturbations, FPNs optimize computational efficiency and memory requirements and hence are specifically suitable for real-time applications. Such advancements facilitate the adoption of new architectures in control systems and thereby optimize overall detection system performance and reliability and promoting the discussion in the literature of engineering on selecting optimal control strategy design (Lin et al., 2017). The analysis of control strategy can be further developed through the examination of the blend of various methods to enhance performance. Hybrid control schemes that bring together the strengths of linear control, MPC, and machine learning have been shown to be effective solutions for adaptive and robust systems with high complexity levels. These use model-based and data-based methods to create a complete control architecture that can operate in dynamic systems. For instance, a hybrid approach would integrate the predictive ability of MPC with self-tuning mechanisms of neural networks in order to improve the system's ability to predict and react to changes. This integration not only improves the strengths of both approaches but also immunity against system behavior unpredictability. By means of intensive comparative study, scholars can more accurately estimate the consequences of such integrations, and ultimately inform subsequent control strategy developments that maximize both robustness and performance (Manzoor et al., 2020; Lin et al., 2017).

## 10.6.3. Lessons Learned from Case Studies

Case study analysis in the context of linear control systems gives invaluable information about how performance and robustness influence each other. It is an essential lesson that both achieving desired performance targets and maintaining robustness to disturbances and uncertainties require a fine balance. For instance, in automated vehicles scenarios, the performance might be quantified by way of response time and path following. However, case studies within the field have indicated that discounting the systems divergent external effects—for example, variable weather or unforeseen road blocks—can lead to performance loss, indeed failure. These findings reinforce the necessity of creating controllers that not only optimize output but also incorporate robust mechanisms to address such unexpected challenges, as reported in the analyses of advanced control applications (Dinh et al., 2021). Besides, case studies of Federated Learning (FL) integration into IoT ecosystems reveal deeper intricacies of linear control systems. Specifically, the findings illustrate how FL can enable collaborative knowledge based on distributed devices without undermining data security and privacy—a increasingly important consideration in the current data era. The versatility of FL makes it applicable to a wide range of applications, from intelligent health care to smart transportation systems, and being effective to optimize decision-making in complex systems. These results suggest that using advanced data processing techniques can optimize the robustness of linear control systems via online tuning based on collective local wisdom. This alignment highlights that performance improvement must be seen in terms of both enhanced immediate productivity but also the ability of systems to learn and adapt, stimulating continual improvement (Lim et al., 2020). Furthermore, case studies highlight the realities of real-world applications, with theoretical models often being faced with unanticipated practical restraints. For instance, metrics such as communication latency in Mobile Edge Computing (MEC) settings can exert an impact on linear control systems' overall performance. Observations gained from such settings claim that although such systems provide excellent performance in laboratory settings, they do not perform well under real-time and varying settings. This discrepancy emphasizes the necessity of multiple re-testing and validation on a variety of real-life scenarios. Through these case studies, researchers and engineers are able to develop more robust systems that not only forecast possible failures but also adaptively counter the performance threats, thus optimizing the reliability and efficiency of linear control operations in different operating conditions.

# SUMMARY

The robustness and performance control theory discussion in linear control, it is imperative to acknowledge the complexity of achieving optimum system performance with varying uncertainties. From the above analysis, it is clear that while control design technical aspects such as the development of intuitive mappings are important, individually they may not decide on system performance outcomes. Most significantly, the study suggests that the natural skills of the operator far surpass the power of the control map as a conceptualization in Steafan et al. (2023). This finding requires further study of the relationship between human skill acquisition and the underlying mechanisms of robotic interfaces, encouraging us to investigate other channels which may help to enhance the performance of the operator as well as the interaction of the system. Besides, the stability of linear control systems is fundamentally due to the ability of these systems to preserve performance integrity in the face of uncertainties and disturbances. The combined examination of discrete singularly perturbed systems justifies the significance of stability analysis and implementation of robust $H\infty$ control methodology. As set forth in Feng et al. (2023), the development of controller conditions that achieve the $H\infty$ performance index represents a tremendous leap in ensuring systems are robust to anticipated as well as unanticipated predicaments. This transformation of control issues into computationally feasible quadratic performance indices reflects progress in integrating advanced mathematical techniques into practicable application, ultimately leading to more reliable system responses under dynamic circumstances. Based on these observations, research directions in the future must thus shift towards improving feedback measures and adaptive learning systems for operators. The control strategy optimization, as noted in the analysis, might require re-tuning to place greater emphasis on other performance contributors such as successful feedback loops and individually customized learning structures. As innovations in robot-user interaction continue to evolve, embracing an all-encompassing view that brings together operator training and effective control design will be central to addressing the multilateral issues provided in modern control systems. Focusing our attention on these areas could greatly enhance the reliability and performance of linear control systems in many applications.

In linear control system performance and robustness discussions, there exists a necessity to synthesize basic results presenting the intricacy of such ideas. Robustness merely defines a system's ability to operate in accordance with requirements irrespective of parameter uncertainties or external disturbances. Performance defines the efficiency and effectiveness of system responses relative to optimal goals. As systems become more and more complex, it becomes of even greater concern to be able to resist disruption and offer peak performance as a benchmark of achievement. This dual weighting of robustness and performance serves to better illuminate control system behavior, driving the development of techniques which can critically assess and optimize system reliability. Additionally, the mutual dependency between robustness and performance forces us to also consider that optimizing one could, of necessity, impact the other. This is a usual phenomenon that any enhancement in robustness incurs a cost in terms of performance measures such as speed or accuracy. This interconnectedness necessitates very elaborate modeling and simulation approaches to grasp the dynamics underlying linear control systems entirely. For instance, recent advances in computational techniques have augmented these efforts so that there can be enhanced predictive power after analyzing system response for varied conditions. Techniques utilized within this analysis include statistical techniques and algorithms permitting increased penetration in system behavior. Combined, these advancements represent a significant advance towards filling gaps in understanding in robustness-performance trade-offs, ultimately informing future research and application in control systems. Ultimately, real-world applications of robust linear control systems illustrate the significant stakes of these determinate points across multiple industries. Manufacturing to healthcare sectors increasingly rely on control systems that can function irrespective of environmental variations and unforeseen challenges. For instance, in the healthcare sector, application of strong control principles has the potential to enhance guaranteed-performance patient monitoring systems under varying patient conditions. As illustrated by Christopher et al. (2022), fighting antimicrobial resistance through customized medical treatment also resonates with robust design principle; it is aimed at a systematic method of minimizing risks and maximizing efficiency. Similarly, in biomedicine, advancements in predictivity modeling, as documented in Jumper et al. (2021), have potential to redefine standards natively linked with robustness and performance. Such ubiquitous cross-disciplinary applications emphasize the necessity to include performance and robustness as better issues in linear control system design.

# REFERENCES


1. Akinrinde, H. A., Ayinde, B., & El-Ferik, S. (2024). Adaptive position control of electrohydraulic servo systems with parameter uncertainty using artificial bee colony optimization algorithm. *Asian Research Journal of Mathematics*. https://www.semanticscholar.org/paper/89d2261e4748335 50339f85ebad2bbd06e823a3b (accessed on 19th September 2025).
2. Aleyaasin, M. (2024). Comparing the performance of robust controllers for vibration suppression in long rotor systems. Acoustics. https://www.semanticscholar.org/paper/2fbb434acead6b33a4351b2d2ed8a2ba05 1b0672 (accessed on 19th September 2025).
3. Apte, A., Joshi, V., Mehta, H., & Walambe, R. (2019). Disturbance-observer-based sensorless control of PMSM using integral state feedback controller. *IEEE Transactions on Power Electronics, 35*, 6082–6090. https://doi.org/10.1109/tpel.2019.2949921 (accessed on 19th September 2025).
4. Baker, S., Bloom, N., & Davis, S. J. (2016). Measuring economic policy uncertainty. The Quarterly Journal of Economics, 131, 1593–1636. https://doi.org/10.1093/qje/qjw024 (accessed on 19th September 2025).
5. Balas, G. J., & Doyle, J. C. (1994). Robustness and performance trade-offs in control design for flexible structures. https://core.ac.uk/download/4869909.pdf (accessed on 19th September 2025).
6. Carecci, D., Ficara, E., Geraci, I., Leva, A., & Ferretti, G. (2025). Plant-wide modelling of a biorefinery: Microalgae for the valorization of digestate in biomethane plants. *Systems and Control Transactions*. https://www.semanticscholar.org/paper/902ada13ff37c1b815d7dae9ad5dbd6e99 738e44 (accessed on 19th September 2025).
7. Chen, Z., Yao, B., & Wang, Q. (2014). μ-synthesis-based adaptive robust control of linear motor driven stages with high-frequency dynamics: A case study. IEEE/ASME Transactions on Mechatronics, 20, 1482–1490. https://doi.org/10.1109/tmech.2014.2369454 (accessed on 19th September 2025).
8. Cochran, J. J., & Cox, L. A. (2011). Wiley encyclopedia of operations research and management science. *Choice Reviews Online, 49*, 49–0052. https://doi.org/10.5860/choice.49-0052 (accessed on 19th September 2025).
9. Coulson, J., Lygeros, J., & Dörfler, F. (2021). Distributionally robust chance constrained data-enabled predictive control. *IEEE Transactions on Automatic Control, 67*, 3289–3304. https://doi.org/10.1109/tac.2021.3097706 (accessed on 19th September 2025).

10. Cox, J., Hein, M. Y., Luber, C. A., Paron, I., Nagaraj, N., & Mann, M. (2014). Accurate proteome-wide label-free quantification by delayed normalization and maximal peptide ratio extraction, termed MaxLFQ. Molecular & Cellular Proteomics, 13, 2513–2526. https://doi.org/10.1074/mcp.m113.031591 (accessed on 19th September 2025).
11. Djari, A., Bouarroudj, N., Houam, Y., & Djabri, R. (2024). Definitions and terms adjustment for linear control systems. *2024 2nd International Conference on Electrical Engineering and Automatic Control (ICEEAC)*. https://www.semanticscholar.org/paper/695d6e16ddfa0a2606e93dedded5620fb71417ed (accessed on 19th September 2025).
12. Feng, S., & Liu, L. (2023). Robust H∞ control for discrete-time singularly perturbed systems with uncertainty and time-delays. 2023 IEEE 18th Conference on Industrial Electronics and Applications (ICIEA). https://www.semanticscholar.org/paper/93aa29f8cb6238d14b3efde4734e3a08054887b1 (accessed on 19th September 2025).
13. Garcia, C. E., Lee, J. H., & Morari, M. (1992). State-space interpretation of model predictive control. https://core.ac.uk/download/4892684.pdf (accessed on 19th September 2025).
14. Ghiloubi, I. B., Abdou, L., & Lahmar, O. (2024). Adjusted linear quadratic regulator-proportional-derivative control of Quanser's three degrees of freedom helicopter based on flower pollination algorithm under external disturbances. *IAES International Journal of Robotics and Automation (IJRA)*. https://www.semanticscholar.org/paper/093ab021f245bef79c7af87d5bda35342f5b9adc (accessed on 19th September 2025).
15. Gunasekaran, J., Deenadayalan, E., Karunakaran, R. T., & Rajendran, D. P. (2025). A modified graphical based tuning and performance analysis of second order LADRC using frequency domain technique for SISO systems. *An International Journal of Optimization and Control: Theories & Applications (IJOCTA)*. https://www.semanticscholar.org/paper/c48f103bcb18af4fb0da1f418759ba0a61d9fa1f (accessed on 19th September 2025).
16. Jamshidifar, H., Khosravani, S., Fidan, B., & Khajepour, A. (2018). Vibration decoupled modeling and robust control of redundant cable-driven parallel robots. *IEEE/ASME Transactions on Mechatronics, 23*, 690–701. https://doi.org/10.1109/tmech.2018.2793578 (accessed on 19th September 2025).
17. Jordan, M. I., & Mitchell, T. M. (2015). Machine learning: Trends, perspectives, and prospects. Science, 349, 255–260. https://doi.org/10.1126/science.aaa8415 (accessed on 19th September 2025).

18. Jumper, J., Taki, K., Pritzel, A., Green, T., Figurnov, M., Ronneberger, O., Tunyasuvunakool, K., et al. (2021). Highly accurate protein structure prediction with AlphaFold. Nature, 596, 583–589. https://doi.org/10.1038/s41586-021-03819-2 (accessed on 19th September 2025).

19. Khosravi, M., Behrunani, V., Myszkorowski, P., Smith, R. S., Rupenyan, A., & Lygeros, J. (2021). Performance-driven cascade controller tuning with Bayesian optimization. IEEE Transactions on Industrial Electronics, 69, 1032–1042. https://doi.org/10.1109/tie.2021.3050356 (accessed on 19th September 2025).

20. Kühne, T. D., Iannuzzi, M., Del Ben, M., Rybkin, V. V., Seewald, P., Stein, F., Laino, T., et al. (2020). CP2K: An electronic structure and molecular dynamics software package–Quickstep: Efficient and accurate electronic structure calculations. The Journal of Chemical Physics, 152. https://doi.org/10.1063/5.0007045 (accessed on 19th September 2025).

21. Lamraoui, H. C., Bouzid, Y., & Nemra, A. (2024). Flatness-based nonlinear internal model control with disturbances compensation via improved extended state observer for remotely operated vehicle. *International Journal of Control, 98*, 1480–1494. https://www.semanticscholar.org/paper/18347d234f7db1fecd35fcc98e7a33b2b560b83a (accessed on 19th September 2025).

22. Larsson, E. G., Edfors, O., Tufvesson, F., & Marzetta, T. L. (2014). Massive MIMO for next generation wireless systems. *IEEE Communications Magazine, 52*, 186–195. https://doi.org/10.1109/mcom.2014.6736761 (accessed on 19th September 2025).

23. Le, V. C., Phan, V. D., Ho, S. P., Duong, D. T., Ta, H. C., & Dang, T. S. (2025). Enhanced performance of permanent magnet linear motors under uncertainties and disturbances using fast nonsingular terminal sliding mode control with neural network. 2025 3rd International Conference on Smart Systems for Applications in Electrical Sciences (ICSSES). https://www.semanticscholar.org/paper/ab8de16ac021bddaf85c50fc6002c03cfef2e539 (accessed on 19th September 2025).

24. Li, J., Wang, M., Liu, M., Zhang, C., & Li, L. (2024). Iterative learning controller based on linear extended state observer for six-degree-of-freedom micro-motion stage. 2024 27th International Conference on Electrical Machines and Systems (ICEMS). https://www.semanticscholar.org/paper/d1817c5f915e465db8f216d4792ff41e9e0d6c77 (accessed on 19th September 2025).

25. Lim, W. Y. B., Luong, N. C., Hoang, D. T., Jiao, Y., Liang, Y. C., Yang,

Q., Niyato, D., et al. (2020). Federated learning in mobile edge networks: A comprehensive survey. IEEE Communications Surveys & Tutorials, 22, 2031–2063. https://doi.org/10.1109/comst.2020.2986024 (accessed on 19th September 2025).

26. Lin, T. Y., Dollár, P., Girshick, R., He, K., Hariharan, B., & Belongie, S. (2017). Feature pyramid networks for object detection. https://doi.org/10.1109/cvpr.2017.106 (accessed on 19th September 2025).
27. Mahfoud, S., Derouich, A., Iqbal, A., & El Ouanjli, N. (2021). ANT-colony optimization-direct torque control for a doubly fed induction motor: An experimental validation. Energy Reports, 8, 81–98. https://doi.org/10.1016/j.egyr.2021.11.239 (accessed on 19th September 2025).
28. Manzoor, T., Xia, Y., Zhai, D. H., & Ma, D. (2020). Trajectory tracking control of a VTOL unmanned aerial vehicle using offset-free tracking MPC. Chinese Journal of Aeronautics, 33, 2024–2042. https://doi.org/10.1016/j.cja.2020.03.003 (accessed on 19th September 2025).
29. Massenio, P. R., Tipaldi, M., Naso, D., & Rizzello, G. (2024). Design of exponentially stabilizers for distributed control systems subject to cyber disconnections. 2024 32nd Mediterranean Conference on Control and Automation (MED). https://www.semanticscholar.org/paper/981e42a227044618780afa968f994565381d7d09 (accessed on 19th September 2025).
30. Mejdi, H., & Ezzedine, T. (2024). Novel dynamic defense strategies in networked control systems under stochastic jamming attacks. *Mathematics*. https://www.semanticscholar.org/paper/d3fd0ffc43676e1c39b331397eba77349cb3124d (accessed on 19th September 2025).
31. Murray, C. J. L., Ikuta, K. S., Sharara, F., Swetschinski, L. R., Robles Aguilar, G., Gray, A. P., Han, C., et al. (2022). Global burden of bacterial antimicrobial resistance in 2019: A systematic analysis. *The Lancet, 399*, 629–655. https://doi.org/10.1016/s0140-6736(21)02724-0 (accessed on 19th September 2025).
32. Nguyen, D. C., Ding, M., Pathirana, P. N., Seneviratne, A., Li, J., & Poor, H. V. (2021). Federated learning for Internet of Things: A comprehensive survey. *IEEE Communications Surveys & Tutorials, 23*, 1622–1658. https://doi.org/10.1109/comst.2021.3075439 (accessed on 19th September 2025).
33. Oboe, R., Ohnishi, K., & Sariyildiz, E. (2019). Disturbance observer-based robust control and its applications: 35th anniversary overview. http://arxiv.org/abs/1902.09032 (accessed on 19th September 2025).

34. Page, M. J., Moher, D., Bossuyt, P. M., Boutron, I., Hoffmann, T., Mulrow, C. D., Shamseer, L., et al. (2021). PRISMA 2020 explanation and elaboration: Updated guidance and exemplars for reporting systematic reviews. BMJ. https://doi.org/10.1136/bmj.n160 (accessed on 19th September 2025).
35. Pauli, P., Berberich, J., & Allgöwer, F. (2022). Robustness analysis and training of recurrent neural networks using dissipative theory. At–Automatisierungstechnik, 70, 730–739. https://www.semanticscholar.org/paper/75c0b49c41e2d7bfebb21d5ec907b391c8068025 (accessed on 19th September 2025).
36. Ponikowski, P., Voors, A. A., Anker, S. D., Bueno, H., Cleland, J. G. F., Coats, A. J. S., Falk, V., et al. (2016). 2016 ESC guidelines for the diagnosis and treatment of acute and chronic heart failure. European Heart Journal, 37, 2129–2200. https://doi.org/10.1093/eurheartj/ehw128 (accessed on 19th September 2025).
37. Sadiq, R., Wang, Z., Chung, C. Y., Gan, D., & Tong, C. (2024). Coordinated robust PID-based damping control of permanent magnet synchronous generators for low-frequency oscillations considering power system operational uncertainties. Journal of Modern Power Systems and Clean Energy. https://www.semanticscholar.org/paper/a6454baebb297a81b7e45638960d62623254a061 (accessed on 19th September 2025).
38. Shangguan, X. C., Zhang, C. K., He, Y., Jin, L., Jiang, L., Spencer, J. W., & Wu, M. (2020). Robust load frequency control for power system considering transmission delay and sampling period. IEEE Transactions on Industrial Informatics, 17, 5292–5303. https://doi.org/10.1109/tii.2020.3026336 (accessed on 19th September 2025).
39. Somers, F., Roos, C., Preda, V., & Sanfedino, F. (2024). H-analysis based algorithms for probabilistic finite-frequency H2-performance analysis of uncertain linear systems. *2024 32nd Mediterranean Conference on Control and Automation (MED)*. https://www.semanticscholar.org/paper/98f93602fba8f69ea575c9da6c6c94dbadd2337d (accessed on 19th September 2025).
40. Stengel, R. F. (1990). Investigation of Air Transportation Technology at Princeton University, 1989–1990. https://core.ac.uk/download/pdf/10427644.pdf (accessed on 19th September 2025).
41. Sun, N., Liang, D., Wu, Y., Chen, Y., Qin, Y., & Fang, Y. (2019). Adaptive control for pneumatic artificial muscle systems with parametric uncertainties and unidirectional input constraints. IEEE Transactions on Industrial Informatics, 16, 969–979. https://doi.org/10.1109/tii.2019.2923715 (accessed on 19th September 2025).

42. Thanmay, K. J., & Harsha, M. V. (2024). Design automation of 2-stage CMOS op-amp using PSO algorithm. International Journal for Research in Applied Science and Engineering Technology. https://www.semanticscholar.org/paper/27b64ede6a22c144e29058dc13c77b316b55c7cb (accessed on 19th September 2025).

43. Vernekar, P. (2025). Nonlinear feedback linearization and LQG/LTR control: A comparative study for a single-machine infinite-bus system. https://www.semanticscholar.org/paper/80f2da9991cb1c75b3180442a1043e5f44d8bb74 (accessed on 19th September 2025).

44. Wang, B., Li, J., Hargrove, L., & Kamavuako, E. (2024). Unravelling influence factors in pattern recognition myoelectric control systems: The impact of limb positions and electrode shifts. *Sensors (Basel, Switzerland), 24.* https://www.semanticscholar.org/paper/df2a3564553f07600f24f820c7b95f4328f9ab73 (accessed on 19th September 2025).

45. Wu, Y., Wang, L., Zhang, J., & Li, F. (2019). Path following control of autonomous ground vehicle based on nonsingular terminal sliding mode and active disturbance rejection control. IEEE Transactions on Vehicular Technology, 68, 6379–6390. https://doi.org/10.1109/tvt.2019.2916982 (accessed on 19th September 2025).

46. Yanarateş, C., & Altan, A. (2024). Compact analysis of the necessity of Padé approximation for delayed continuous-time models in LQR, H-infinity and root locus control strategies. *Black Sea Journal of Engineering and Science.* https://www.semanticscholar.org/paper/69d33c3e39256201f79cd8ff8df35e3a3b0d7637 (accessed on 19th September 2025).

47. Yang, T., Sun, N., Chen, H., & Fang, Y. (2019). Neural network-based adaptive antiswing control of an underactuated ship-mounted crane with roll motions and input dead zones. IEEE Transactions on Neural Networks and Learning Systems, 31, 901–914. https://doi.org/10.1109/tnnls.2019.2910580 (accessed on 19th September 2025).

48. Zehra, A., Farman, M., Bonny, T., Sambas, A., Al Nassan, W., Nisar, K. S., Voon, B. W. N., et al. (2025). Dynamics of 4D supply chain system with fractal fractional derivatives insight of stability analysis and ANN prediction. *Scientific Reports, 15.* https://www.semanticscholar.org/paper/1ba9925f9158fa6b2d21233f6c8127495f9e087a (accessed on 19th September 2025).

49. Zhu, J. Y., Park, T., Isola, P., & Efros, A. A. (2017). Unpaired image-to-image translation using cycle-consistent adversarial networks. https://doi.org/10.1109/iccv.2017.244 (accessed on 19th September 2025).

# INDEX

## D

## E



## F

## G



## H

## I



## K



## L



## M

## N



## O

## P

## Q



## R



## S

## T

## U



## V



## W



## Z